Dynamic Stiffness and Substructures

A.Y.T. Leung

Dynamic Stiffness and Substructures

With 72 Figures

Springer-Verlag
London Berlin Heidelberg New York
Paris Tokyo Hong Kong
Barcelona Budapest

Andrew Y.T. Leung, MSc, PhD, CEng, FRAeS
Department of Civil and Structural Engineering, University of Hong Kong, Pokfulam Road, Hong Kong

ISBN 3-540-19807-5 Springer-Verlag Berlin Heidelberg New York
ISBN 0-387-19807-5 Springer-Verlag New York Berlin Heidelberg

British Library Cataloguing in Publication Data
Leung, Andrew Y.T.
Dynamic Stiffness and Substructures
I. Title
624.1
ISBN 3-540-19807-5

Library of Congress Cataloging-in-Publication Data
Leung, A.Y.T.
Dynamic stiffness and substructures / A.Y.T. Leung.
p. cm.
Includes bibliographical references and index.
ISBN 0-387-19807-5 (alk. paper)
1. Structural analysis (Engineering) – Matrix methods. 2. Modal analysis.
3. Finite element method. I. Title.
TA642.L48 1993 93-15461
624.1'71 – dc20 CIP

Printed in Great Britian

Typeset by Asco Trade Typesetting Ltd., Hong Kong
Printed by Antony Rowe Ltd., Chippenham, Wiltshire
69/3830-543210 Printed on acid-free paper

Contents

Preface

This book is concerned with the dynamic response of structures in the frequency domain. The stiffness (and mass) matrices are frequency dependent and are called dynamic stiffness matrices. Natural frequencies are obtained by equating the determinant of the dynamic stiffness matrix to zero and the natural modes by inverse iteration. In general, an infinite number of natural modes can be found by a small number of degrees of freedom. If one generalizes the frequency to complex number, response corresponding to exponentially evolving harmonic force can also be studied. The computational algorithms to find all the desired modes are discussed in detail. For complicated members, when the analytical solutions are not possible, the dynamic stiffness matrices are alternatively established by the methods of dynamic substructures. Parametrically excited and non-conservatively excited systems are included. The extension of the methods to multifrequency and chaotic responses is a currently active research area.

This book summarizes the author's experience, together with the knowledge of scientists from various countries, including the research groups of W.H. Wittrick, F.W. Williams and A. Simpson in the UK, B. Akkeson in Sweden, and W. Pilkey in the USA.

In conclusion, grateful acknowledgement is due to all those who have contributed in any way towards the successful completion of the book: in particular, the Croucher Foundation and the Hong Kong Research Grants Council (RGC), for providing the necessary financial support.

The author also wishes to acknowledge the work of the publishers.

Acknowledgements

The author is grateful to publishers for the permission to use parts of the papers published in the following journals:

TH Richards & AYT Leung 1977. An accurate method in structural vibration analysis. J Sound Vib 55, 363–376

AYT Leung 1978. An accurate method of dynamic condensation in structural analysis. Int J Num Meth Engng 12, 1705–1716

AYT Leung 1979. An accurate method of dynamic substructuring with simplified computation. Int J Num Meth Engng 14, 1241–1256

AYT Leung 1979. Accelerated convergence of dynamic flexibility in series form. Engng Struct 1, 203–206

AYT Leung 1980. Dynamic of periodic structures. J Sound Vib 72, 451–467

AYT Leung 1982. On a dynamic substructure method. Letter to Editor. Int J Num Meth Engng 18, 629–630

AYT Leung 1983. Fast modal analysis for continuous systems. J Sound Vib 87, 449–467

AYT Leung 1983. Fast modal response method for structures. Int J Num Meth Engng 19, 1435–1451

AYT Leung 1985. Dynamic stiffness method for exponentially varying harmonic excitation of continuous systems. J Sound Vib 98, 337–347

AYT Leung 1985. Structural response to exponentially varying harmonic excitations. J Earthquake Engng Struct Dyn 13, 677–681

AYT Leung 1986. Steady state response of undamped systems to excitations expressed as polynomials in time. J Sound Vib 106, 145–151

AYT Leung 1987. A simple method for exponentially modulated random excitation. J Sound Vib 112, 273–282

AYT Leung 1987. Dynamic stiffness and response analysis. Dyn Stabil Syst 2, 125–137

AYT Leung 1987. Inverse iteration for damped natural vibration. J Sound Vib 118, 193–198

AYT Leung 1988. Inverse iteration for the quadratic eigenvalue problem. J Sound Vib 124, 249–267

AYT Leung 1988. Direct method for the steady state response of structures. J Sound Vib 124, 135–139

AYT Leung 1988. A simple dynamic substructure method. J Earthquake Engng Struct Dyn 16, 827–837

AYT Leung 1988. Dynamic stiffness analysis of follower force. J Sound Vib 126, 533–543

AYT Leung 1988. Damped dynamic substructures. Int J Num Meth Engng 26, 2355–2365

AYT Leung 1989. Multilevel dynamic substructures. Int J Num Meth Engng 28, 181–192

AYT Leung 1989. Stability boundaries for parametrically excited systems by dynamic stiffness. J Sound Vib 131, 265–273

AYT Leung 1989. Dynamic stiffness and nonconservative modal analysis. Int J Analyt Experim Modal Analysis. 4, 77–82

AYT Leung 1990. Nonconservative dynamic substructures. Dyn Stabil Sys 5, 47–57

AYT Leung 1991. Natural shape functions of a compressed Vlasov element. Thin Walled Struct 11, 431–438

AYT Leung 1991. Lambda matrix flexibility. J Sound Vib 148, 521–531

AYT Leung 1991. Dynamic substructure response. J Sound Vib 149, 83–90

AYT Leung 1991. Dynamic stiffness analysis of follower moments. Microcomputers Civil Engng 6, 229–236

AYT Leung 1992. Dynamic stiffness for lateral buckling. Computers and Struct 42, 321–325

AYT Leung 1992. Dynamic stiffness for thin-walled structures. Thin Walled Struct. 14, 209–222

AYT Leung 1993. An algorithm for matrix polynomial eigenproblems. J Sound Vib. 158(2), 363–368

Chapter 1

Harmonic Analysis

An assemblage of coupled objects possessing inertia and elasticity is called a mechanical system. A mechanical system possesses inertia and elasticity. When a small disturbance is applied to a system, the propagation of this small disturbance through the medium of the system is called vibration. Most machines and engineering structures experience vibration in varying degrees.

If the medium of a system vibrates sinusoidally in time, the vibration is called harmonic. The geometric state (displacements, strains, etc.) of a system at any instance is called a configuration. When a harmonic vibration exists, the configuration of the system will repeat itself in equal intervals of time. The time elapsed during which the motion repeats itself is called the period, and the motion completed during a period is referred to as a cycle. The number of complete cycles in a unit time is known as the frequency of vibration, while the peak value of motion is called the amplitude. The set of parameters needed to specify a configuration is called the set of generalized coordinates, and the time rates of change of generalized coordinates are called generalized velocities.

The vibration of a system is generally non-linear in nature. However, if there exists an equilibrium configuration of the system, i.e. a configuration in which the system can remain permanently at rest, or about which the system undergoes a prescribed steady state motion, we can expand all the non-linearities (geometry, material, etc.) in Taylor series about the equilibrium configuration in terms of the generalized coordinates and their time derivatives. When the vibration of a system is not far away from its equilibrium configuration and when the disturbances are small, we can study the vibration approximately using the first two terms of such Taylor series, and we say the system is linearized.

Vibrations of linear systems fall into two general classes, free and forced. Free vibration takes place when a system vibrates under the action of forces inherent in the system itself, with no externally applied forces present. When a system described by a finite number of generalized coordinates is subjected to arbitrary initial conditions, the free vibration is periodic with several frequency components. However, among these will be some simple harmonic motions called principal modes, or natural modes of vibration. These are characterized by a certain distribution of amplitude over the body, in which each point in the body undergoes harmonic motion of common frequency (the natural frequency) with all points passing through their equilibrium configuration simultaneously.

Vibration that takes place under the excitation of external forces is called forced vibration. When the exciting force is harmonic, the forced vibration takes place at the frequency of the excitation (independent of the natural frequencies). When the frequency of the exciting forces is coincident with one of the natural frequencies of the system, a condition of resonance is encountered and a dangerously large amplitude may result. Consequently, the calculation of natural frequencies is of interest in all types of vibrating systems.

Vibration systems are all more or less subject to damping because energy is dissipated by friction and other resistances. Since no external energy is supplied during free vibrations, the motion will diminish with time, and is said to be damped. On the other hand, forced vibration may be maintained at constant amplitude with the required energy supplied by an external force. For the convenience of study, we shall assume no damping effect to be present unless otherwise stated.

The behaviour of an oscillatory system may be examined according to the type of excitation to which the system is subjected. Excitation forces may be classified as harmonic, periodic, non-periodic and stochastic, the latter occurring when the applied forces are statistical. Since a periodic excitation can be expressed in terms of Fourier integrals, the time variables are transformed to harmonic frequency spectra, and the free vibration of a system can be considered as harmonic motion vibrating with its natural frequencies. Therefore, we will study the harmonic vibration of a mechanical system in great detail, then general free vibrations and the other types of forced vibrations will be treated as applications of the harmonic vibrations.

Two types of external forces are of interest. One acts in the same direction as a coordinate and does work. The other acts in a direction orthogonal to the associated coordinate and does no work. One example is the constant axial force acting along the neutral axis of a straight beam. If transverse vibration is considered, the constant axial force does no work; it merely changes the stiffness of the beam. The resulting vibration is not classified as forced vibration. An axial force with constant direction is conservative. If, however, the axial force changes in direction according to the vibration configuration, then it is non-conservative and is called a follower force. Finally, if the axial force is no longer constant in magnitude, but rather, periodic in time, then, the beam is said to be parametrically excited.

The types of vibration to be considered are summarized in Fig. 1.0.1.

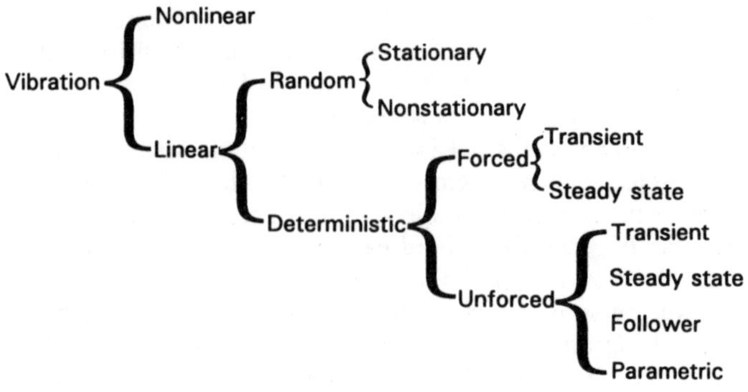

Fig. 1.0.1

1.1. Steady State

Consider the single degree of freedom system

$$\ddot{u} + 2\zeta\omega\dot{u} + \omega^2 u = f(t) \tag{1.1.1}$$

with initial conditions $u(0)$ and $\dot{u}(0)$. The response in terms of the Duhamel integral $u_d(t)$ is given traditionally by

$$u(t) = u_d(t) + g(t)u(0) + h(t)\dot{u}(0) \tag{1.1.2}$$

where

$$\begin{aligned} u_d(t) &= \int_0^t h(t-\tau)f(\tau)\,d\tau, \qquad h(t) = e^{-\zeta\omega t}\omega_d^{-1}\sin\omega_d t \\ g(t) &= e^{-\zeta\omega t}\cos\omega_d t + \zeta\omega h(t), \qquad \omega_d^2 = (1-\zeta^2)\omega^2 \end{aligned} \tag{1.1.3}$$

The range of integration of the Duhamel integral is from zero to t because the Duhamel integral is the forced response when the system is assumed to be at rest initially (at $t = 0$). However, if the excitation is assumed to have existed for a long time, the steady state response is given by

$$u_s(t) = \int_{-\infty}^t h(t-\tau)f(\tau)\,d\tau \tag{1.1.4}$$

Obviously, $u_d(0) = \dot{u}_d(0) = 0$, but $u_s(0)$ and $\dot{u}_s(0)$ need not be zero, so for uniqueness of solution,

$$u_d(t) = u_s(t) - h(t)\dot{u}_s(0) - g(t)u_s(0) \tag{1.1.5}$$

to account for the initial conditions. The solution in terms of the steady state response is thus

$$u(t) = u_s(t) + g(t)[u(0) - u_s(0)] + h(t)[\dot{u}(0) - \dot{u}_s(0)] \tag{1.1.6}$$

The steady state solution (1.1.4) can be obtained in the following alternative manner without explicit integration. Let

$$f(t) = (f_0 + f_1 t + f_2 t^2 + f_3 t^3)e^{\alpha t} \tag{1.1.7}$$

where f_j, $j = 0, 1, 2, 3$, are real constants and $\alpha = \beta + i\nu$ is complex. The real and imaginary parts of the solution $u_s(t)$ obtained in what follows correspond, respectively, to the excitations of the functions $(f_0 + f_1 t + f_2 t^2 + f_2 t^3)e^{\beta t}\cos\nu t$ and $(f_0 + f_1 t + f_2 t^2 + f_2 t^3)e^{\beta t}\sin\nu t$. An example of excitation in the form of Eq. (1.1.7) is earthquake ground motion [1].

Assume the steady state to be

$$u_s(t) = (u_0 + u_1 t + u_2 t^2 + u_3 t^3)e^{\alpha t} \tag{1.1.8}$$

where u_j, $j = 0, 1, 2, 3$, are complex constants to be determined. Substitute Eqs (1.1.7) and (1.1.8) into Eq. (1.1.1) and compare similar terms,

$$\begin{bmatrix} D(\alpha) & 2\zeta(\omega+\alpha) & 2 & \\ & D(\alpha) & 4\zeta(\omega+\alpha) & 6 \\ & & D(\alpha) & 6\zeta(\omega+\alpha) \\ & & & D(\alpha) \end{bmatrix} \begin{Bmatrix} u_0 \\ u_1 \\ u_2 \\ u_3 \end{Bmatrix} = \begin{Bmatrix} f_0 \\ f_1 \\ f_2 \\ f_3 \end{Bmatrix} \tag{1.1.9}$$

where $D(\alpha) = \alpha^2 + 2\alpha\zeta\omega + \omega^2$. The coefficients u_j are determined immediately by back substitution since Eq. (1.1.9) is in upper triangular form already.

Generally, if

$$f(t) = e^{\alpha t} \sum_{j=0}^{n} f_j t^j \tag{1.1.10}$$

then the steady state response is given by

$$u_s(t) = e^{\alpha t} \sum_{j=0}^{n} u_j t^j \tag{1.1.11}$$

where

$$\begin{gathered} u_j = [f_j - 2j(\zeta\omega + \alpha)u_{j+1} - j(j+1)u_{j+2}]/(\alpha^2 + 2\alpha\zeta\omega + \omega^2) \\ j = n, n-1, \ldots, 0, \qquad u_{n+1} = u_{n+2} = 0 \end{gathered} \tag{1.1.12}$$

If the excitation is piecewise continuous and expressed in form (1.1.10) in each time interval $t_k < t < t_{k+1}$, then Eq. (1.1.6) must be modified to account for the discontinuity. From Eq. (1.1.6),

$$\begin{Bmatrix} u(t) \\ \dot{u}(t) \end{Bmatrix} = \begin{Bmatrix} u_s(t) \\ \dot{u}_s(t) \end{Bmatrix} + \begin{bmatrix} g(t) & h(t) \\ \dot{g}(t) & \dot{h}(t) \end{bmatrix} \begin{Bmatrix} u(0) - u_s(0) \\ \dot{u}(0) - \dot{u}_s(0) \end{Bmatrix} \tag{1.1.13}$$

where

$$\begin{Bmatrix} \dot{g}(t) \\ \dot{h}(t) \end{Bmatrix} = \begin{bmatrix} 0 & -\omega^2 \\ 1 & -2\zeta\omega \end{bmatrix} \begin{Bmatrix} g(t) \\ h(t) \end{Bmatrix} \tag{1.1.14}$$

When these are rewritten in the time interval $t_k < t < t_{k+1}$,

$$\begin{Bmatrix} u(t_{k+1}) \\ u(t_{k+1}) \end{Bmatrix} = \begin{Bmatrix} u_s(t_{k+1}) \\ u_s(t_{k+1}) \end{Bmatrix} + \begin{bmatrix} g(\delta) & h(\delta) \\ \dot{g}(\delta) & \dot{h}(\delta) \end{bmatrix} \begin{Bmatrix} u(t_k) - u_s(t_k) \\ \dot{u}(t_k) - \dot{u}_s(t_k) \end{Bmatrix} \tag{1.1.15}$$

where $\delta = t_{k+1} - t_k$. Therefore, the response at time t_{k+1} is completely defined by the response at the previous time t_k.

Example 1.1.1

In harmonic vibration, $f(t) = F\,e^{i\Omega t}$, where Ω is the excitation frequency, and one assumes a steady state solution of the form $u_s(t) = U\,e^{i\Omega t}$, where U is a complex constant to be determined. Equation (1.1.1) gives

$$u_s(t) = \frac{F\,e^{i\Omega t}}{(\omega^2 - \Omega^2) + i2\zeta\omega\Omega} = \frac{F}{\rho} e^{i(\Omega t - \theta)},$$

where the polar form of the denominator $\rho e^{i\theta}$ is given by

$$\rho^2 = (\omega^2 - \Omega^2)^2 - 4\zeta^2\omega^2\Omega^2$$

and

$$\theta = \tan^{-1}[2\zeta\omega\Omega/(\omega^2 - \Omega^2)]$$

These are plotted against the excitation frequency Ω in Fig. 1.1.1.

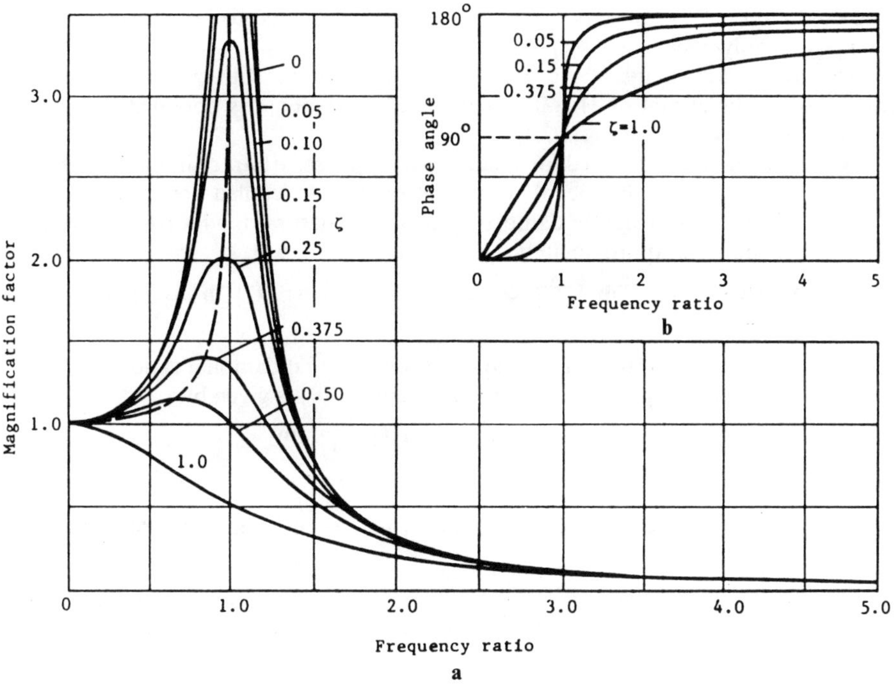

Fig. 1.1.1. **a** Magnification factor against frequency ratio. **b** Phase angle factor against frequency ratio

1.2. Multiple Degrees of Freedom

The governing equation for a p-degree of freedom system is

$$[\mathbf{M}]\{\ddot{\mathbf{u}}\} + [\mathbf{C}]\{\dot{\mathbf{u}}\} + [\mathbf{K}]\{\mathbf{u}\} = [\mathbf{F}]\{\mathbf{f}(t)\} \tag{1.2.1}$$

where $[\mathbf{F}]$ is a constant $m \times p$ matrix characterizing the superposition of the m forcing function components in $\{\mathbf{f}(t)\}$. As in references [2–4], only the steady state solution will be discussed here.

Let the excitation be expressed as

$$\{\mathbf{f}(t)\} = \mathrm{e}^{\alpha t} \sum_{j=0}^{n} \{\mathbf{f}_j\} t^j = \mathrm{e}^{\beta t}(\cos vt + \mathrm{i} \sin vt) \sum_{j=0}^{n} \{\mathbf{f}_j\} t^j \tag{1.2.2}$$

where $\alpha = \beta + \mathrm{i}v$ is complex and $\{\mathbf{f}_j\}$ are real, $j = 0, 1, 2, \ldots, n$. Assume that the steady state solution has the similar form,

$$\{\mathbf{u}_s(t)\} = \mathrm{e}^{\alpha t} \sum_{j=0}^{n} \{\mathbf{u}_j\} t^j \tag{1.2.3}$$

where the $\{\mathbf{u}_j\}$ are complex coefficient vectors to be determined. Substitute Eqs (1.2.2) and (1.2.3) into Eq. (1.2.1), and compare similar terms,

$$\{\mathbf{u}_j\} = [\mathbf{D}]^{-1}\{\mathbf{F}\mathbf{f}_j - j[\mathbf{C} + 2\alpha\mathbf{M}]\mathbf{u}_{j+1} - j(j+1)\mathbf{u}_{j+2}\} \tag{1.2.4}$$

where $[\mathbf{D}] = [\alpha^2\mathbf{M} + \alpha\mathbf{C} + \mathbf{K}]$ and $\{\mathbf{u}_{n+1}\} = \{\mathbf{u}_{n+2}\} = \{\mathbf{0}\}$. The real and imaginary parts of the steady state response (1.2.3) correspond to real and imaginary parts of the excitation (1.2.2) respectively. Note that $[\mathbf{D}]$ is required to be inverted (or decomposed) only once.

An obvious advantage of the present method over the Ritz or Lanczos vector methods for the excitation (1.2.2) is that the exact response is obtained in a finite number of matrix operations (1.2.4).

While the method is tailored for steady state response, arbitrary initial conditions can also be considered by means of natural modes. A detailed derivation can be found in Sect. 1.3. For lightly damped systems, the natural modes $[\mathbf{\Phi}]$ satisfy the following conditions approximately:

$$[\mathbf{\Phi}]^{\mathrm{T}}[\mathbf{M}][\mathbf{\Phi}] = [\mathbf{I}], \quad [\mathbf{\Phi}]^{\mathrm{T}}[\mathbf{K}][\mathbf{\Phi}] = \mathrm{diag}[\omega_i^2]; \quad [\mathbf{\Phi}]^{\mathrm{T}}[\mathbf{C}][\mathbf{\Phi}] = \mathrm{diag}[2\omega_i\zeta_i] \tag{1.2.5}$$

Here ω_i is the ith natural frequency and ζ_i the ith modal damping ratio. The transient response due to initial conditions $\{\mathbf{u}(0)\}$ and $\{\dot{\mathbf{u}}(0)\}$ is given by

$$\{\mathbf{u}(t)\} = [\mathbf{g}(t)]\{\mathbf{u}(0)\} + [\mathbf{h}(t)]\{\dot{\mathbf{u}}(0)\} \tag{1.2.6}$$

where

$$[\mathbf{g}(t)] = [\mathbf{\Phi}]\,\mathrm{diag}[g_i(t)][\mathbf{\Phi}]^{\mathrm{T}}[\mathbf{M}] \quad \text{and} \quad [\mathbf{h}(t)] = [\mathbf{\Phi}]\,\mathrm{diag}[h_i(t)][\mathbf{\Phi}]^{\mathrm{T}}[\mathbf{M}]$$

in which $g_i(t) = \mathrm{e}^{-\zeta_i\omega_i t}\cos\lambda_i t + \zeta_i\omega_i h_i(t)$ and $h_i(t) = \mathrm{e}^{-\zeta_i\omega_i t}\lambda_i^{-1}\sin\lambda_i t$, where $\lambda_i^2 = (1-\zeta_i^2)\omega_i^2$. Since the system is linear, the total response is obtained by the superposition of Eqs (1.2.3) and (1.2.6) with the necessary modification of the initial conditions:

$$\{\mathbf{u}(t)\} = \{\mathbf{u}_s(t)\} + [\mathbf{g}(t)]\{\mathbf{u}(0) - \mathbf{u}_s(0)\} + [\mathbf{h}(t)]\{\mathbf{u}(0) - \mathbf{u}_s(0)\} \tag{1.2.7}$$

For heavily damped systems, the complex modes λ_i, $\{\phi_i\}$, $i = 1, 2, \ldots, 2N$, are required to satisfy the orthonormality conditions

$$[\mathbf{\Lambda}][\mathbf{\Phi}]^{\mathrm{T}}[\mathbf{M}][\mathbf{\Phi}] + [\mathbf{\Phi}]^{\mathrm{T}}[\mathbf{M}][\mathbf{\Phi}][\mathbf{\Lambda}] + [\mathbf{\Phi}]^{\mathrm{T}}[\mathbf{C}][\mathbf{\Phi}] = [\mathbf{I}] \tag{1.2.8}$$

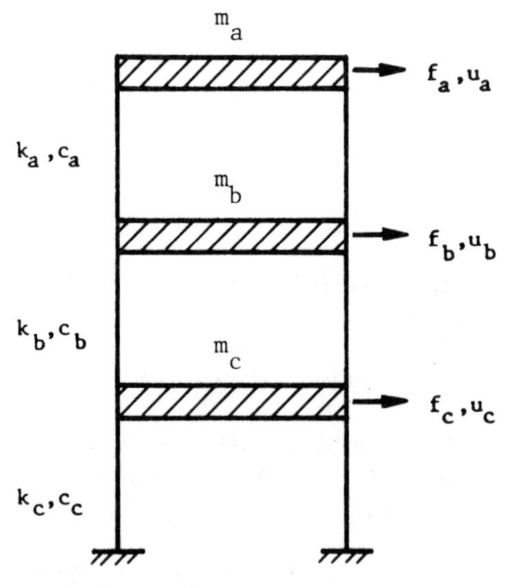

Fig. 1.2.1. A discrete model for a three storey system

where $[\Lambda] = \text{diag}[\lambda_i]$, $[\Phi] = \text{row}[\phi_i]$ and N is the order of the system. The matrices $[\mathbf{g}(t)]$ and $[\mathbf{h}(t)]$ in Eqs (1.2.6) and (1.2.7) now have the following forms

$$[\mathbf{g}(t)] = [\mathbf{\Phi}]\exp[\Lambda t][\Lambda\mathbf{\Phi}^{\mathrm{T}}\mathbf{M} + \mathbf{\Phi}^{\mathrm{T}}\mathbf{C}]; \qquad [\mathbf{h}(t)] = [\mathbf{\Phi}]\exp[\Lambda t][\mathbf{\Phi}^{\mathrm{T}}\mathbf{M}]$$

Example 1.2.1

Consider the system shown in Fig. 1.2.1. Let

$$m_{\mathrm{a}} = 1\,\mathrm{kg}, \quad k_{\mathrm{a}} = 600\,\mathrm{N\,m^{-1}}, \quad c_{\mathrm{a}} = 6\,\mathrm{N\,s\,m^{-1}}$$
$$m_{\mathrm{b}} = 1.5\,\mathrm{kg}, \quad k_{\mathrm{b}} = 1200\,\mathrm{N\,m^{-1}}, \quad c_{\mathrm{b}} = 12\,\mathrm{N\,s\,m^{-1}}$$
$$m_{\mathrm{c}} = 2\,\mathrm{kg}, \quad k_{\mathrm{c}} = 1800\,\mathrm{N\,m^{-1}}, \quad c_{\mathrm{c}} = 18\,\mathrm{N\,s\,m^{-1}}$$

and $f_{\mathrm{a}}(t) = 2\mathrm{e}^{-0.2t}\sin 2t$, $f_{\mathrm{b}}(t) = 2\mathrm{e}^{-0.1t}\cos t$, $f_{\mathrm{c}}(t) = \mathrm{e}^{-0.3t}\sin t$, then the following equation is to be solved for $\{\mathbf{u}_j\}$:

$$[\alpha_j^2\mathbf{M} + \alpha_j\mathbf{C} + \mathbf{K}]\{\mathbf{u}_j\}\mathrm{e}^{\alpha_j t} = \{\mathbf{f}_j\}\mathrm{e}^{\alpha_j t}$$

where $\alpha_1 = 2\mathrm{i} - 0.2$, $\alpha_2 = \mathrm{i} - 0.1$, $\alpha_3 = \mathrm{i} - 0.3$ and

$$[\mathbf{M}] = \begin{bmatrix} 1 & & \\ & 1.5 & \\ & & 2 \end{bmatrix}, \quad [\mathbf{C}] = \begin{bmatrix} 6 & -6 & 0 \\ -6 & 18 & -12 \\ 0 & -12 & 30 \end{bmatrix}, \quad [\mathbf{K}] = \begin{bmatrix} 600 & -600 & 0 \\ -600 & 1800 & -1200 \\ 0 & -1200 & 3000 \end{bmatrix}$$

$$\{\mathbf{f}_1\} = [2 \quad 0 \quad 0]^{\mathrm{T}}, \qquad \{\mathbf{f}_2\} = [0 \quad 1 \quad 0]^{\mathrm{T}}, \qquad \{\mathbf{f}_3\} = [0 \quad 0 \quad 1]^{\mathrm{T}}$$

The solutions are

$$\{\mathbf{u}_1\} = \begin{Bmatrix} 6.2256 \\ 2.8455 \\ 1.1412 \end{Bmatrix} + \mathrm{i}\begin{Bmatrix} -0.10548 \\ -0.045359 \\ -0.017641 \end{Bmatrix}\mathrm{mm}$$

$$\{\mathbf{u}_2\} = \begin{Bmatrix} 1.3979 \\ 1.3956 \\ 0.55860 \end{Bmatrix} + \mathrm{i}\begin{Bmatrix} -0.012502 \\ -0.012924 \\ -0.0051022 \end{Bmatrix}\mathrm{mm}$$

$$\{\mathbf{u}_3\} = \begin{Bmatrix} 0.56040 \\ 0.55954 \\ 0.55846 \end{Bmatrix} + \mathrm{i}\begin{Bmatrix} -0.0035379 \\ -0.0040860 \\ -0.0047698 \end{Bmatrix}\mathrm{mm}$$

and finally, the steady state response is given by (see Fig. 1.2.2)

$$\{\mathbf{u}(t)\} = \mathrm{Im}\{\mathrm{e}^{\alpha_1 t}\mathbf{u}_1\} + \mathrm{Re}\{\mathrm{e}^{\alpha_2 t}\mathbf{u}_2\} + \mathrm{Im}\{\mathrm{e}^{\alpha_3 t}\mathbf{u}_3\} \tag{1.2.9}$$

It is noted that the steady state solution (1.2.9) implies the following initial conditions:

$$\{\mathbf{u}(0)\} = \mathrm{Im}\{\mathbf{u}_1\} + \mathrm{Re}\{\mathbf{u}_2\} + \mathrm{Im}\{\mathbf{u}_3\}$$
$$= \{1.2889 \quad 1.3462 \quad 0.5362\}\,\mathrm{mm}$$

$$\{\dot{\mathbf{u}}(0)\} = \mathrm{Im}\{\alpha_1\mathbf{u}_1\} + \mathrm{Re}\{\alpha_2\mathbf{u}_2\} + \mathrm{Im}\{\alpha_3\mathbf{u}_3\}$$
$$= \{12.906 \quad 6.134 \quad 2.795\}\,\mathrm{mm\,s^{-1}}$$

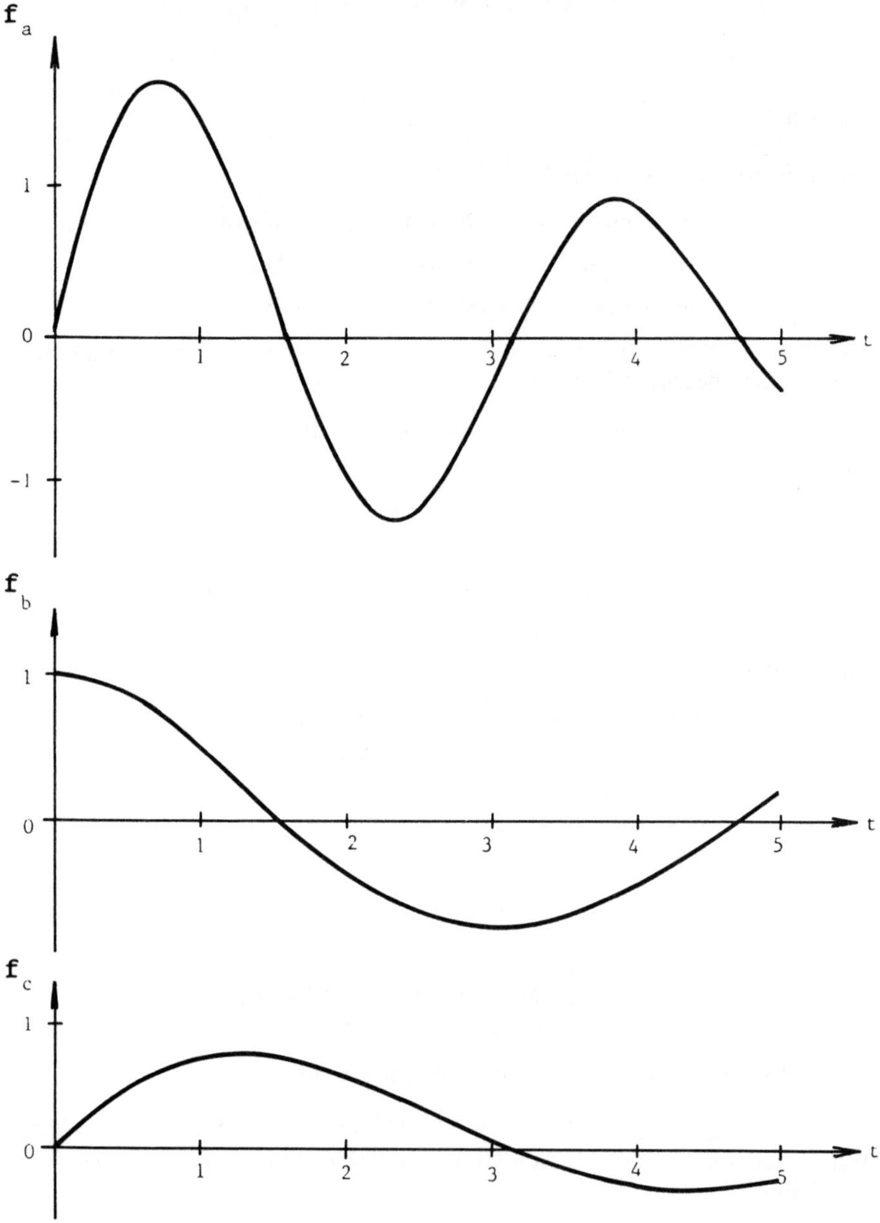

a

Fig. 1.2.2. **a** Time histories for the excitations. **b** Time histories for the responses

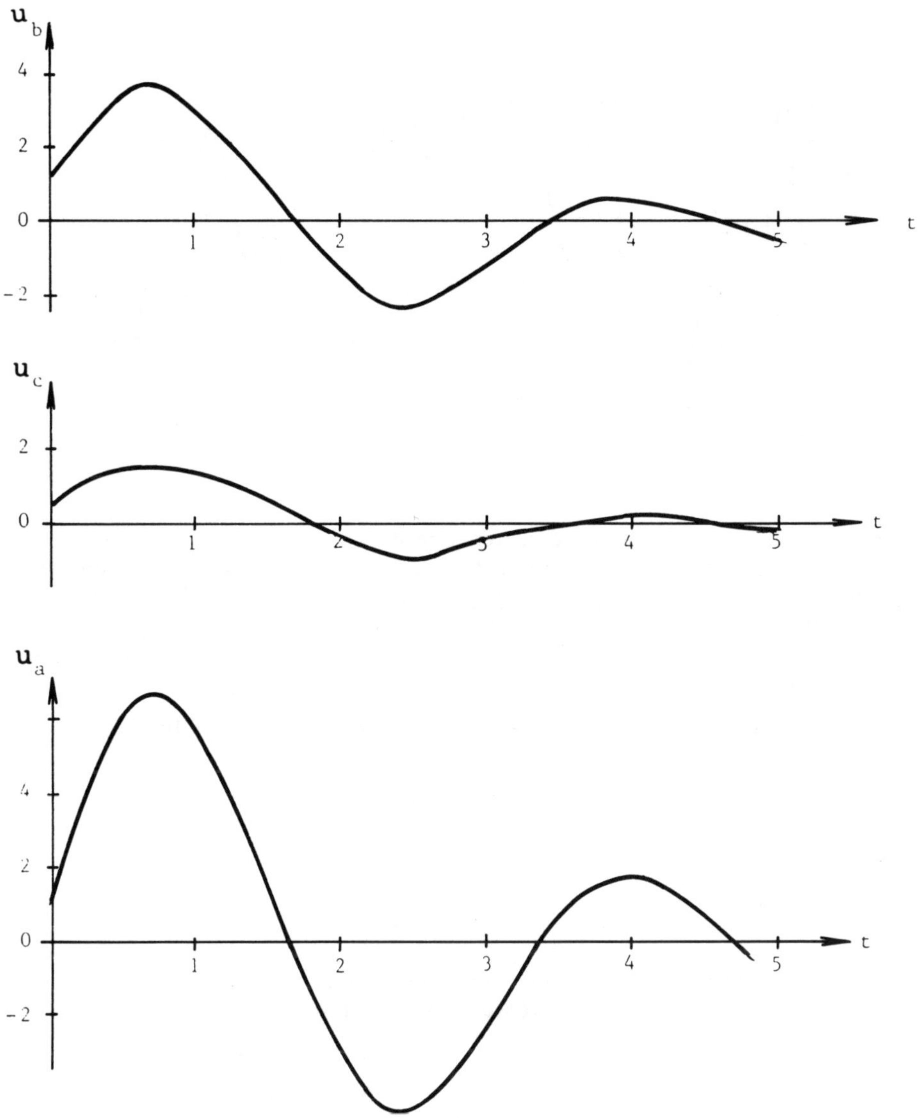

b

Fig. 1.2.2 (*continued*)

The steady state response is obtained as if the forcing terms had been applied since time $t = -\infty$; therefore, nonzero initial conditions are generated for a real situation. If "at rest" initial conditions had been imposed, modal analysis would be necessary. Again, the required steady state response can also be obtained by modal analysis. The present method does not assume any particular form of the damping matrix [**C**], e.g. proportional.

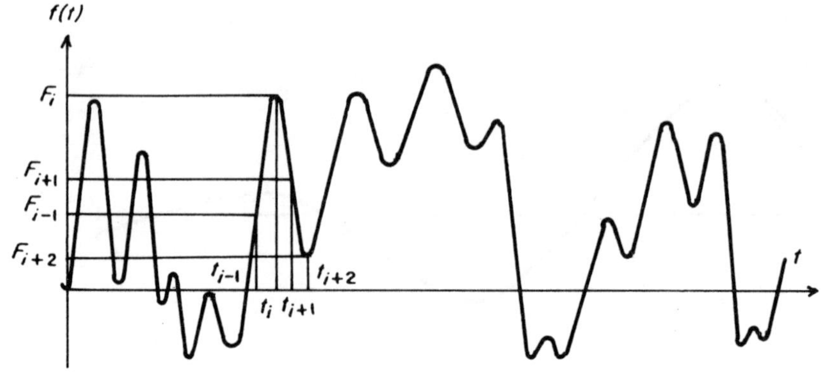

Fig. 1.2.3. Piecewise linear discretization of a time history

Example 1.2.2

Consider the piecewise linear forcing function shown in Fig. 1.2.3, as can be found in typical field measurements. Suppose the nodal force and displacement vectors are given as $\{\mathbf{f}(t)\} = \{\mathbf{f}_0\} + \{\mathbf{f}_1\}\tau$ and $\{\mathbf{q}_s(t)\} = \{\mathbf{q}_0\} + \{\mathbf{q}_1\}\tau$, where $\tau = t - t_i$, $\{\mathbf{f}_0\} = \{\mathbf{F}_i\} = \{\mathbf{f}(t_i)\}$ and $\{\mathbf{f}_1\} = (\{\mathbf{F}_{i+1}\} - \{\mathbf{F}_i\})/(t_{i+1} - t_i)$. The steady state response is obtained immediately from $[\mathbf{K}]\{\mathbf{q}_0\} = \{\mathbf{f}_0\}$ and $[\mathbf{K}]\{\mathbf{q}_1\} = \{\mathbf{f}_1\}$. Rewrite Eq. (1.2.7) in the form

$$\{\mathbf{q}(t)\} = \{\mathbf{q}_s(t)\} + [\mathbf{g}(t)]\{\mathbf{q}(0) - \mathbf{q}_s(0)\} + [\mathbf{h}(t)]\{\dot{\mathbf{q}}(0) - \dot{\mathbf{q}}_s(0)\}$$

where

$$[\mathbf{g}(t)] = \sum_j \{\boldsymbol{\Phi}_j\}\{\boldsymbol{\Phi}_j\}^{\mathrm{T}}[\mathbf{M}]\cos\omega_j t$$

and

$$[\mathbf{h}(t)] = \sum_j \{\boldsymbol{\Phi}_j\}\{\boldsymbol{\Phi}_j\}^{\mathrm{T}}[\mathbf{M}]\omega_j^{-1}\sin\omega_j t$$

are the indical response and impulsive response matrices respectively.

We arrive at the following recurrence formula for response at time $t = t_{i+1}$:

$$\begin{Bmatrix} \mathbf{q}(t_{i+1}) \\ \dot{\mathbf{q}}(t_{i+1}) \end{Bmatrix} = \begin{Bmatrix} \mathbf{q}_0 + \mathbf{q}_1\delta \\ \mathbf{q}_1 \end{Bmatrix} + \begin{bmatrix} \mathbf{g}(\delta) & \mathbf{h}(\delta) \\ \dot{\mathbf{g}}(\delta) & \dot{\mathbf{h}}(\delta) \end{bmatrix} \begin{Bmatrix} \mathbf{q}(t_i) - \mathbf{q}_0 \\ \dot{\mathbf{q}}(t_i) - \mathbf{q}_1 \end{Bmatrix}$$

where $\delta = t_{i+1} - t_i$. If the natural modes are given, the above equation can be programmed easily in a recursive manner.

Example 1.2.3

When the forcing function is due to a vibrating environment, $\{\mathbf{f}(t)\}$ may be narrow band. It can be desirable to use only the recorded peak and trough values (see Fig. 1.2.4). The nodal force and displacement vectors are then expressed as

$$\begin{aligned} &\{\mathbf{f}(t)\} = \{\mathbf{f}_0\} + \{\mathbf{f}_1\}\tau + \{\mathbf{f}_2\}\tau^2 + \{\mathbf{f}_3\}\tau^3 \\ \text{and} \quad &\{\mathbf{q}(t)\} = \{\mathbf{q}_0\} + \{\mathbf{q}_1\}\tau + \{\mathbf{q}_2\}\tau^2 + \{\mathbf{q}_3\}\tau^3 \end{aligned} \tag{1.2.10}$$

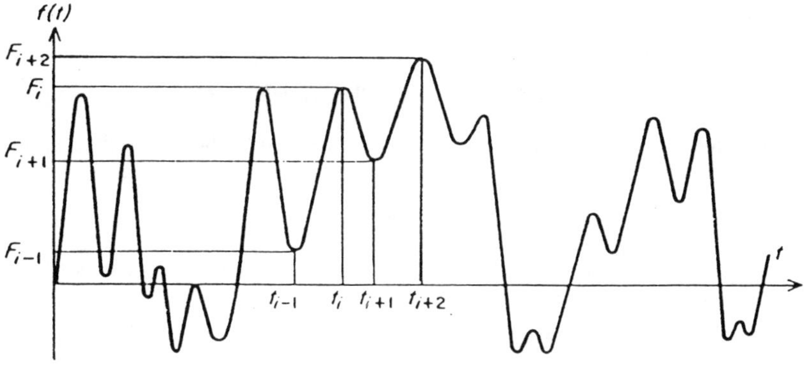

Fig. 1.2.4. Piecewise cubic discretization of a time history

where again $\tau = t - t_i$ and $\delta = t_{i+1} - t_i$. In terms of the values at the discrete time, $\{\mathbf{f}(t_i)\} = \{\mathbf{F}_i\}$,

$$\{\mathbf{f}(t)\} = [1 - 3(\tau/\delta)^2 + 2(\tau/\delta)^3]\{\mathbf{F}_i\} + [3(\tau/\delta)^2 - 2(\tau/\delta)^3]\{\mathbf{F}_{i+1}\} \quad (1.2.11)$$

Comparing Eqs (1.2.10) and (1.2.11), we obtain $\{\mathbf{f}_0\} = \{\mathbf{F}_i\}$, $\{\mathbf{f}_1\} = \{0\}$, $\{\mathbf{f}_2\} = 3\{\mathbf{F}_{i+1} + \mathbf{F}_i\}/\delta^2$ and $\{\mathbf{f}_3\} = 2\{\mathbf{F}_i - \mathbf{F}_{i+1}\}/\delta^3$. The coefficient vectors of the response are obtained from the equations $[\mathbf{K}]\{\mathbf{q}_3\} = \{\mathbf{f}_3\}$, $[\mathbf{K}]\{\mathbf{q}_2\} = \{\mathbf{f}_2\}$, $[\mathbf{K}]\{\mathbf{q}_1\} = -6[\mathbf{M}]\{\mathbf{q}_3\}$ and $[\mathbf{K}]\{\mathbf{q}_0\} = \{\mathbf{f}_0\} - 2[\mathbf{M}]\{\mathbf{q}_2\}$. As these equations have the same coefficient matrix, the solutions for $\{\mathbf{q}_3\}$, $\{\mathbf{q}_2\}$, $\{\mathbf{q}_1\}$ and $\{\mathbf{q}_0\}$ are quite simple. Finally from Eq. (1.2.7), we arrive at the following recurrence formula:

$$\begin{Bmatrix} \mathbf{q}(t_{i+1}) \\ \dot{\mathbf{q}}(t_{i+1}) \end{Bmatrix} = \begin{Bmatrix} \mathbf{q}_0 + \mathbf{q}_1\delta + \mathbf{q}_2\delta^2 + \mathbf{q}_3\delta^3 \\ \mathbf{q}_1\delta + 2\mathbf{q}_2\delta + 3\mathbf{q}_3\delta^2 \end{Bmatrix} + \begin{bmatrix} \mathbf{g}(\delta) & \mathbf{h}(\delta) \\ \dot{\mathbf{g}}(\delta) & \dot{\mathbf{h}}(\delta) \end{bmatrix} \begin{Bmatrix} \mathbf{q}(t_i) - \mathbf{q}_0 \\ \dot{\mathbf{q}}(t_i) - \mathbf{q}_1 \end{Bmatrix}$$

1.3. Modal Analysis and Acceleration

Consider the governing matrix equation of a system which has been derived from approximate methods such as the finite element method,

$$[\mathbf{M}]\{\ddot{\mathbf{q}}\} + [\mathbf{C}]\{\dot{\mathbf{q}}\} + [\mathbf{K}]\{\mathbf{q}\} = \{\mathbf{Q}\} \quad (1.3.1)$$

with initial conditions

$$\{\mathbf{q}(0)\} = \{\mathbf{q}_0\}, \qquad \{\dot{\mathbf{q}}(0)\} = \{\dot{\mathbf{q}}_0\} \quad (1.3.2)$$

where $[\mathbf{M}]$, $[\mathbf{C}]$ and $[\mathbf{K}]$ are the $n \times n$ mass, damping and stiffness matrices, respectively, and $\{\mathbf{q}\}$ and $\{\mathbf{Q}\}$ are the time dependent generalized displacement and force vectors, respectively. If

$$[\mathbf{C}] = 2\beta[\mathbf{M}] + 2\gamma[\mathbf{K}] \quad (1.3.3)$$

where β and γ are damping constants, Eq. (1.3.1) can be uncoupled by means of a transformation using the modal matrix $[\mathbf{\Phi}]$ of the undamped system. The modal matrix $[\mathbf{\Phi}]$ is a collection of the first m normal modes $\{\phi_i\}$ of the system

$$[\mathbf{M}]\{\phi_i\}\omega_i^2 = [\mathbf{K}]\{\phi_i\} \quad \text{or} \quad [\mathbf{M}][\mathbf{\Phi}][\mathbf{\Omega}^2] = [\mathbf{K}][\mathbf{\Phi}], \qquad i = 1, 2, \ldots, m \quad (1.3.4)$$

where $[\mathbf{\Phi}] = \{\phi_1 \quad \phi_2 \ldots \phi_m\}$, $[\mathbf{\Omega}^2] = \mathrm{diag}[\omega_1^2, \omega_2^2 \ldots \omega_m^2]$ and $\omega_1^2 \leq \omega_2^2 \leq \cdots \leq \omega_m^2$. The modes are normalized so that

$$[\mathbf{\Phi}]^{\mathrm{T}}[\mathbf{M}][\mathbf{\Phi}] = [\mathbf{I}], \qquad [\mathbf{\Phi}]^{\mathrm{T}}[\mathbf{K}][\mathbf{\Phi}] = [\mathbf{\Omega}^2] \tag{1.3.5}$$

Now, the generalized displacements $\{\mathbf{q}\}$ are transformed to the modal displacements $\{\mathbf{p}\}$ by

$$\{\mathbf{q}\} = [\mathbf{\Phi}]\{\mathbf{p}\} \tag{1.3.6}$$

Substituting Eq. (1.3.6) into Eq. (1.3.1), premultiplying by $[\mathbf{\Phi}]^{\mathrm{T}}$ and making use of the orthogonality conditions (1.3.5) yields

$$\{\ddot{\mathbf{p}}\} + 2[\beta\mathbf{I} + \gamma\mathbf{\Omega}^2]\{\dot{\mathbf{p}}\} + [\mathbf{\Omega}^2]\{\mathbf{p}\} = \{\mathbf{P}\} \tag{1.3.7a}$$

or

$$\ddot{p}_i + 2(\beta + \gamma\omega_i^2)\dot{p}_i + \omega_i^2 p_i = P_i \tag{1.3.7b}$$

where $\{\mathbf{P}\} = \{\mathbf{\Phi}]^{\mathrm{T}}\{\mathbf{Q}\}$ is a vector of modal forces. Equations (1.3.7) are to be solved for $\{\mathbf{p}\}$ with the initial conditions

$$\{\mathbf{p}_0\} = [\mathbf{\Phi}]^{\mathrm{T}}[\mathbf{M}]\{\mathbf{q}_0\} \quad \text{and} \quad \{\dot{\mathbf{p}}_0\} = [\mathbf{\Phi}]^{\mathrm{T}}[\mathbf{M}]\{\dot{\mathbf{q}}_0\} \tag{1.3.8}$$

Let

$$\lambda_i^2 = \omega_i^2 - \beta^2 - \gamma^2\omega_1^4 \quad \text{and} \quad \mu_i = \beta + \gamma\omega_i^2 \tag{1.3.9}$$

then the solution for Eqs (1.3.7) subjected to the initial conditions (1.3.8) are

$$p_i = \int_0^t h_i(t-\tau)P_i(\tau)\,\mathrm{d}\tau + g_i p_{i0} + h_i\dot{p}_{i0} \tag{1.3.10}$$

where $g_i(t) = \mathrm{e}^{-\mu_i t}\cos\lambda_i t + \mu_i h_i(t)$ and $h_i(t) = \lambda_i^{-1}\mathrm{e}^{-\mu_i t}\sin\lambda_i t$.
From Eq. (1.3.6)

$$\{\mathbf{q}\} = \{\mathbf{q}_c\} + \{\bar{\mathbf{q}}\} \tag{1.3.11}$$

where

$$\{\mathbf{q}_c\} = [\mathbf{\Phi}]\int_0^t [h_i(t-\tau)][\mathbf{\Phi}]^{\mathrm{T}}\{\mathbf{Q}(\tau)\}\,\mathrm{d}\tau \tag{1.3.12}$$

represents the convolution integral, and

$$\{\bar{\mathbf{q}}\} = [\mathbf{\Phi}][g_i(t)][\mathbf{\Phi}]^{\mathrm{T}}[\mathbf{M}]\{\mathbf{q}_0\} + [\mathbf{\Phi}][h_i(t)][\mathbf{\Phi}]^{\mathrm{T}}[\mathbf{M}]\{\dot{\mathbf{q}}_0\} \tag{1.3.13}$$

represents the initial transient.

It can be seen from Eqs (1.3.12) and (1.3.13) that all the terms involve at least λ_i^{-1} which is dependent on ω_i^{-1} in the limit. The convergence rate with respect to the number of natural modes may be very slow.

It should be noted that the values of β and γ may be changed from mode to mode as the solution depends on λ_i and μ_i, given by Eq. (1.3.9), rather than β and γ directly. It should also be noted from Eq. (1.3.9) that $\lambda_i^2 = \omega_i^2(1 - \gamma^2\omega_i^2) - \beta^2$, and λ_i will become imaginary if $\omega_i > 1/\gamma$ for small β. Therefore, it is necessary either to keep γ small or make γ decreasing with respect to the normal modes.

For viscously damped system, Eqs (1.3.7) have the form

$$\ddot{p}_i + 2\zeta_i\omega_i\dot{p}_i + \omega_i^2 p_i = P_i \tag{1.3.14}$$

where ζ_i is the modal damping coefficient, and λ_i^2 and μ_i become $\lambda_i^2 = (1 - \zeta_i^2)\omega_i$ and $\mu_i = \zeta_i\omega_i$, respectively.

The initial transient given by Eq. (1.3.13) converges very slowly with respect to the number of natural modes, and can be improved in the following manner.

The initial displacement $\{\mathbf{q}_0\}$ can be considered as being produced by the fictitious initial force $\{\mathbf{F}_0\}$, so that $[\mathbf{K}]\{\mathbf{q}_0\} = \{\mathbf{F}_0\}$. Therefore,

$$[\mathbf{\Phi}][\mathbf{M}]\{\mathbf{q}_0\} = [\mathbf{\Phi}][\mathbf{M}][\mathbf{K}]^{-1}\{\mathbf{F}_0\} \tag{1.3.15}$$

and from Eq. (1.3.4)

$$[\mathbf{K}]^{-1} = [\mathbf{\Phi}][\mathbf{\Omega}^{-2}][\mathbf{\Phi}]^{\mathrm{T}} \tag{1.3.16}$$

Equation (1.3.15) thus becomes

$$[\mathbf{\Phi}]^{\mathrm{T}}[\mathbf{M}]\{\mathbf{q}_0\} = [\mathbf{\Phi}]^{\mathrm{T}}[\mathbf{M}][\mathbf{\Phi}][\mathbf{\Omega}^{-2}][\mathbf{K}]^{-1}[\mathbf{\Phi}]^{\mathrm{T}}\{\mathbf{F}_0\} = [\mathbf{\Omega}^{-2}][\mathbf{\Phi}]^{\mathrm{T}}\{\mathbf{F}_0\} \tag{1.3.17a}$$

and similarly

$$[\mathbf{\Phi}]^{\mathrm{T}}[\mathbf{M}]\{\dot{\mathbf{q}}_0\} = [\mathbf{\Omega}^{-2}][\mathbf{\Phi}]^{\mathrm{T}}\{\dot{\mathbf{F}}_0\} \tag{1.3.17b}$$

Substituting Eqs (1.3.17) into (1.3.13) gives

$$\{\bar{\mathbf{q}}\} = [\mathbf{\Phi}][\omega_i^{-2}g_i(t)][\mathbf{\Phi}]^{\mathrm{T}}[\mathbf{F}_0] + [\mathbf{\Phi}][\omega_i^{-2}h_i(t)][\mathbf{\Phi}]^{\mathrm{T}}[\dot{\mathbf{F}}_0] \tag{1.3.18}$$

and it can be seen by comparing Eqs (1.3.12) and (1.3.18) that the convergence rate is greatly improved by the newly appeared term ω_i^{-2}. The effect of higher mode response is included by the operation represented in Eq. (1.3.15).

The convergence rate of the convolution integral can be achieved by a different approach. Carry out integration by parts four times on the integral Eq. (1.3.12), arriving at

$$\{\mathbf{q}_c\} = \{\mathbf{q}_e\} + \{\mathbf{q}_f\} + \{\mathbf{q}_r\} \tag{1.3.19}$$

where

$$\begin{aligned}\{\mathbf{q}_e\} = &-[\mathbf{\Phi}][\rho_i^{-2}e^{-\mu_i t}(\cos\lambda_i t + \mu_i\lambda_i^{-1}\sin\lambda_i t)][\mathbf{\Phi}]^{\mathrm{T}}\{\mathbf{Q}_0\}\\ &+ [\mathbf{\Phi}][\rho_i^{-4}e^{-\mu_i t}\{2\mu_i\cos\lambda_i t + (\mu_i^2 - \lambda_i^2)\lambda_i^{-1}\sin\lambda_i t)][\mathbf{\Phi}]^{\mathrm{T}}\{\dot{\mathbf{Q}}_0\}\\ &-[\mathbf{\Phi}][\rho_i^{-6}e^{-\mu_i t}\{(3\mu_i^2 - \lambda_i^2)\cos\lambda_i t + \mu_i(\mu_i^2 - 3\lambda_i^2)\lambda_i^{-1}\sin\lambda_i t][\mathbf{\Phi}]^{\mathrm{T}}\{\ddot{\mathbf{Q}}_0\}\\ &+[\mathbf{\Phi}][\rho_i^{-8}e^{-\mu_i t}\{4\mu_i(\mu_i^2-\lambda_i^2)\cos\lambda_i t+(\mu_i^4-6\mu_i^2\lambda_i^2+\lambda_i^4)\lambda_i^{-1}\sin\lambda_i t\}][\mathbf{\Phi}]^{\mathrm{T}}\{\dddot{\mathbf{Q}}_0\}\end{aligned} \tag{1.3.20}$$

represents the exponentially decaying oscillating parts,

$$\begin{aligned}\{\mathbf{q}_f\} = &[\mathbf{\Phi}][\rho_i^{-2}][\mathbf{\Phi}]^{\mathrm{T}}\{\mathbf{Q}\} - 2[\mathbf{\Phi}][\rho_i^{-4}\mu_i][\mathbf{\Phi}]^{\mathrm{T}}\{\dot{\mathbf{Q}}\}\\ &+ [\mathbf{\Phi}][\rho_i^{-6}(3\mu_i - \lambda_i^2)][\mathbf{\Phi}]^{\mathrm{T}}\{\ddot{\mathbf{Q}}\} - 4[\mathbf{\Phi}][\rho_i^{-8}\mu_i(\mu_i^2 - \lambda_i^2)][\mathbf{\Phi}]^{\mathrm{T}}\{\dddot{\mathbf{Q}}\}\end{aligned} \tag{1.3.21}$$

represents the force function parts, and

$$\begin{aligned}\{\mathbf{q}_r\} = [\mathbf{\Phi}][\rho_i^{-8}]\int_0^t &[e^{-\mu_i(t-\tau)}4\mu_i(3\mu_i^2 - \lambda_i^2)\cos\lambda_i(t-\tau)\\ &\times \lambda_i^{-1}(\mu_i^4 - 6\mu_i^2\lambda_i^2 + \lambda_i^4)\sin\lambda_i(t-\tau)][\mathbf{\Phi}]^{\mathrm{T}}\{\mathbf{Q}^{\mathrm{iv}}\}\,d\tau\end{aligned} \tag{1.3.22}$$

represents the remaining convolution integral. As before, dots denote derivatives with respect to t, $\{\mathbf{Q}_0\} = \{\mathbf{Q}(0)\}$, $\{\dot{\mathbf{Q}}_0\} = \{\dot{\mathbf{Q}}(0)\}$, etc., are the initial values of the forcing functions and their time derivatives, and

$$\rho_i^2 = \lambda_i^2 + \mu_i^2 = (1 + 2\beta\gamma)\omega_i^2 \tag{1.3.23}$$

Although the formulae become more complicated, the computations involved are quite simple. The reward is that the convergence rate is greatly improved. In Eq. (1.3.20), as ρ_i^2 is proportional to ω_i^2, $\{\mathbf{q}_e\}$ converges as fast as ω_i^{-2}. However, $\{\mathbf{q}_e\}$ decays exponentially and is not affected by the forcing functions except at the beginning. $\{\mathbf{q}_e\}$ becomes unimportant as time goes on and accurate computation is needed only at the beginning.

In Eq. (1.3.21), all the terms have closed form representation:

$$\begin{aligned}
[\mathbf{A}_1] &= [\mathbf{\Phi}][\rho_i^{-2}][\mathbf{\Phi}]^{\mathrm{T}} = (1 + 2\beta\gamma)^{-1}[\mathbf{\Phi}][\omega_i^{-2}][\mathbf{\Phi}]^{\mathrm{T}} = (1 + 2\beta\gamma)^{-1}[\mathbf{K}]^{-1} \\
[\mathbf{A}_2] &= [\mathbf{\Phi}][2\mu_i\rho_i^{-4}][\mathbf{\Phi}]^{\mathrm{T}} = 2(1 + 2\beta\gamma)^{-2}[\beta\mathbf{K}^{-1}\mathbf{M}\mathbf{K}^{-1} + \gamma\mathbf{K}^{-1}] \\
[\mathbf{A}_3] &= [\mathbf{\Phi}][(3\mu_i^2 - \lambda_i^2)\rho_i^{-6}][\mathbf{\Phi}]^{\mathrm{T}} \\
&= (1 + 2\beta\gamma)^{-3}[4\beta^2(\mathbf{K}^{-1}\mathbf{M})^2\mathbf{K}^{-1} + (6\beta\gamma - 1)^{-1}\mathbf{M}\mathbf{K}^{-1} + 4\gamma^2\mathbf{K}^{-1}] \\
[\mathbf{A}_4] &= [\mathbf{\Phi}][(4\mu_i^3 - 4\mu_i\lambda_i^2)\rho_i^{-8}][\mathbf{\Phi}]^{\mathrm{T}} = 4(1 + 2\beta\gamma)^{-4}[4\beta^3(\mathbf{K}^{-1}\mathbf{M})^3\mathbf{K}^{-1} \\
&\quad + (4\beta\gamma - 1)\beta(\mathbf{K}^{-1}\mathbf{M})^2\mathbf{K}^{-1} + (4\beta\gamma - 1)\gamma\mathbf{K}^{-1}\mathbf{M}\mathbf{K}^{-1} + 2\gamma^3\mathbf{K}^{-1}]
\end{aligned} \tag{1.3.24}$$

These are independent of time and can be calculated once and for all, therefore

$$\{\mathbf{q}_f\} = [\mathbf{A}_1]\{\mathbf{Q}\} - [\mathbf{A}_2]\{\dot{\mathbf{Q}}\} + [\mathbf{A}_3]\{\ddot{\mathbf{Q}}\} - [\mathbf{A}_4]\{\dddot{\mathbf{Q}}\} \tag{1.3.25}$$

Finally, $\{\mathbf{q}_r\}$ is computed by numerical integration. The expression, however, contains ω_i^{-5} so the convergence rate is extremely fast, and one or two terms will give very accurate results.

Example 1.3.1

Consider a simply supported beam with Young's modulus E, mass density ρ, cross-sectional area A, second moment of area I, and length l. The natural frequencies are $\omega_n^2 = (n\pi)^4 EI/\rho A l^4$. If the active coordinate is at midspan, the normal modes are given by

$$\Phi_n = \sqrt{\frac{2}{\rho A l}}\sin\left(\frac{n\pi}{2}\right)$$

The condensed stiffness and mass are, respectively,

$$K_0 = \frac{48\,EI}{l^2}, \qquad M_0 = \frac{17\,\rho A l}{35}$$

and it can be checked that

$$K_0^{-1} = \sum_{\text{odd}} \phi_n^2\omega_n^{-2} = \frac{2}{\rho A l}\,\frac{\rho A l^4}{EI}\,\frac{1}{\pi^4}\sum_{\text{odd}} n^{-4} = \frac{l^3}{48\,EI}$$

$$K_0^{-1}M_0K_0^{-1} = \sum_{\text{odd}} \phi_n^2\omega_n^{-4} = \frac{17\,\rho A l^7}{80\,640(EI)^2}$$

If the forcing functions are $f_1(t) = tN$ and $f_2(t) = t^3N$ acting at the midspan, and

Table 1.3.1. Initial conditions for beam

	$f_1 = t$	$f_2 = t^3$
q_0	0	0
$\dot{q}_0$	0.02083	0

Table 1.3.2. Linear response for beam (0.1% errors in brackets; $\beta = 0.100$; unit = 0.01 mm)

Time (s)	Fast method 1 mode	Conventional method					
		1 mode	3 modes	5 modes	7 modes	9 modes	Exact
0.10	35(4)	32(83)	34(14)	34(4)	34(1)	34(0)	34
0.20	223(1)	217(28)	223(4)	223(1)	223(0)	223(0)	224
0.30	580(0)	571(15)	579(2)	580(0)	580(0)	580(0)	580
0.40	971(0)	958(12)	969(1)	970(0)	970(0)	971(0)	971
0.50	1231(0)	1216(12)	1229(1)	1230(0)	1231(0)	1231(0)	1231
0.60	1319(0)	1301(13)	1316(2)	1318(0)	1318(0)	1319(0)	1319
0.70	1344(0)	1323(15)	1341(2)	1343(0)	1344(0)	1344(0)	1344
0.80	1470(0)	1446(16)	1466(2)	1469(0)	1470(0)	1470(0)	1470
0.90	1769(0)	1742(15)	1765(2)	1768(0)	1769(0)	1769(0)	1770
1.00	2157(0)	2126(13)	2152(2)	2155(0)	2156(0)	2156(0)	2156
1.10	2471(0)	2438(13)	2466(2)	2470(0)	2471(0)	2471(0)	2472
1.20	2621(0)	2585(13)	2615(2)	2619(0)	2620(0)	2620(0)	2621
1.30	2660(0)	2621(14)	2654(2)	2658(0)	2659(0)	2660(0)	2660
1.40	2742(0)	2700(15)	2736(2)	2740(0)	2741(0)	2742(0)	2742
1.50	2978(0)	2933(15)	2970(2)	2975(0)	2977(0)	2977(0)	2978
1.60	3340(0)	3292(14)	3333(2)	3338(0)	3339(0)	3340(0)	3340
1.70	3690(0)	3638(13)	3681(2)	3687(0)	3688(0)	3689(0)	3690
1.80	3901(0)	3847(13)	3892(2)	3898(0)	3900(0)	3900(0)	3901
1.90	3974(0)	3917(14)	3966(2)	3972(0)	3973(0)	3974(0)	3975
2.00	4032(0)	3972(14)	4023(2)	4029(0)	4031(0)	4032(0)	4032
2.10	4209(0)	4146(15)	4199(2)	4206(0)	4208(0)	4208(0)	4209
2.20	4530(0)	4464(14)	4519(2)	4526(0)	4528(0)	4529(0)	4530
2.30	4892(0)	4823(14)	4881(2)	4889(0)	4891(0)	4892(0)	4892
2.40	5159(0)	5086(14)	5147(2)	5155(0)	5157(0)	5158(0)	5159
2.50	5279(0)	5204(14)	5267(2)	5275(0)	5277(0)	5278(0)	5279
2.60	5335(0)	5256(14)	5322(2)	5331(0)	5333(0)	5334(0)	5335
2.70	5462(0)	5381(14)	5449(2)	5458(0)	5460(0)	5461(0)	5462
2.80	5733(0)	5648(14)	5719(2)	5728(0)	5731(0)	5732(0)	5733
2.90	6087(0)	6000(14)	6073(2)	6083(0)	6085(0)	6086(0)	6087
3.00	6395(0)	6304(14)	6380(2)	6390(0)	6393(0)	6394(0)	6395
3.10	6568(0)	6475(14)	6553(2)	6563(0)	6566(0)	6567(0)	6568
3.20	6641(0)	6544(14)	6625(2)	6636(0)	6638(0)	6639(0)	6640
3.30	6735(0)	6636(14)	6720(2)	6731(0)	6733(0)	6734(0)	6737
3.40	6953(0)	6850(14)	6936(2)	6948(0)	6950(0)	6951(0)	6953
3.50	7282(0)	7177(14)	7265(2)	7277(0)	7280(0)	7281(0)	7282
3.60	7614(0)	7505(14)	7597(2)	7609(0)	7612(0)	7613(0)	7614
3.70	7838(0)	7727(14)	7820(2)	7832(0)	7836(0)	7837(0)	7838
3.80	7943(0)	7829(14)	7925(2)	7938(0)	7941(0)	7942(0)	7943
3.90	8023(0)	7905(14)	8004(2)	8017(0)	8020(0)	8021(0)	8023
4.00	8192(0)	8072(14)	8173(2)	8186(0)	8190(0)	8191(0)	8192

$EI = \rho A = l = 1$, the response with initial conditions listed in Table 1.3.1 are given in Tables 1.3.2 and 1.3.3 for f_1 and f_2. The damping constants are $\gamma = 0$ and $\beta = 0.1$, 0.001, respectively. The results are listed in numerical form rather than graphical as the graphical presentations cannot show the differences. The convergent rates are illustrated and the percentage errors are included for comparison purpose. (It

Table 1.3.3. Cubic response for beam (0.1% errors in brackets; $\beta = 0.001$; unit = 0.1 mm)

Time (s)	Fast method 1 mode	Conventional method					
		1 mode	3 modes	5 modes	7 modes	9 modes	Exact
0.10	0(0)	0([a])	0([a])	0(93)	0(87)	0(84)	0
0.20	0(0)	0(78)	0(15)	0(7)	0(4)	0(4)	0
0.30	2(0)	2(39)	2(6)	2(2)	2(1)	2(0)	2
0.40	7(0)	7(26)	7(4)	7(1)	7(0)	7(0)	7
0.50	18(0)	18(20)	18(3)	18(1)	18(0)	18(0)	18
0.60	37(0)	36(17)	37(2)	37(0)	37(0)	37(0)	37
0.70	63(0)	62(16)	63(2)	63(0)	63(0)	63(0)	63
0.80	98(0)	96(15)	98(2)	98(0)	98(0)	98(0)	98
0.90	141(0)	139(15)	141(2)	141(0)	141(0)	141(0)	141
1.00	195(0)	192(15)	195(2)	195(0)	195(0)	195(0)	195
1.10	262(0)	258(15)	261(2)	262(0)	262(0)	262(0)	262
1.20	344(0)	339(15)	343(2)	344(0)	344(0)	344(0)	344
1.30	442(0)	435(14)	441(2)	441(0)	441(0)	441(0)	442
1.40	555(0)	547(14)	554(2)	555(0)	555(0)	555(0)	555
1.50	685(0)	675(14)	684(2)	685(0)	685(0)	685(0)	685
1.60	833(0)	821(14)	831(2)	832(0)	833(0)	833(0)	833
1.70	1001(0)	986(14)	999(2)	1000(0)	1001(0)	1001(0)	1001
1.80	1191(0)	1173(14)	1188(2)	1190(0)	1191(0)	1191(0)	1191
1.90	1405(0)	1384(14)	1401(2)	1404(0)	1404(0)	1405(0)	1405
2.00	1642(0)	1618(14)	1638(2)	1641(0)	1642(0)	1642(0)	1642
2.10	1904(0)	1876(14)	1900(2)	1903(0)	1903(0)	1904(0)	1904
2.20	2191(0)	2159(14)	2186(2)	2189(0)	2190(0)	2190(0)	2191
2.30	2505(0)	2468(14)	2499(2)	2503(0)	2504(0)	2504(0)	2505
2.40	2848(0)	2807(14)	2842(2)	2846(0)	2847(0)	2848(0)	2848
2.50	3223(0)	3176(14)	3215(2)	3221(0)	3222(0)	3222(0)	3223
2.60	3629(0)	3576(14)	3621(2)	3627(0)	3628(0)	3629(0)	3629
2.70	4068(0)	4008(14)	4058(2)	4065(0)	4066(0)	4067(0)	4068
2.80	4539(0)	4472(14)	4528(2)	4535(0)	4537(0)	4538(0)	4539
2.90	5044(0)	4970(14)	5032(2)	5040(0)	5042(0)	5043(0)	5044
3.00	5586(0)	5504(14)	5573(2)	5582(0)	5584(0)	5585(0)	5586
3.10	6166(0)	6076(14)	6152(2)	6162(0)	6164(0)	6165(0)	6166
3.20	6786(0)	6688(14)	6771(2)	6781(0)	6784(0)	6785(0)	6786
3.30	7446(0)	7338(14)	7429(2)	7441(0)	7444(0)	7445(0)	7446
3.40	8146(0)	8028(14)	8127(2)	8140(0)	8144(0)	8145(0)	8146
3.50	8888(0)	8759(14)	8867(2)	8881(0)	8885(0)	8886(0)	8888
3.60	9673(0)	9533(14)	9651(2)	9666(0)	9670(0)	9672(0)	9673
3.70	10505(0)	10352(14)	10480(2)	10497(0)	10501(0)	10503(0)	10505
3.80	11383(0)	11218(14)	11357(2)	11375(0)	11380(0)	11381(0)	11383
3.90	12310(0)	12131(14)	12281(2)	12301(0)	12306(0)	12308(0)	12310
4.00	13284(0)	12091(14)	13253(2)	13274(0)	13280(0)	13282(0)	13284

[a] Overflows.

Table 1.3.4. Initial conditions for plate

	$f_1 = t$	$f_2 = t^3$
q_0	0	0
$\dot{q}_0$	0.01160	0

Table 1.3.5. Linear response for plate (0.1% errors in brackets; $\beta = 0.001$; unit = 0.001 mm)

Times (s)	Fast method 1 mode	Conventional method					
		1 mode	5 modes	11 modes	19 modes	29 modes	Exact
0.05	146(51)	79([a])	129([a])	140(87)	146(52)	149(35)	154
0.10	682(5)	548([a])	634(75)	659(38)	669(23)	675(15)	686
0.15	1647(4)	1446([a])	1563(46)	1599(24)	1615(14)	1623(9)	1639
0.20	2695(2)	2428([a])	2586(38)	2636(19)	2657(12)	2668(8)	2689
0.25	3407(0)	3074([a])	3283(37)	3344(19)	3370(11)	3383(7)	3410
0.30	3664(2)	3264([a])	3518(41)	3592(21)	3623(13)	3639(8)	3672
0.35	3756(0)	3289([a])	3576(47)	3663(24)	3699(15)	3718(10)	3756
0.40	4121(2)	3587([a])	3907(49)	4006(25)	4047(15)	4068(10)	4111
0.45	4952(0)	4351([a])	4717(46)	4830(24)	4876(14)	4900(9)	4949
0.50	6024(0)	5357([a])	5774(42)	5897(22)	5949(13)	5976(8)	6030
0.55	6895(0)	6161([a])	6619(40)	6756(21)	6813(12)	6842(8)	6901
0.60	7304(0)	6503([a])	6992(42)	7142(21)	7204(13)	7236(8)	7301
0.65	7405(1)	6537([a])	7063(45)	7224(23)	7292(14)	7326(9)	7397
0.70	7627(0)	6693([a])	7268(47)	7443(24)	7515(14)	7552(9)	7628
0.75	8292(0)	7291([a])	7915(46)	8101(23)	8179(14)	8219(9)	8300
0.80	9324(0)	8256([a])	8916(44)	9116(22)	9199(13)	9242(9)	9328
0.85	10316(0)	9182([a])	9873(42)	10085(21)	10173(13)	10218(8)	10310
0.90	10900(0)	9699([a])	10432(42)	10655(21)	10749(13)	10796(8)	10894
0.95	11071(0)	9803([a])	10587(44)	10824(22)	10923(13)	10973(9)	11076
1.00	11197(0)	9862([a])	10692(45)	10940(23)	11043(14)	11096(9)	11205
1.05	11685(0)	10284([a])	11146(46)	11407(23)	11516(14)	11572(9)	11686
1.10	12618(0)	11150([a])	12045(44)	12318(23)	12432(14)	12491(9)	12610
1.15	13679(0)	12144([a])	13085(43)	13371(22)	13490(13)	13551(9)	13676
1.20	14436(0)	12835([a])	13827(42)	14125(22)	14249(13)	14313(8)	14443
1.25	14731(0)	13062([a])	14095(43)	14405(22)	14535(13)	14601(9)	14737
1.30	14819(0)	13084([a])	14148(45)	14471(23)	14606(14)	14675(9)	14815
1.35	15142(0)	13340([a])	14441(45)	14775(23)	14915(14)	14987(9)	15133
1.40	15931(0)	14062([a])	15213(45)	15560(23)	15706(14)	15780(9)	15932
1.45	16998(0)	15062([a])	16261(43)	16621(22)	16772(13)	16849(9)	17006
1.50	17906(0)	15904([a])	17139(42)	17510(22)	17666(13)	17746(9)	17908
1.55	18361(0)	16292([a])	17558(43)	17943(22)	18104(13)	18187(9)	18355
1.60	18475(0)	16339([a])	17647(44)	18043(22)	18210(13)	18295(9)	18468
1.65	18667(0)	16464([a])	17823(45)	18233(23)	18404(14)	18492(9)	18671
1.70	19284(0)	17015([a])	18419(45)	18841(23)	19017(14)	19108(9)	19292
1.75	20294(0)	17958([a])	19395(44)	19829(22)	20011(13)	20104(9)	20293
1.80	21310(0)	18908([a])	20378(43)	20825(22)	21012(13)	21108(9)	21301
1.85	21942(0)	19472([a])	20989(43)	21447(22)	21640(13)	21738(9)	21938
1.90	22142(0)	19606([a])	21173(44)	21646(22)	21843(13)	21944(9)	22150
1.95	22252(0)	19649([a])	21257(44)	21741(23)	21943(14)	22046(9)	22258
2.00	22694(0)	20024([a])	21662(45)	22160(23)	22368(14)	22474(9)	22690

[a] Overflows.

should be noted that, in practice, Timoshenko beam theory must be employed for higher beam modes.)

Example 1.3.2

A simply supported square plate with flexural rigidity D, mass density ρ, thickness h and side a is studied. The natural frequencies are given by

Table 1.3.6. Cubic response for plate (0.1% errors in brackets; $\beta = 0.001$; unit = 0.01 mm)

Time (s)	Fast method 1 mode	Conventional method					
		1 mode	5 modes	11 modes	19 modes	29 modes	Exact
0.05	0(25)	0([a])	0([a])	0([a])	0([a])	0(84)	0
0.10	0(0)	0([a])	0([a])	0(86)	0(52)	0(35)	0
0.15	2(0)	1([a])	1([a])	2(53)	2(32)	2(21)	2
0.20	6(0)	4([a])	5(74)	5(38)	5(23)	5(15)	6
0.25	13(0)	11([a])	13(59)	13(30)	13(18)	13(12)	13
0.30	26(0)	23([a])	25(52)	26(27)	26(16)	26(11)	26
0.35	45(0)	39([a])	42(49)	44(25)	44(15)	44(10)	45
0.40	69(0)	60([a])	65(47)	67(24)	68(15)	68(10)	69
0.45	99(0)	87([a])	94(47)	97(24)	98(14)	98(9)	99
0.50	137(0)	120([a])	130(46)	133(24)	135(14)	135(9)	137
0.55	183(0)	161([a])	175(46)	179(24)	181(14)	182(9)	183
0.60	241(0)	212([a])	229(46)	235(23)	237(14)	238(9)	241
0.65	308(0)	272([a])	294(45)	301(23)	304(14)	305(9)	308
0.70	388(0)	342([a])	370(45)	378(23)	382(14)	384(9)	388
0.75	478(0)	422([a])	456(45)	467(23)	471(14)	474(9)	478
0.80	581(0)	513([a])	555(45)	568(23)	573(14)	576(9)	581
0.85	698(0)	616([a])	667(45)	682(23)	688(14)	692(9)	698
0.90	831(0)	733([a])	793(45)	811(23)	819(14)	823(9)	831
0.95	979(0)	865([a])	935(44)	957(23)	966(14)	970(9)	979
1.00	1145(0)	1011([a])	1093(44)	1118(23)	1129(14)	1134(9)	1145
1.05	1327(0)	1173([a])	1268(44)	1296(23)	1308(14)	1314(9)	1327
1.10	1527(0)	1349([a])	1458(44)	1491(23)	1505(14)	1512(9)	1527
1.15	1746(0)	1543([a])	1667(44)	1705(23)	1721(14)	1729(9)	1746
1.20	1985(0)	1754([a])	1896(44)	1939(23)	1957(14)	1966(9)	1985
1.25	2246(0)	1985([a])	2145(44)	2194(23)	2214(14)	2224(9)	2246
1.30	2528(0)	2235([a])	2416(44)	2470(23)	2493(14)	2505(9)	2528
1.35	2834(0)	2505([a])	2707(44)	2768(23)	2794(14)	2807(9)	2834
1.40	3161(0)	2795([a])	3021(44)	3089(23)	3117(14)	3132(9)	3161
1.45	3513(0)	3106([a])	3357(44)	3432(23)	3464(14)	3480(9)	3513
1.50	3891(0)	3440([a])	3717(44)	3801(23)	3836(13)	3854(9)	3891
1.55	4295(0)	3798([a])	4104(44)	4196(23)	4235(13)	4254(9)	4295
1.60	4726(0)	4180([a])	4516(44)	4618(22)	4660(13)	4682(9)	4726
1.65	5186(0)	4586([a])	4955(44)	5067(22)	5113(13)	5137(9)	5186
1.70	5673(0)	5017([a])	5421(44)	5543(22)	5594(13)	5620(9)	5673
1.75	6189(0)	5474([a])	5914(44)	6047(22)	6103(13)	6131(9)	6189
1.80	6736(0)	5958([a])	6437(44)	6582(22)	6642(13)	6673(9)	6736
1.85	7315(0)	6470([a])	6990(44)	7147(22)	7213(13)	7246(9)	7315
1.90	7927(0)	7011([a])	7574(44)	7745(22)	7816(13)	7852(9)	7927
1.95	8571(0)	7582([a])	8191(44)	8375(22)	8452(13)	8491(9)	8571
2.00	9250(0)	8182([a])	8839(44)	9037(22)	9121(13)	9163(9)	9250

[a] Overflows.

$$\omega_{mn}^2 = D\pi^4 \frac{(m^2 + n^2)^2}{\rho h a^4}$$

and the normal modes at the centre point of the plate are

$$\phi_{mn} = \frac{2\sin(m\pi/2)\sin(n\pi/2)}{\sqrt{\rho a^2 h}}$$

The condensed stiffness and mass from a finite element analysis are found to be

$$K_0 = \frac{86.201\,19D}{a^2} \quad \text{and} \quad M_0 = 0.196\,457\,31\,\rho h a^2$$

If the forcing functions are $f_1(t) = tN$ and $f_2(t) = t^3N$, respectively, and if $D = \rho h = a = 1$, the response with initial conditions listed in Table 1.3.4 are computed and given in Table 1.3.5 for f_1 and in Table 1.3.6 for f_2. The damping constants are taken as $\beta = 0.001$, 0.01 and $\gamma = 0$. It is seen from the results that the classical modal analysis converges very slowly. However, the present method gives nearly exact results using just one natural mode.

References

1. MD Trifunac 1971. Response envelope spectrum and interpretation of strong earthquake ground motion. Bull Seismic Soc Am 61(2)
2. EL Wilson, MW Yuan, JM Dickens 1982. Dynamic analysis by direction superposition of Ritz vectors. Earthquake Engng Struct Dyn 10, 813–821
3. B Nour-Omid, RW Clough 1985. Dynamics analysis of structures using Lanczos coordinates. Earthquake Engng Struct Dyn 12, 565–577
4. B Nour-Omid, RW Clough 1985. Block Lanczos method for dynamic analysis of structures. Earthquake Engng Struct Dyn 13, 271–275

Chapter 2

Finite Elements and Continuum Elements

Since our method is very similar to the finite element method, it is worth giving a brief introduction to these methods as applied to elastic systems. If one takes the exact solution of the governing equations for the vibrating member as shape functions, a continuum element results. After deriving the element matrices for straight beams and plates, we prove that the mass matrix for a continuum element can be obtained simply by differentiating the dynamic stiffness matrix with respect to the frequency. We then relate finite elements and continuum elements by means of Simpson's hypothesis. Therefore, many useful frequency algorithms for the finite element method are made available to the continuum element models.

2.1. Formulation

Any of the energy principles may be used as a basis for numerical analysis by the finite element method. The finite element discretization implies a division of the total volume *vol* into subvolumes or subdomains denoting finite elements. The functions chosen to represent approximate displacement and stress fields are specified within each element, and conditions imposed on certain parameters at interelement boundaries provide the necessary continuity requirement of field functions.

In the case of the standard displacement method, the displacement field is assumed to be

$$\{\mathbf{u}(x, y, z)\} = [\boldsymbol{\phi}(x, y, z)]\{\boldsymbol{\alpha}\}$$

where $[\boldsymbol{\phi}(x, y, z)]$ is the vector of chosen modes of displacement and $\{\boldsymbol{\alpha}\}$ is a vector of constants to be determined by nodal displacements. At any node i, the vector of displacement components is given by

$$\{\mathbf{q}_i\} = \{\mathbf{u}(x_i, y_i, z_i)\} = [\boldsymbol{\phi}(x_i, y_i, z_i)]\{\boldsymbol{\alpha}\}$$

where (x_i, y_i, z_i) are the coordinates of the node. If all displacement components of the n nodes of the element are arranged in a vector $\{\mathbf{q}\}$, then

$$\{\mathbf{q}\} = [\boldsymbol{\Phi}]\{\boldsymbol{\alpha}\}$$

where the constant matrix $[\boldsymbol{\Phi}]$ is given by

$$[\boldsymbol{\Phi}] = \begin{bmatrix} \boldsymbol{\phi}(x_1, y_1, z_1) \\ \boldsymbol{\phi}(x_2, y_2, z_2) \\ \vdots \\ \boldsymbol{\phi}(x_n, y_n, z_n) \end{bmatrix}$$

and the displacement field is expressed in terms of nodal displacements,

$$\{\mathbf{u}(x, y, z)\} = [\boldsymbol{\phi}(x, y, z)][\boldsymbol{\Phi}]^{-1}\{\mathbf{q}\} = [\mathbf{N}(x, y, z]\{\mathbf{q}\} \tag{2.1.1}$$

where $[\mathbf{N}(x, y, z)] = [\boldsymbol{\phi}(x, y, z)][\boldsymbol{\Phi}]^{-1}$. The strain field is obtained from the kinematic relation as

$$\{\boldsymbol{\varepsilon}(x, y, z)\} = [\mathbf{B}(x, y, z)]\{\mathbf{q}\} \tag{2.1.2}$$

For vibration analysis, if the external force can be expressed by a potential V, the most convenient energy principle is Hamilton's principle, which states that among all admissible displacements which satisfy the prescribed geometrical constraints and the prescribed condition at the limits $t = t_1$ and $t = t_2$, the functional

$$\int_{t_1}^{t_2} \left[T - U - \int_{vol} W \, \mathrm{d}\,vol \right] \mathrm{d}t \tag{2.1.3}$$

is stationary.

Now the kinetic energy and strain energy are given by

$$T = \frac{1}{2} \int_{vol} \{\dot{\mathbf{u}}\}^{\mathrm{T}} [\boldsymbol{\rho}] \{\dot{\mathbf{u}}\} \, \mathrm{d}\,vol$$

and

$$U = \frac{1}{2} \int_{vol} \{\boldsymbol{\varepsilon}\}^{\mathrm{T}} [\mathbf{C}] \{\boldsymbol{\varepsilon}\} \, \mathrm{d}\,vol \tag{2.1.4}$$

where $[\boldsymbol{\rho}]$ is the inertia matrix and $[\mathbf{C}]$ is the matrix of elastic constants. From Eqs (2.1.1) and (2.1.2), we have

$$\delta(\tfrac{1}{2}\{\dot{\mathbf{q}}\}^{\mathrm{T}}[\mathbf{M}]\{\dot{\mathbf{q}}\} + \tfrac{1}{2}\{\mathbf{q}\}^{\mathrm{T}}[\mathbf{K}]\{\mathbf{q}\} - \{\mathbf{q}\}^{\mathrm{T}}\{\mathbf{Q}\}) = 0 \tag{2.1.5}$$

where the mass matrix $[\mathbf{M}]$ and stiffness matrix $[\mathbf{K}]$ are given by

$$[\mathbf{M}] = \int [\mathbf{N}]^{\mathrm{T}} [\boldsymbol{\rho}] [\mathbf{N}] \, \mathrm{d}\,vol$$

and

$$[\mathbf{K}] = \int [\mathbf{B}]^{\mathrm{T}} [\mathbf{C}] [\mathbf{B}] \, \mathrm{d}\,vol \tag{2.1.6}$$

respectively, and $\{\mathbf{Q}\}$ is the load vector resulting from the volume integral in expression (2.1.3).

Since the kinetic energy of the system is the summation of kinetic energies associated with the individual elements,

$$T = \frac{1}{2} \sum_{\substack{\text{all}\\ \text{elements}}} \int_{\substack{\text{vol of}\\ \text{element}}} \{\dot{\mathbf{u}}_e\}^T [\boldsymbol{\rho}_e] \{\dot{\mathbf{u}}_e\} \, d\,vol \tag{2.1.7a}$$

and similarly

$$U = \frac{1}{2} \sum_{\substack{\text{all}\\ \text{elements}}} \int_{\substack{\text{vol of}\\ \text{element}}} \{\boldsymbol{\varepsilon}_e\}^T [\mathbf{C}_e] \{\boldsymbol{\varepsilon}_e\} \, d\,vol \tag{2.1.7b}$$

where the subscript e denotes quantities referred to the individual elements. Applying the requirement of stationary energy (2.1.3) and with reference to Eqs (2.1.1) and (2.1.2), we have

$$\delta\left(\sum_{\substack{\text{all}\\ \text{elements}}} \frac{1}{2}\{\dot{\mathbf{q}}_e\}^T[\mathbf{M}_e]\{\dot{\mathbf{q}}_e\} + \sum_{\substack{\text{all}\\ \text{elements}}} \frac{1}{2}\{\mathbf{q}_e\}^T[\mathbf{K}_e]\{\mathbf{q}_e\} - \sum_{\substack{\text{all}\\ \text{elements}}} \{\mathbf{q}_e\}^T\{\mathbf{Q}_e\} \right) = 0 \tag{2.1.8}$$

Now, if all the coordinate vectors $\{\mathbf{q}_e\}$ are transformed to a common coordinate vector base $\{\mathbf{q}\}$ by

$$\{\mathbf{q}_e\} = [\mathbf{n}_e]\{\mathbf{q}\}$$

then we have

$$\delta\left[\frac{1}{2}\{\dot{\mathbf{q}}\}^T\left(\sum_{\substack{\text{all}\\ \text{elements}}} [\overline{\mathbf{M}}_e] \right)\{\dot{\mathbf{q}}\} + \frac{1}{2}\{\mathbf{q}\}^T\left(\sum_{\substack{\text{all}\\ \text{elements}}} [\overline{\mathbf{K}}_e] \right)\{\mathbf{q}\} - \{\mathbf{q}\}^T\left(\sum_{\substack{\text{all}\\ \text{elements}}} \{\overline{\mathbf{Q}}_e\} \right)\right] = 0 \tag{2.1.9}$$

where

$$\begin{aligned} [\overline{\mathbf{M}}_e] &= [\mathbf{n}_e]^T[\mathbf{M}_e][\mathbf{n}_e] \\ [\overline{\mathbf{K}}_e] &= [\mathbf{n}_e]^T[\mathbf{K}_e][\mathbf{n}_e] \\ \{\overline{\mathbf{Q}}_e\} &= [\mathbf{n}_e]^T\{\mathbf{Q}_e\} \end{aligned} \tag{2.1.10}$$

Comparing Eqs (2.1.5) and (2.1.9), we have, for the system,

$$[\mathbf{M}] = \sum_{\substack{\text{all}\\ \text{elements}}} [\overline{\mathbf{M}}_e]$$

$$[\mathbf{K}] = \sum_{\substack{\text{all}\\ \text{elements}}} [\overline{\mathbf{K}}_e]$$

$$\{\mathbf{Q}\} = \sum_{\substack{\text{all}\\ \text{elements}}} \{\overline{\mathbf{Q}}_e\}$$

Equations (2.1.10) are used to assemble the system equations of motion. If we perform the variation of Eq. (2.1.5), we have

$$[\mathbf{M}]\{\ddot{\mathbf{q}}\} + [\mathbf{K}]\{\mathbf{q}\} = \{\mathbf{Q}\} \tag{2.1.11}$$

which is the governing equation of motion in matrix form.

2.2. Bar Elements

2.2.1. Stiffness Matrix

Consider a bar element with several nodes and one degree of freedom at each node (Fig. 2.2.1). In dynamic analysis, the shape functions for each bar can be either frequency independent or frequency dependent. The first type is the same as that for static analysis, and an obvious choice involves Lagrange polynomials of the first degree, i.e.

$$N_1 = 1 - \xi, \qquad N_2 = \xi \tag{2.2.1}$$

where $\xi = x/l$. These shape functions satisfy the governing differential equation for a bar without distributed loads, and the associated natural element will yield exact results in static analysis. For the dynamic case, the above statement is no longer true and the frequencies produced by this natural element and its corresponding consistent mass matrix are only approximate.

The stiffness matrix of the bar element can be computed via Eqs (2.1.2) and (2.1.6),

$$[\mathbf{B}] = \frac{\partial[\mathbf{N}]}{\partial x} = \frac{1}{l}[-1 \quad 1]$$

$$[\mathbf{k}] = \int_{vol} [\mathbf{B}]^{\mathrm{T}}[\mathbf{D}][\mathbf{B}]\,\mathrm{d}\,vol = \int_0^l EA[\mathbf{B}]^{\mathrm{T}}[\mathbf{B}]\,\mathrm{d}x = \int_0^1 EAl[\mathbf{B}]^{\mathrm{T}}[\mathbf{B}]\,\mathrm{d}\xi$$

$$= \frac{EA}{l}\int_0^1 \begin{bmatrix} 1 & -1 \\ -1 & 1 \end{bmatrix} \mathrm{d}\xi = \frac{EA}{l}\begin{bmatrix} 1 & -1 \\ -1 & 1 \end{bmatrix} \tag{2.2.2}$$

2.2.2. Consistent Mass Matrix Formulation

Following Eq. (2.1.6), the consistent mass matrix based on the natural shape functions is

$$[\mathbf{m}] = \int_{vol} [\mathbf{N}]^{\mathrm{T}}[\boldsymbol{\rho}][\mathbf{N}]\,\mathrm{d}\,vol = \int_0^1 \rho Al \begin{bmatrix} \xi^2 & \xi(1-\xi) \\ \xi(1-\xi) & \xi^2 \end{bmatrix} \mathrm{d}\xi = \frac{\rho Al}{6} = \begin{bmatrix} 2 & 1 \\ 1 & 2 \end{bmatrix} \tag{2.2.3}$$

If only one element is used for the longitudinal vibration of a fixed-free bar, u_1 is equal to zero and just one degree of freedom u_2 remains in the frequency equation of $([\mathbf{k}] - \omega^2[\mathbf{m}])\{\mathbf{u}\} = \{\mathbf{0}\}$, with the result that only one natural frequency can be computed:

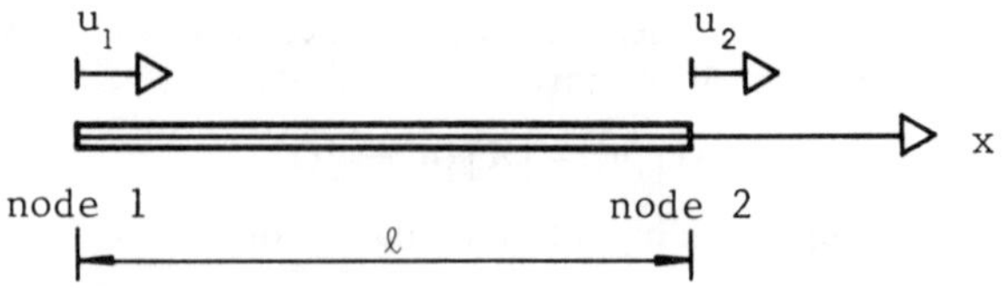

Fig. 2.2.1. A bar element

Table 2.2.1. Longitudinal vibration of a fixed–free bar – ω/ω_{exact} ratio

n	Frequency number									
	1	2	3	4	5	6	7	8	9	10
1	1.103									
2	1.026	1.195								
3	1.012	1.103	1.200							
4	1.006	1.058	1.154	1.191						
5	1.004	1.037	1.103	1.181	1.282					
6	1.003	1.026	1.072	1.137	1.195	1.273				
7	1.002	1.019	1.053	1.103	1.161	1.200	1.266			
8	1.002	1.015	1.041	1.079	1.128	1.177	1.201	1.259		
9	1.001	1.012	1.032	1.063	1.103	1.148	1.188	1.200	1.254	
10	1.001	1.009	1.026	1.051	1.084	1.123	1.163	1.195	1.198	1.250

$$\frac{EA}{l}u_2 - \omega^2\frac{\rho Al}{6}(2u_2) = 0$$

$$\omega^2 = \frac{3E}{\rho l^2} \quad \text{or} \quad \omega = 1.7321\sqrt{\frac{E}{\rho l^2}}$$

The computed natural frequency, when compared with the exact value of $\omega = 1.5708\sqrt{E/\rho l^2}$, is 10.3% higher. The accuracy can be improved by increasing the number of elements, and the results from such an exercise can be found in Table 2.2.1.

From Table 2.2.1, the following general conclusions are illustrated regarding natural vibration analysis using consistent mass matrices:

1. An n-degree-of-freedom mathematical model can only predict up to a maximum of n natural modes and frequencies.
2. The predicted natural frequencies are higher than the true ones, thus forming an upper bound solution.
3. The accuracy of the predicated frequencies tends to decrease with the increase in frequency number.

2.2.3. Continuous Mass Matrix Formulation

Alternatives to the natural shape functions are the frequency-dependent shape functions which are solutions of the governing equation for harmonic oscillation

$$EAu'' + \omega^2\rho Au = 0 \tag{2.2.4}$$

with boundary conditions $u(0) = u_1$, and $u(l) = u_2$. The solution can similarly be written as

$$u(x) = N_1u_1 + N_2u_2 \tag{2.2.5a}$$

but with

$$\begin{aligned} N_1 &= \cos\psi\xi - \cot\psi\sin\psi\xi \\ N_2 &= \csc\psi\sin\psi\xi \end{aligned} \tag{2.2.5b}$$

where

$$\psi^2 = \omega^2 \rho l^2/E$$

Once again, in accordance with Eqs (2.1.1) and (2.1.6), the mass matrix and stiffness matrix can be derived:

$$[\mathbf{m}] = \frac{\rho A l}{2\psi} = \begin{bmatrix} \psi \csc^2\psi - \cot\psi & \csc\psi - \psi\csc\psi\cot\psi \\ \csc\psi - \psi\csc\psi\cot\psi & \psi\csc^2\psi - \cot\psi \end{bmatrix} \tag{2.2.6a}$$

and

$$[\mathbf{k}] = \frac{EA\psi}{l} = \begin{bmatrix} \cot\psi & -\csc\psi \\ -\csc\psi & \cot\psi \end{bmatrix} + \omega^2[\mathbf{m}] \tag{2.2.6b}$$

Since the shape functions are mathematically exact solutions of the governing equation for the free vibration of a bar, the element matrices can be used to predict an infinite number of modes accurately with a minimum number of elements. As an example, the frequency equation of the previously mentioned fixed–free bar formed with only one element is

$$\cot\psi = 0$$

which gives an infinite number of solutions

$$\psi = (2m+1)\pi/2, \qquad m = 1, 2, \ldots, \infty$$

If $m = 1$, then $\psi = 1.5708$, which is the exact solution of the fundamental frequency.

Now, the amplitudes of the nodal forces for a vibrating bar are given by

$$f_1 = -EAu'(0) \quad \text{and} \quad f_2 = EAu'(l)$$

which together with the exact displacement-function solution given by Eq. (2.2.5), will produce the dynamic stiffness relation

$$\begin{Bmatrix} f_1 \\ f_2 \end{Bmatrix} = \frac{EA\psi}{l}\begin{bmatrix} \cot\psi & -\csc\psi \\ -\csc\psi & \cot\psi \end{bmatrix}\begin{Bmatrix} u_1 \\ u_2 \end{Bmatrix}$$

In other words, with very little effort, the dynamic stiffness matrix has been established as

$$[\mathbf{d}(\omega)] = \frac{EA\psi}{l}\begin{bmatrix} \cot\psi & -\csc\psi \\ -\csc\psi & \cot\psi \end{bmatrix}$$

It can be easily verified that

$$[\mathbf{d}(\omega)] = [\mathbf{k}(\omega)] - \omega^2[\mathbf{m}(\omega)]$$

2.2.4. Other Mass Matrix Models

The generation of the element mass matrix involves only integration of the shape functions, in contrast with the stiffness matrix where differentiations of the shape functions are necessary. If shape functions are approximate, then differentiation reduces the accuracy. Therefore, in general, the mass matrix is more accurate than the stiffness matrix, and this means that some simplifications on the formulation of the mass matrix will not affect the numerical accuracy unduly. One such approach, called the lumped mass model, produces a mass matrix with diagonal components only. This leads to a diagonal system mass matrix and hence a significant reduction in the overall solution cost.

There are two kinds of lumped mass models. One lumps the element mass in equal shares at all the nodes, while the other lumps the element mass in proportional shares to the diagonal components of the consistent mass matrix. The former is called the lumped mass method and the latter the diagonal mass method. In the case of longitudinal vibration of a uniform bar, both methods produce the same matrix

$$[\mathbf{m}_{\mathrm{l}}] = \rho A l \begin{bmatrix} 0.5 & 0 \\ 0 & 0.5 \end{bmatrix} \tag{2.2.7}$$

Due to the fact that the coupling of inertia terms is ignored, the lumped mass model generally predicts lower values for natural frequencies. Therefore, it is thought that the average of the consistent and the lumped mass model may give better results, and this gives rise to the so-called average mass matrix, which, again for the uniform bar case, is

$$[\mathbf{m}_{\mathrm{a}}] = \frac{\rho A l}{26} \begin{bmatrix} 9 & 4 \\ 4 & 9 \end{bmatrix} \tag{2.2.8}$$

By using 30 finite elements, the percentage errors in various natural frequencies computed using the three models are listed in Table 2.2.2 for a fixed–fixed bar. It can be concluded that the average mass model is indeed much more accurate after checking through the results. Leung [1] has derived explicitly the error ratio $\delta = \psi/\psi_{\mathrm{exact}}$ as follows:

Consistent
$$\delta_{\mathrm{c}} = \frac{\varepsilon^2}{12} + \frac{\varepsilon^4}{360} - \frac{17\varepsilon^6}{30\,240} + \cdots \tag{2.2.9a}$$

Lumped
$$\delta_{\mathrm{l}} = \frac{-\varepsilon^2}{12} + \frac{\varepsilon^4}{360} - \frac{\varepsilon^6}{20\,160} + \cdots \tag{2.2.9b}$$

Average
$$\delta_{\mathrm{a}} = \frac{-\varepsilon^4}{240} + \frac{\varepsilon^6}{6048} + \cdots \tag{2.2.9c}$$

where $\varepsilon = j\pi/n$, j is the mode number and n the number of elements involved. Because the continuous model is exact, Przemieniecki [2] attempted to approximate the shape functions by taking the first two terms of the expansion of the exact model's powers of the frequency squared. The resulting matrices are

Table 2.2.2. Percentage errors for frequencies for a fixed–fixed bar

Mode j	Consistent mass	Lumped mass	Average mass
1	0.092	−0.091	−0.000
2	0.366	−0.365	−0.001
3	0.825	−0.820	−0.004
4	1.471	−1.454	−0.013
5	2.305	−2.264	−0.032
6	3.331	−3.247	−0.066
10	9.427	−8.810	−0.521
15	21.585	−18.947	−2.732
20	37.956	−31.608	−8.811

$$[\mathbf{k}] = \frac{AE}{l}\begin{bmatrix} 1 & -1 \\ -1 & 1 \end{bmatrix} + \frac{\omega^4\rho^2 A l^3}{360\,E}\begin{bmatrix} 8 & 7 \\ 7 & 8 \end{bmatrix}$$

$$[\mathbf{m}] = \frac{\rho A l}{6}\begin{bmatrix} 2 & 1 \\ 1 & 2 \end{bmatrix} + \frac{\omega^2\rho^2 A l^3}{180\,E}\begin{bmatrix} 8 & 7 \\ 7 & 8 \end{bmatrix}$$

and the error ratio is

$$\delta_{\mathrm{p}} = \frac{\varepsilon^4}{120} - \frac{7\varepsilon^6}{5800} + \cdots$$

which is in fact not better that the average mass model, although more computational efforts are involved.

A simple proportionality relationship between the series terms for $[\mathbf{k}]$ and $[\mathbf{m}]$ of the form

$$(n+1)\mathbf{k}_{n+1} = n\mathbf{m}_n$$

has been demonstrated by Fergusson and Pilkey [3], where $\mathbf{k}_{n+1}$ is the $(n+1)$th term in the stiffness matrix expansion, and $\mathbf{m}_n$ the nth term in the corresponding mass matrix expansion. This result is essentially a power series equivalent of Leung's theorem given in Sect. 2.7. The authors also present a formula by which the $2n$th mass matrix term may be calculated using only the first n terms in the shape function expansion.

2.2.5. Torsion

Since the governing equations for longitudinal and torsional vibrations are in exactly the same form, i.e.

$$EAu'' + \omega^2\rho A u = 0 \qquad (2.2.10a)$$

$$GJ\alpha'' + \omega^2\rho I_0\alpha = 0 \qquad (2.2.10b)$$

where GJ is the torsional rigidity, I_0 is the polar moment of cross-sectional area about the neutral axis and α is the torsional rotation about the neutral axis, all formulae for the torsion-element matrices can be obtained from the bar case directly, by replacing EA by GJ, A by I_0, u by α and axial force by torque. The following are examples for torsional vibration equations. For the consistent mass model,

$$\frac{GJ}{l}\begin{bmatrix} 1 & -1 \\ -1 & 1 \end{bmatrix}\begin{Bmatrix} \alpha_1 \\ \alpha_2 \end{Bmatrix} - \alpha^2\frac{\rho I_0 l}{6}\begin{bmatrix} 2 & 1 \\ 1 & 2 \end{bmatrix}\begin{Bmatrix} \alpha_1 \\ \alpha_2 \end{Bmatrix} = \begin{Bmatrix} T_1 \\ T_2 \end{Bmatrix}$$

and for the continuous mass model,

$$\frac{GJ}{l}\begin{bmatrix} \cot v & -\csc v \\ -\csc v & \cot v \end{bmatrix}\begin{Bmatrix} \alpha_1 \\ \alpha_2 \end{Bmatrix} = \begin{Bmatrix} T_1 \\ T_2 \end{Bmatrix}, \qquad v^2 = \frac{\rho I_0\omega^2 l^2}{GJ}$$

2.3. Beam Elements

The beam element, as shown in Fig. 2.3.1, is the simplest bending element and one of the most widely used finite elements in structural dynamics. Beam-element matrices have been standardized and incorporated into most structural engineering computer packages. Research is still being carried out to reduce other complex elements such as plates and shells into equivalent beams by certain energy criteria, and therefore special attention should be paid to the formulation and use of beam elements.

2.3.1. Thin Uniform Beam Element

Unlike the axial deformation problem, the bending of a beam element requires that the nodal rotations as well as the nodal displacements be compatible between adjacent elements. For an element with two nodes, four coefficients are required for a polynomial displacement function

$$v(x) = a_0 + a_1 x + a_2 x^2 + a_3 x^3 \tag{2.3.1}$$

where a_i are to be determined from the conditions $v(0) = v_1$, $v'(0) = \phi_1$, $v(l) = v_2$, $v'(l) = \phi_2$ as shown in Fig. 2.3.1. Solving for the a_i and substituting back into Eq. (2.3.1), we arrive at

$$\begin{aligned} v(x) &= [1 - 3\xi^2 + 2\xi^3,\ \xi l(1 - 2\xi + \xi^2),\ 3\xi^2 - 2\xi^3,\ \xi l(\xi^2 - \xi)] \begin{Bmatrix} v_1 \\ \phi_1 \\ v_2 \\ \phi_2 \end{Bmatrix} \\ &= [\mathbf{N}]\{\mathbf{u}_e\} \end{aligned} \tag{2.3.2}$$

The shape functions are in fact Hermite polynomials, which could have been used directly. For a thin uniform beam, bending deformation is predominant and other effects can be neglected in the stiffness formulation.

Following the standard finite element procedure given in Sect. 2.1, for a beam element with length l and the second moment of area I, we have

$$\{\boldsymbol{\varepsilon}\} = -\frac{d^2 v}{dx^2} = [\mathbf{B}]\{\mathbf{u}_e\}$$

$$\{\boldsymbol{\sigma}\} = [\mathbf{D}]\{\boldsymbol{\varepsilon}\} = EI\{\boldsymbol{\varepsilon}\}$$

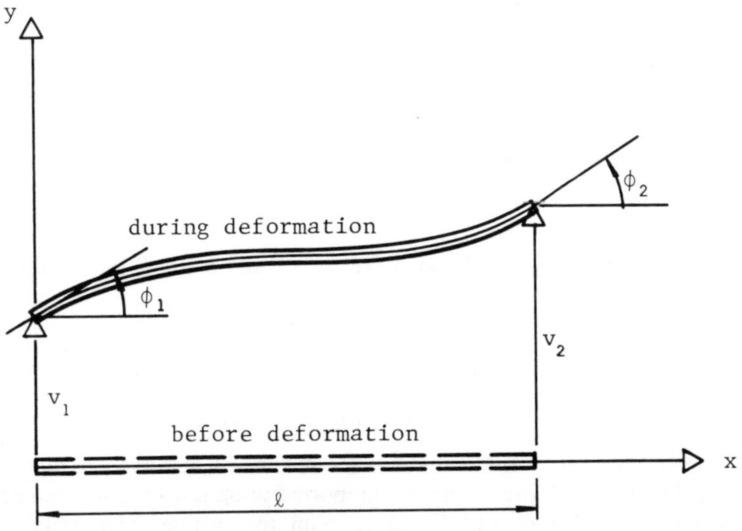

Fig. 2.3.1. A beam element

$$[\mathbf{k}] = \int_0^l [\mathbf{B}]^T[\mathbf{D}][\mathbf{B}]\,dx = EI\int_0^l [\mathbf{B}]^T[\mathbf{B}]\,dx$$

$$= \frac{EI}{l^3}\begin{bmatrix} 12 & & & \text{sym.} \\ 6l & 4l^2 & & \\ -12 & -6l & 12 & \\ 6l & 2l^2 & -6l & 4l^2 \end{bmatrix} \tag{2.3.3}$$

Since [N] satisfies the static governing equation,

$$EIv^{iv} = 0$$

the element is a natural element. The consistent mass matrix is

$$[\mathbf{m}] = \int_{vol} [\mathbf{N}]^T[\boldsymbol{\rho}][\mathbf{N}]\,d\,vol = \rho Al\int_0^l [\mathbf{N}]^T[\mathbf{N}]\,dx$$

$$= \frac{\rho Al}{420}\begin{bmatrix} 156 & & & \text{sym.} \\ 22l & 4l^2 & & \\ 54 & 13l & 156 & \\ -13l & -3l^2 & -22l & 4l^2 \end{bmatrix} \tag{2.3.4}$$

The lumped mass matrix is obtained by assigning equal shares of the element mass to nodal translation degrees of freedom (d.o.f.)

$$[\mathbf{m}_l]\,dx = \rho Al\begin{bmatrix} 0.5 & & & \text{sym.} \\ 0 & 0 & & \\ 0 & 0 & 0.5 & \\ 0 & 0 & 0 & 0 \end{bmatrix} \tag{2.3.5}$$

and the diagonal mass matrix by proportional shares of the diagonal components corresponding to translational d.o.f. and rotational d.o.f., respectively, of the consistent mass matrix

$$[\mathbf{m}_d] = \frac{\rho Al}{420}\begin{bmatrix} 210 & & & 0 \\ & l^2 & & \\ & & 210 & \\ 0 & & & l^2 \end{bmatrix} \tag{2.3.6}$$

Finally, taking the average mass matrix $[\mathbf{m}_a] = \frac{1}{2}([\mathbf{m}_l] + [\mathbf{m}_d])$, we obtain

$$[\mathbf{m}_a] = \frac{\rho Al}{420}\begin{bmatrix} 183 & & & \text{sym.} \\ 11l & 2.5l^2 & & \\ 27 & 6.5l & 183 & \\ -6.5l & -1.5l^2 & -11l & 2.5l^2 \end{bmatrix} \tag{2.3.7}$$

The convergence behaviour of the consistent, lumped, and average mass models are depicted in Fig. 2.3.2 where a simply supported beam is taken as an example and the results given as $\lambda_i^4 = \omega_i^2\rho Al^4/EI$. Once again, the average mass model is found to produce the best results.

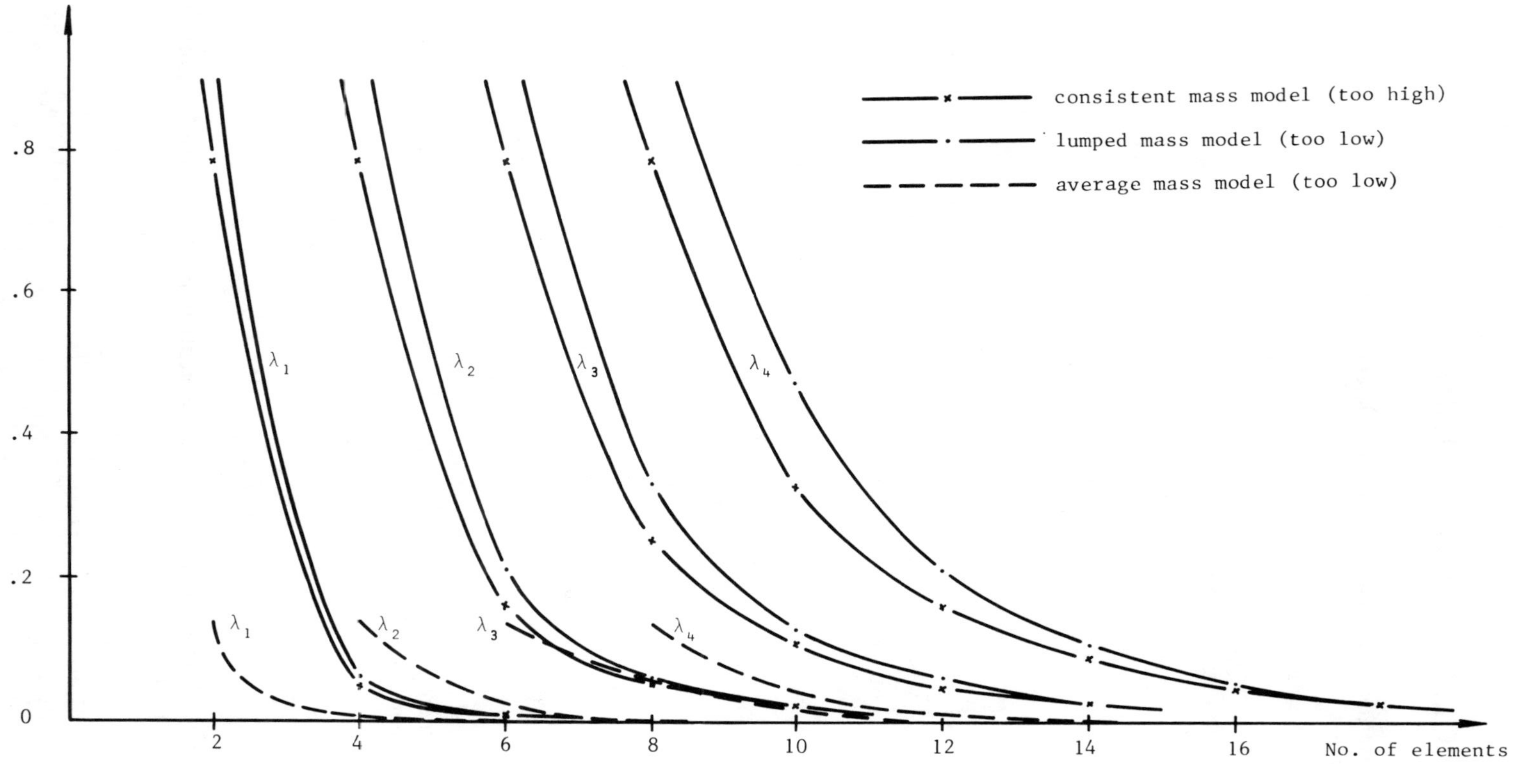

Fig. 2.3.2. Convergence of the consistent, lumped, and averaged mass models

2.4. Continuous Mass Model

The governing equation for harmonic lateral vibration of a thin beam with frequency ω is

$$\frac{EI\partial^4 v}{\partial x^4} = \rho A\omega^2 v$$

The solution associated with the boundary conditions

$$v(0) = v_1, \qquad v'(0) = \phi_1, \qquad v(l) = v_2, \qquad v'(l) = \phi_2$$

is

$$v(x) = [\mathbf{N}(x)]\{\mathbf{u}_e\}$$

where

$$[\mathbf{N}] = \begin{Bmatrix} \cos\lambda\xi \\ \sin\lambda\xi \\ \cosh\lambda\xi \\ \sinh\lambda\xi \end{Bmatrix}^{\mathrm{T}} \begin{bmatrix} 1/2 - F_4/2\lambda^2 & F_2 l/2\lambda^2 & -F_3/2\lambda^2 & F_1 l/2\lambda^2 \\ -F_6/2\lambda^3 & l/2\lambda + F_4 l/2\lambda^3 & -F_5/2\lambda^3 & -F_3 l/2\lambda^3 \\ 1/2 + F_4/2\lambda^2 & -F_2 l/2\lambda^2 & F_3/2\lambda^2 & -F_1 l/2\lambda^2 \\ F_6/2\lambda^2 & l/2\lambda - F_4 l/2\lambda^3 & F_5/2\lambda^3 & F_3 l/2\lambda^3 \end{bmatrix} \tag{2.4.1}$$

in which $\lambda^4 = \omega^2\rho A l^4/El$, and the frequency functions F_i are defined by

$$F_1 = -\lambda(\sinh\lambda - \sin\lambda)/\delta$$

$$F_2 = -\lambda(\cosh\lambda\sin\lambda - \sinh\lambda\cos\lambda)/\delta$$

$$F_3 = -\lambda^2(\cosh\lambda - \cos\lambda)/\delta$$

$$F_4 = \lambda^2(\sinh\lambda\sin\lambda)/\delta$$

$$F_5 = \lambda^3(\sinh\lambda + \sin\lambda)/\delta$$

$$F_6 = -\lambda^3(\cosh\lambda\sin\lambda + \sinh\lambda\cos\lambda)/\delta$$

$$\delta = \cosh\lambda\cos\lambda - 1$$

The detailed derivations of these functions can be found in Kolousek [4].

The dynamic stiffness $[\mathbf{d}(\omega)]$ relates the end-shears and moments to the end-displacements in which the end-forces are

$$V_1 = EI\frac{\partial^3 v(0)}{\partial x^3} \qquad M_1 = -EI\frac{\partial^2 v(0)}{\partial x^2}$$

$$V_2 = -EI\frac{\partial^3 v(l)}{\partial x^3} \qquad M_2 = EI\frac{\partial^2 v(l)}{\partial x^2}$$

Therefore by carrying out appropriate differentiations of the shape functions given in Eq. (2.4.1), one can write

$$\begin{Bmatrix} V_1 \\ M_1 \\ V_2 \\ M_2 \end{Bmatrix} = \frac{EI}{l^3}\begin{bmatrix} F_6 & -F_4 l & F_5 & F_3 l \\ -F_4 l & F_2 l^2 & -F_3 l & F_1 l^2 \\ F_5 & -F_3 l & F_6 & F_4 l \\ F_3 l & F_1 l^2 & F_4 l & F_2 l^2 \end{bmatrix}\{\mathbf{u}_e\} = [\mathbf{d}]\{\mathbf{u}_e\} \tag{2.4.2}$$

and the mass matrix is

$$[\mathbf{m}] = \rho A l \begin{bmatrix} G_6 & -G_4 l & G_5 & G_3 l \\ -G_4 l & G_2 l^2 & -G_3 l & G_1 l^2 \\ G_5 & -G_3 l & G_6 & G_4 l \\ G_3 l & G_1 l^2 & G_4 l & G_2 l^2 \end{bmatrix}$$

where the frequency functions G_i are defined by

$$\begin{aligned} G_1 &= (F_1 F_2 - F_3 - F_1)/4\lambda^4 \\ G_2 &= (F_1^2 - F_2)/4\lambda^4 \\ G_3 &= -(F_1 F_4 + 2F_3)/4\lambda^4 \\ G_4 &= -(F_1 F_3 + 2F_4)/4\lambda^4 \\ G_5 &= (F_3 F_4 - 3F_5)/4\lambda^4 \\ G_6 &= (F_3^2 - 3F_6)/4\lambda^4 \end{aligned} \tag{2.4.3}$$

These formulae are exact so that, when the natural modes of a cantilever are required, one-element analysis gives the frequency equation

$$\det \begin{bmatrix} F_6 & F_4 l \\ F_4 l & F_2 l^2 \end{bmatrix} = 0$$

or

$$1 + \cos\lambda \cos\lambda = 0$$

The solution of the frequency equation yields

$$\lambda = 1.875, 4.694, \ldots (2r-1)\pi/2, \ldots$$

which are indeed exact solutions, and once more it has been demonstrated that one-element analysis can predict an arbitrary number of modes exactly.

If the effects of a constant axial force P (tension is positive), shear deformation and rotatory inertia are all included then the governing equations for harmonic oscillation are [5]

$$\begin{aligned} \phi^{\text{iv}} + \frac{b^2 \Delta \phi''}{(1 - s^2 p^2)} - \frac{b^2(1 - b^2 r^2 s^2)\phi}{(1 - s^2 p^2)} &= 0 \\ v^{\text{iv}} + \frac{b^2 \Delta v''}{(1 - s^2 p^2)} - \frac{b^2(1 - b^2 r^2 s^2)v}{(1 - s^2 p^2)} &= 0 \end{aligned} \tag{2.4.4}$$

The slope due to bending ϕ and the lateral deflection v are related by

$$v'' + \frac{b^2 s^2 v}{(1 - s^2 p^2)} - \frac{l\phi'}{(1 - s^2 p^2)} = 0 \tag{2.4.5}$$

where $b^2 = \rho A l^4 \omega^2/EI$, $r^2 = I_0/Al^2$, $s^2 = EI/GA_s l^2$, $p^2 = -Pl^4/EI$ and $\Delta = (p^2/b^2) + r^2(1 - s^2 p^2) + s^2$.

It is assumed here that $b^2 r^2 s^2 < 1$, since $b^2 r^2 s^2 = \rho\omega^2 I_0/GA > 1$ implies very high frequencies of vibration which are beyond the applicability of Timoshenko's theory. The solutions of Eqs (2.4.4) are

$$\begin{aligned} v &= C_1 \cosh \alpha\xi + C_2 \sinh \alpha\xi + C_3 \cos \beta\xi + C_4 \sin \beta\xi \\ \phi &= C_1' \sinh \alpha\xi + C_2' \cosh \alpha\xi + C_3' \sin \beta\xi + C_4' \cos \beta\xi \end{aligned} \tag{2.4.6}$$

where

$$\left.\begin{matrix}\alpha \\ \beta\end{matrix}\right\} = b\{\mp\Delta + [\Delta^2 + 4(1 - s^2p^2)(1 - b^2r^2s^2)/b^2]^{1/2}\}^{1/2}/[2(1 - s^2p^2)]^{1/2}$$

The four constants C_i' can be eliminated through the relationship given by Eq. (2.4.5). This results in

$$C_1' = hC_1, \quad C_2' = hC_2, \quad C_3' = -eC_3, \quad C_4' = eC_4$$

where

$$h = [(1 - s^2p^2)\alpha^2 + b^2s^2]/\alpha l$$
$$e = [(1 - s^2p^2)\beta^2 - b^2s^2]/\beta l$$

The remaining four constants, C_i, are determined by the boundary conditions,

$$v(0) = v_1, \quad \phi(0) = \phi_1, \quad v(l) = v_2, \quad \phi(l) = \phi_2$$

With the displacement functions now given in terms of the end displacements, it is a simple matter to form the dynamic stiffness matrix through the following relationships:

1. $M = EI\phi'$, the generalized stress–strain relationship for a Timoshenko beam.
2. $V = M' + \rho V' - \rho I_0 \phi$, the moment equilibrium of all forces acting on an element of length dx.

Finally, taking $V_1 = V(0)$, $M_1 = -M(0)$, $V_2 = -V(l)$, $M_2 = M(l)$

$$\begin{Bmatrix} V_1 \\ M_1 \\ V_2 \\ M_2 \end{Bmatrix} = \frac{EI}{l^3} \begin{bmatrix} F_6 & -F_4 l & F_5 & F_3 l \\ -F_4 l & F_2 l^2 & -F_3 l & F_1 l^2 \\ F_5 & -F_3 l & F_6 & F_4 l \\ F_3 l & F_1 l^2 & F_4 l & F_2 l^2 \end{bmatrix} \begin{Bmatrix} v_1 \\ \phi_1 \\ v_2 \\ \phi_2 \end{Bmatrix} \tag{2.4.7}$$

or

$$\{\mathbf{f}_e\} = [\mathbf{d}(\omega)]\{\mathbf{u}_s\}$$

The frequency functions are

$$\begin{aligned} F_1 &= (\alpha + \eta\beta)(\eta \sinh \alpha - \sin \beta)/\delta \\ F_2 &= (\alpha + \eta\beta)(\sin \beta \cosh \alpha - \eta \cos \beta \sinh \alpha)/\delta \\ F_3 &= \eta(1 - s^2p^2)(\alpha^2 + \beta^2)(\cosh \alpha - \cos \beta)/\delta \\ F_4 &= el[(\alpha - \eta\beta)(1 - \cosh \alpha \cos \beta) - (\beta + \eta\alpha) \sin \rho \sinh \alpha]/\delta \\ F_5 &= -b^2(\alpha + \eta\beta)(\eta \sinh \alpha + \sin \beta)/\alpha\beta\delta \\ F_6 &= b^2(\alpha + \eta\beta)(\sinh \alpha \cos \beta + \eta \sin \beta \cosh \alpha)/\alpha\beta\delta \end{aligned} \tag{2.4.8}$$

2.5. Rectangular Plate

A rectangular plate with two opposite edges simply supported and with the other two edges connected to other structures by prescribed displacement patterns will be

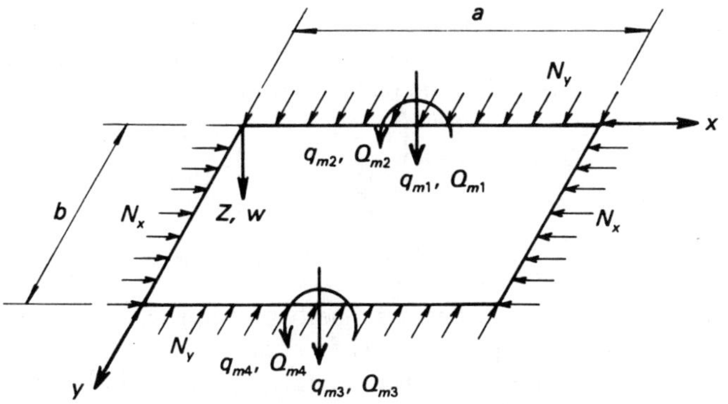

Fig. 2.5.1. Rectangular plate with $x = 0$ and $x = a$ simply supported

discussed in this section. Distributed coordinates on the edges will be used in this example.

To satisfy the boundary conditions of two opposite edges being simply supported, the displacement pattern of the plate may be written as

$$w(x, y) = \sum_{m=1}^{N} Y_m(y) \sin \frac{m\pi x}{a} \tag{2.5.1}$$

where N is the number of terms taken, a and b are the dimensions as shown in Fig. 2.5.1, and $Y_m(y)$ are the functions to be determined in order to satisfy the governing equation of vibration.

The generalized displacements q_{mi}, $m = 1, 2, \ldots, N$ and $i = 1, 2, 3, 4$, are defined by

$$\begin{aligned} w(x, 0) &= \sum_{m=1}^{N} q_{m1} \sin \frac{m\pi x}{a} \\ w(x, b) &= \sum_{m=1}^{N} q_{m2} \sin \frac{m\pi x}{a} \\ \frac{\partial w(x, 0)}{\partial y} &= \sum_{m=1}^{N} q_{m3} \sin \frac{m\pi x}{a} \\ \frac{\partial w(x, b)}{\partial y} &= \sum_{m=1}^{N} q_{m4} \sin \frac{m\pi x}{a} \end{aligned} \tag{2.5.2}$$

and the generalized forces Q_{mi} are defined by

$$\begin{aligned} Q_y(x, 0) &= \sum_{m=1}^{N} Q_{m1} \sin \frac{m\pi x}{a} \\ M_y(x, 0) &= \sum_{m=1}^{N} Q_{m2} \sin \frac{m\pi x}{a} \\ -Q_y(x, b) &= \sum_{m=1}^{N} Q_{m3} \sin \frac{m\pi x}{a} \\ -M_y(x, b) &= \sum_{m=1}^{N} Q_{m4} \sin \frac{m\pi x}{a} \end{aligned} \tag{2.5.3}$$

where Q_y and M_y are the Kirchhoff shear and the bending moment of the plate along the line $y = \text{constant}$ [6]. The generalized forces are related to the displacements through the conditions of equilibrium on the edges $y = 0$ and $y = b$,

$$Q_y(x, y) = -D\left[\frac{\partial^3 w}{\partial y^3} + (2 - v)\frac{\partial^3 w}{\partial x^2 \partial y}\right] \tag{2.5.4}$$

$$M_y(x, y) = -D\left[\frac{\partial^2 w}{\partial y^2} + v\frac{\partial^2 w}{\partial x^2}\right] \tag{2.5.5}$$

where $D = Eh^3/12(1 - v^2)$ is the flexural rigidity of the plate, h is the thickness and v is Poisson's ratio. Before applying Eqs (2.5.2)–(2.5.5) to find the dynamic stiffness matrix, one must first calculate the functions $Y_m(y)$ in Eq. (2.5.1). If the loadings are harmonic with frequency ω, the governing equation is given by

$$D\nabla^4 w - \rho h\omega^2 w + N_x\frac{\partial^2 w}{\partial x^2} + N_y\frac{\partial^2 w}{\partial y^2} = p(x, y) \tag{2.5.6}$$

where N_x and N_y are the compressive in-plane loads in the x and y directions respectively, ∇^4 is the biharmonic operator in rectangular coordinates, and $p(x, y)$ is the downward distributed load intensity, being represented as

$$p(x, y) = \sum_{m=1}^{N} p_m \sin\frac{m\pi x}{a} \qquad \text{per unit area} \tag{2.5.7}$$

Substituting Eq. (2.5.1) into (2.5.6) gives

$$\sum_{m=1}^{N} \sin\frac{m\pi x}{a}\left\{\left(\frac{m\pi}{a}\right)^4 Y_m - 2\left(\frac{m\pi}{a}\right)^2 Y_m'' + Y_m^{\text{iv}}\right.$$
$$\left. - \frac{\rho h\omega^2}{D} Y_m - \frac{N_x}{D}\left(\frac{m\pi}{a}\right)^2 Y_m + \frac{N_y}{D} Y_m'' - \frac{p_m}{D} Y_m\right\} = 0$$

Multiplication by $\sin(n\pi x/a)$, and integration from $x = 0$ to $x = a$, where n is a positive integer, and use of the orthogonality of sine functions, yields

$$Y_m^{\text{iv}} - 2\left\{\left(\frac{m\pi}{a}\right)^2 - \frac{N_y}{2D}\right\} Y_m'' + \left\{\left(\frac{m\pi}{a}\right)^4 - \frac{\rho h\omega^2}{D} - \frac{N_x}{D}\left(\frac{m\pi}{a}\right)^2 - \frac{p_m}{D}\right\} Y_m = 0$$
$$m = 1, 2, \ldots \tag{2.5.8}$$

The associated boundary conditions for these fourth order differential equations are obtained from Eq. (2.5.2) as

$$Y_m(0) = q_{m1}, \quad Y_m(b) = q_{m3}, \quad Y_m'(0) = q_{m2}, \quad Y_m'(b) = q_{m4} \tag{2.5.9}$$

The auxiliary roots of Eqs (2.5.8) are obtained by letting $Y_m = \mathrm{e}^{\sigma y}$

$$\sigma^2 = \left[\left(\frac{m\pi}{a}\right)^2 - \frac{N_y}{2D}\right]$$
$$\pm\left\{\left[\left(\frac{m\pi}{a}\right)^2 - \left(\frac{N_y}{2D}\right)\right]^2 - \left(\frac{m\pi}{a}\right)^4 + \frac{\rho h\omega^2}{D} + \frac{N_x}{D}\left(\frac{m\pi}{a}\right)^2 + \frac{p_m}{D}\right\}^{1/2} \tag{2.5.10}$$

$Y_m(y)$ will have four different forms of solution depending on whether σ^2 is positive, negative or complex. These four cases are studied as follows.

Case 1. When all four roots are real, they take the form $\pm\sigma_1$, $\pm\sigma_2$, and the general solution is

$$Y_m(y) = A\cosh\frac{\sigma_2 y}{b} + B\sinh\frac{\sigma_2 y}{b} + C\cosh\frac{\sigma_1 y}{b} + D\sin\frac{\sigma_1 y}{b} \tag{2.5.11}$$

where A, B, C, D are integration constants and are determined from the boundary conditions (2.5.9) as

$$\begin{aligned}
A &= \left(\frac{\sigma_1^2 - F_4}{\sigma_1^2 - \sigma_2^2}\right)q_{m1} + \left(\frac{-F_2}{\sigma_1^2 - \sigma_2^2}\right)bq_{m2} + \left(\frac{-F_3}{\sigma_1^2 - \sigma_2^2}\right)q_{m3} + \left(\frac{F_1}{\sigma_1^2 - \sigma_2^2}\right)bq_{m4} \\
B &= \left(\frac{F_6}{\sigma_1^2 - \sigma_2^2}\right)\frac{q_{m1}}{\sigma_2} + \left(\frac{\sigma_2^2 - F_4}{\sigma_1^2 - \sigma_2^2}\right)\frac{bq_{m2}}{\sigma_2} + \left(\frac{F_5}{\sigma_1^2 - \sigma_2^2}\right)\frac{q_{m3}}{\sigma_2} + \left(\frac{F_3}{\sigma_1^2 - \sigma_2^2}\right)\frac{bq_{m4}}{\sigma_2} \\
C &= q_{m1} - A, \qquad D = \frac{bq_{m2}}{\sigma_1} - \frac{B\sigma_2}{\sigma_1}
\end{aligned} \tag{2.5.12}$$

where the frequency functions are given by

$$\begin{aligned}
F_1 &= -(\sigma_2\sinh\sigma_1 - \sigma_1\sinh\sigma_2)(\sigma_1^2 - \sigma_2^2)/\delta \\
F_2 &= -(\sigma_1\cosh\sigma_1\sinh\sigma_2 - \sigma_2\sinh\sigma_1\cosh\sigma_2)(\sigma_1^2 - \sigma_2^2)/\delta \\
F_3 &= -\sigma_1\sigma_2(\sigma_1^2 - \sigma_2^2)(\cosh\sigma_1 - \cosh\sigma_2)/\delta \\
F_4 &= \sigma_1\sigma_2[(\sigma_1^2 + \sigma_2^2)(\cosh\sigma_1\cosh\sigma_2 - 1) - 2\sigma_1\sigma_2\sinh\sigma_1\sinh\sigma_2]/\delta \\
F_5 &= \sigma_1\sigma_2(\sigma_1^2 - \sigma_2^2)(\sigma_1\sinh\sigma_1 - \sigma_2\sinh\sigma_2)/\delta \\
F_6 &= -\sigma_1\sigma_2(\sigma_1^2 - \sigma_2^2)(-\sigma_2\cosh\sigma_1\sinh\sigma_2 + \sigma_1\sinh\sigma_1\cosh\sigma_2)/\delta \\
\delta &= 2\sigma_1\sigma_2(\cosh\sigma_1\cosh\sigma_2 - 1) - (\sigma_1^2 + \sigma_2^2)\sinh\sigma_1\sinh\sigma_2
\end{aligned} \tag{2.5.13}$$

Case 2. When there are two real and two pure imaginary roots, $\pm\sigma_1$, $\pm i\sigma_2$, then the general solution has the form

$$Y_m(y) = A\cos\frac{\sigma_2 y}{b} + B\sin\frac{\sigma_2 y}{b} + C\cosh\frac{\sigma_1 y}{b} + D\sinh\frac{\sigma_1 y}{b} \tag{2.5.14}$$

where A, B, C, D are integration constants depending on the boundary conditions (2.5.9), and are found as

$$\begin{aligned}
A &= \left(\frac{\sigma_1^2 - F_4}{\sigma_1^2 + \sigma_2^2}\right)q_{m1} + \left(\frac{F_2}{\sigma_1^2 + \sigma_2^2}\right)bq_{m2} + \left(\frac{-F_3}{\sigma_1^2 + \sigma_2^2}\right)q_{m3} + \left(\frac{F_1}{\sigma_1^2 + \sigma_2^2}\right)bq_{m4} \\
B &= \left(\frac{-F_6}{\sigma_1^2 + \sigma_2^2}\right)\frac{q_{m1}}{\sigma_2} + \left(\frac{\sigma_2^2 + F_4}{\sigma_1^2 + \sigma_2^2}\right)\frac{bq_{m2}}{\sigma_2} + \left(\frac{-F_5}{\sigma_1^2 + \sigma_2^2}\right)\frac{q_{m3}}{\sigma_2} + \left(\frac{-F_3}{\sigma_1^2 + \sigma_2^2}\right)\frac{bq_{m4}}{\sigma_2} \\
C &= q_{m1} - A, \qquad D = \frac{bq_{m2}}{\sigma_1} - \frac{B\sigma_2}{\sigma_1}
\end{aligned} \tag{2.5.15}$$

where the frequency functions are given by

$$
\begin{aligned}
F_1 &= -(\sigma_2 \sinh \sigma_1 - \sigma_1 \sin \sigma_2)(\sigma_1^2 + \sigma_2^2)/\delta \\
F_2 &= -(\sigma_1 \cosh \sigma_1 \sin \sigma_2 - \sigma_2 \sinh \sigma_1 \cos \sigma_2)(\sigma_1^2 + \sigma_2^2)/\delta \\
F_3 &= -\sigma_1 \sigma_2 (\sigma_1^2 + \sigma_2^2)(\cosh \sigma_1 - \cos \sigma_2)/\delta \\
F_4 &= \sigma_1 \sigma_2 [(\sigma_1^2 - \sigma_2^2)(\cosh \sigma_1 \cos \sigma_2 - 1) + 2\sigma_1 \sigma_2 \sinh \sigma_1 \sin \sigma_2]/\delta \qquad (2.5.16) \\
F_5 &= \sigma_1 \sigma_2 (\sigma_1^2 + \sigma_2^2)(\sigma_2 \sin \sigma_2 + \sigma_1 \sinh \sigma_1)/\delta \\
F_6 &= -\sigma_1 \sigma_2 (\sigma_1^2 + \sigma_2^2)(\sigma_2 \cosh \sigma_1 \sin \sigma_2 + \sigma_1 \sinh \sigma_1 \cos \sigma_2)/\delta \\
\delta &= 2\sigma_1 \sigma_2 (\cosh \sigma_1 \cos \sigma_2 - 1) - (\sigma_1^2 - \sigma_2^2) \sinh \sigma_1 \sin \sigma_2
\end{aligned}
$$

Case 3. When all four roots are purely imaginary, $\pm i\sigma_1$, $\pm i\sigma_2$, then the general solution has the form

$$
Y_m(y) = A \cos \frac{\sigma_2 y}{b} + B \sin \frac{\sigma_2 y}{b} + C \cos \frac{\sigma_1 y}{b} + D \sin \frac{\sigma_1 y}{b} \qquad (2.5.17)
$$

where A, B, C, D are integration constants depending on the boundary conditions (2.5.9), and are found to be

$$
\begin{aligned}
A &= \left(\frac{-\sigma_2^2 - F_4}{\sigma_1^2 - \sigma_2^2}\right) q_{m1} + \left(\frac{F_2}{\sigma_1^2 - \sigma_2^2}\right) b q_{m2} + \left(\frac{-F_3}{\sigma_1^2 - \sigma_2^2}\right) q_{m3} + \left(\frac{F_1}{\sigma_1^2 - \sigma_2^2}\right) b q_{m4} \\
B &= \left(\frac{-F_6}{\sigma_1^2 - \sigma_2^2}\right) \frac{q_{m1}}{\sigma_2} + \left(\frac{\sigma_2^2 - F_4}{\sigma_1^2 - \sigma_2^2}\right) \frac{b q_{m2}}{\sigma_2} + \left(\frac{-F_5}{\sigma_1^2 - \sigma_2^2}\right) \frac{q_{m3}}{\sigma_2} + \left(\frac{-F_3}{\sigma_1^2 - \sigma_2^2}\right) \frac{b q_{m4}}{\sigma_2} \\
C &= q_{m1} - A, \qquad D = \frac{b q_{m2}}{\sigma_1} - \frac{B \sigma_2}{\sigma_1} \qquad (2.5.18)
\end{aligned}
$$

where the frequency functions are given by

$$
\begin{aligned}
F_1 &= -(\sigma_2 \sin \sigma_1 - \sigma_1 \sin \sigma_2)(\sigma_1^2 - \sigma_2^2)/\delta \\
F_2 &= -(\sigma_1 \cos \sigma_1 \sin \sigma_2 - \sigma_2 \sin \sigma_1 \cos \sigma_2)(\sigma_1^2 - \sigma_2^2)/\delta \\
F_3 &= -\sigma_1 \sigma_2 (\sigma_1^2 - \sigma_2^2)(\cos \sigma_1 - \cos \sigma_2)/\delta \\
F_4 &= \sigma_1 \sigma_2 [(\sigma_1^2 + \sigma_2^2)(\cos \sigma_1 \cos \sigma_2 - 1) + 2\sigma_1 \sigma_2 \sinh \sigma_1 \sin \sigma_2]/\delta \qquad (2.5.19) \\
F_5 &= \sigma_1 \sigma_2 (\sigma_1^2 - \sigma_2^2)(\sigma_2 \sin \sigma_2 - \sigma_1 \sin \sigma_1)/\delta \\
F_6 &= -\sigma_1 \sigma_2 (\sigma_1^2 - \sigma_2^2)(\sigma_2 \cos \sigma_1 \sin \sigma_2 - \sigma_1 \sin \sigma_1 \cos \sigma_2)/\delta \\
\delta &= 2\sigma_1 \sigma_2 (\cos \sigma_1 \cos \sigma_2 - 1) - (\sigma_1^2 + \sigma_2^2) \sin \sigma_1 \sin \sigma_2
\end{aligned}
$$

Case 4. When all four roots are complex, $\sigma_2 \pm i\sigma_1$, $-\sigma_2 \pm i\sigma_1$, then the general solution has the form

$$
\begin{aligned}
Y_m(y) = {} & A \cos \frac{\sigma_1 y}{b} \cosh \frac{\sigma_2 y}{b} + B \cos \frac{\sigma_1 y}{b} \sinh \frac{\sigma_2 y}{b} \\
& + C \sin \frac{\sigma_1 y}{b} \cosh \frac{\sigma_2 y}{b} + D \sin \frac{\sigma_1 y}{b} \sinh \frac{\sigma_2 y}{b} \qquad (2.5.20)
\end{aligned}
$$

where A, B, C, D are integration constants depending on the boundary conditions (2.5.9), and they are found to be

$$
\left.\begin{aligned}
A &= q_{m1} \\
B &= \{q_{m1}(\sigma_1\sigma_2 \sin\sigma_1 \cos\sigma_1 + \sigma_2^2 \sinh\sigma_1 \cosh\sigma_1) + bq_{m2}(\sigma_2 \sin^2\sigma_1) \\
&\quad - q_{m3}\sigma_1(\sigma_1 \cos\sigma_1 \sinh\sigma_2 + \sigma_2 \sin\sigma_1 \cosh\sigma_2) + bq_{m4}(\sigma_1 \sin\sigma_1 \sinh\sigma_2)\}/\delta \\
C &= \{-q_{m1}(\sigma_1\sigma_2 \sinh\sigma_1 \cosh\sigma_1 + \sigma_2^2 \sin\sigma_1 \cos\sigma_1) - bq_{m2}(\sigma_2 \sinh^2\sigma_1) \\
&\quad + q_{m3}\sigma_2(\sigma_1 \cos\sigma_1 \sinh\sigma_2 + \sigma_2 \sin\sigma_1 \cosh\sigma_2) - bq_{m4}(\sigma_2 \sin\sigma_1 \sinh\sigma_2)\}/\delta \\
D &= \{q_{m1}\sigma_1\sigma_2(\sin^2\sigma_1 + \sinh^2\sigma_2) + bq_{m2}(\sigma_1 \sinh\sigma_2 \cosh\sigma_2 - \sigma_2 \sin\sigma_1 \cos\sigma_2) \\
&\quad - q_{m3}(\sigma_1^2 + \sigma_2^2) \sin\sigma_1 \sinh\sigma_2 + bq_{m4}(\sigma_2 \sin\sigma_1 \cosh\sigma_2 - \sigma_1 \cos\sigma_1 \sinh\sigma_2)\}/\delta \\
\delta &= \sigma_2^2 \sin^2\sigma_1 - \sigma_1^2 \sinh^2\sigma_2
\end{aligned}\right\} \tag{2.5.21}
$$

where the frequency functions are given by

$$
\left.\begin{aligned}
F_1 &= -2\sigma_1\sigma_2(\sigma_2 \sin\sigma_1 \cosh\sigma_2 - \sigma_1 \cos\sigma_1 \sinh\sigma_2)/\delta \\
F_2 &= -2\sigma_1\sigma_2(\sigma_1 \sinh\sigma_2 \cosh\sigma_2 - \sigma_2 \sin\sigma_1 \cos\sigma_1)/\delta \\
F_3 &= 2\sigma_1\sigma_2(\sigma_1^2 + \sigma_2^2)(\sin\sigma_1 \sinh\sigma_2)/\delta \\
F_4 &= (\chi_2\sigma_2 \sin^2\sigma_1 + \chi_1\sigma_1 \sinh^2\sigma_2)/\delta \\
F_5 &= 2\sigma_1\sigma_2(\sigma_1^2 + \sigma_2^2)(\sigma_1 \sinh\sigma_2 \cos\sigma_1 + \sigma_2 \sin\sigma_1 \cosh\sigma_2)/\delta \\
F_6 &= -2\sigma_1\sigma_2(\sigma_1^2 + \sigma_2^2)(\sigma_1 \sinh\sigma_2 \cosh\sigma_2 + \sigma_2 \cos\sigma_1 \sin\sigma_1)/\delta \\
\delta &= \sigma_2^2 \sin^2\sigma_1 - \sigma_1^2 \sinh^2\sigma_2 \\
\chi_1 &= \sigma_1^3 - \sigma_1\sigma_2^2 + (2-\nu)\sigma_1 \\
\chi_2 &= \sigma_2^3 - \sigma_1^2\sigma_2 + (2-\nu)\sigma_2
\end{aligned}\right\} \tag{2.5.22}
$$

Having determined the functions $Y_m(y)$ explicitly in terms of the generalized displacements q_{mi}, the differentiation in Eqs (2.5.4) and (2.5.5) is carried out and Eqs (2.5.3) are invoked, making use of the orthogonality of sine functions. This provides the desired relationship between the generalized forces and generalized displacements.

After some simplification, the dynamic stiffness relations for all these cases have the form

$$
\begin{Bmatrix} Q_{m1} \\ Q_{m2} \\ Q_{m3} \\ Q_{m4} \end{Bmatrix} = \frac{D}{b^3} \begin{bmatrix} F_6 & -F_4 b & F_5 & F_3 b \\ -F_4 b & F_2 b^2 & -F_3 b & F_1 b^2 \\ F_5 & -F_3 b & F_6 & F_4 b \\ F_3 b & F_1 b^2 & F_4 b & F_2 b^2 \end{bmatrix} \begin{Bmatrix} q_{m1} \\ q_{m2} \\ q_{m3} \\ q_{m4} \end{Bmatrix} \quad m = 1, 2, \ldots, N \tag{2.5.23}
$$

where the frequency functions F_i have different forms for the four cases and should be calculated under the individual heading according to the nature of the auxiliary roots.

2.6. Interaction Between Beams and Plates

It is a common engineering practice to stiffen a plate system using beams. The effects of a stiffening beam are threefold: axial, flexural and torsional. In the following analysis, the flexural and torsional effects are considered separately.

The governing equation of a beam in flexural vibration is

$$EI\frac{\partial^4 w}{\partial x^4} - \rho A_0\frac{\partial^2 w}{\partial t^2} + N_x\frac{\partial^2 w}{\partial x^2} = v \tag{2.6.1}$$

where N_x is the axial compressive force and v is the distributed transverse load per unit length along the beam. For harmonic excitation of a simply supported beam,

$$v = \sum_{m=1}^{N} v_m \sin\frac{m\pi x}{a} e^{i\omega t}$$

and

$$w = \sum_{m=1}^{N} w_m \sin\frac{m\pi x}{a} e^{i\omega t} \tag{2.6.2}$$

From Eqs (2.6.1) and (2.6.2), we have

$$\sum_{m=1}^{N} \sin\frac{m\pi x}{a}\left[EI\left(\frac{m\pi}{a}\right)^4 w_m - \rho A_0\omega^2 w_m - N_x\left(\frac{m\pi}{a}\right)^2 w_m - v_m\right] = 0$$

Multiplying the entire equation by $\sin(n\pi x/a)$ and integrating from over $x = 0$ to $x = a$, gives

$$v_m = \left[EI\left(\frac{m\pi}{a}\right)^4 - \rho A_0\omega^2 - N_x\left(\frac{m\pi}{a}\right)^2\right] w_m \tag{2.6.3}$$

which is the stiffness relation required.

The torsional effect is derived as follows. The differential equation governing the torsional vibration of a beam, when the shear centre coincides with the mass centre of cross sectional area, is

$$GJ\frac{\partial^2\theta}{\partial x^2} - \rho I_0\frac{\partial^2 w}{\partial t^2} + T = 0 \tag{2.6.4}$$

where T is the torsional moment acting on the beam per unit length, ρ is the mass density, I_0 is the polar moment of inertia, and GJ is the torsional rigidity.

For harmonic oscillation of a simply supported beam, writing

$$T = \sum_{m=1}^{N} T_m \sin\frac{m\pi x}{a} e^{i\omega t}$$

$$\theta = \sum_{m=1}^{N} \theta_m \sin\frac{m\pi x}{a} e^{i\omega t} \tag{2.6.5}$$

and Eq. (2.6.4) becomes

$$\sum_{m=1}^{N} \sin\frac{m\pi x}{a}\left[-GJ\left(\frac{m\pi}{a}\right)^2\theta_m + \rho I_0\omega^2\theta_m - T_m\right] = 0$$

Multiplying the entire equation by $\sin(n\pi x/a)$ and integrating from $x = 0$ to $x = a$, gives

$$T_m = \left[GJ\left(\frac{m\pi}{a}\right)^2 - \rho I_0\omega^2\right]\theta_m \tag{2.6.6}$$

which is the stiffness relation required.

When the beam member is on an edge of a folded plate, then the generalized displacements of the beam, w_m and θ_m, will correspond to the generalized displacements of the plate, either q_{m1} and q_{m2}, or q_{m3} and q_{m4} respectively, depending on which edge of the plate the beam is situated along.

2.7. Leung's Theorem

The dynamic stiffness matrix is useful for finding the natural modes and the mass matrix is useful for finding the modal mass so that a response analysis can be carried out. The dynamic stiffness matrix $[\mathbf{D}(\omega)]$ of a system vibrating harmonically at frequency ω relates the amplitudes of the response displacements $\{\mathbf{q}\}$ to those of exciting forces $\{\mathbf{D}\}$ according to

$$[\mathbf{D}(\omega)]\{\mathbf{q}\} = \{\mathbf{Q}\} \tag{2.7.1}$$

It will now be proved that the mass matrix is related to the dynamic stiffness matrix by Leung's theorem (after Simpson [7])

$$[\mathbf{M}(\omega)] = -\frac{\partial[\mathbf{D}(\omega)]}{\partial\omega^2} \tag{2.7.2}$$

Let the body of interest be excited independently by two systems of nodal forces $\{\mathbf{Q}_1\}\mathrm{e}^{\mathrm{i}\omega_1 t}$ and $\{\mathbf{Q}_2\}\mathrm{e}^{\mathrm{i}\omega_2 t}$. The steady state responses are $\{\mathbf{u}_1\}\mathrm{e}^{\mathrm{i}\omega_1 t}$ and $\{\mathbf{u}_2\}\mathrm{e}^{\mathrm{i}\omega_2 t}$ respectively. The following relations are obvious:

$$\begin{aligned} \{u_1\} &= [\mathbf{N}(\omega_1)]\{\mathbf{q}_1\}, \qquad & \{u_2\} &= [\mathbf{N}(\omega_2)]\{\mathbf{q}_2\} \\ [\mathbf{D}(\omega_1)]\{\mathbf{q}_1\} &= \{\mathbf{Q}_1\}, \qquad & [\mathbf{D}(\omega_2)]\{\mathbf{q}_2\} &= \{\mathbf{Q}_2\} \end{aligned} \tag{2.7.3}$$

The reciprocal theorem states that the work done by the first set of forces (including inertia) acting through the second set of displacements is equal to the work done by the second set of forces acting through the first set of displacements. That is,

$$\left(\{\mathbf{q}_1\}^{\mathrm{T}}\{\mathbf{Q}_2\} + \omega_2^2\int\{\mathbf{u}_1\}^{\mathrm{T}}[\boldsymbol{\rho}]\{\mathbf{u}_2\}\,\mathrm{d}\,vol\right)\mathrm{e}^{\mathrm{i}(\omega_1+\omega_2)t}$$
$$= \left(\{\mathbf{q}_2\}^{\mathrm{T}}\{\mathbf{Q}_1\} + \omega_1^2\int\{\mathbf{u}_2\}^{\mathrm{T}}[\boldsymbol{\rho}]\{\mathbf{u}_1\}\,\mathrm{d}\,vol\right)\mathrm{e}^{\mathrm{i}(\omega_1+\omega_2)t}$$

where the inertia forces are

$$-\frac{\mathrm{d}^2}{\mathrm{d}t^2}[\boldsymbol{\rho}]\{\mathbf{u}_1\}\exp(\mathrm{i}\omega_1 t) = \omega_1^2[\boldsymbol{\rho}]\{\mathbf{u}_1\}\mathrm{e}^{\mathrm{i}\omega_1 t}$$

and

$$\omega_2^2[\boldsymbol{\rho}]\{\mathbf{u}_1\}\mathrm{e}^{\mathrm{i}\omega_2 t}$$

for the two systems, respectively. By means of the relations (2.7.1),

$$\left(\{\mathbf{q}_1\}^{\mathrm{T}}[\mathbf{D}(\omega_2)]\{\mathbf{q}_2\} + \omega_2^2\{\mathbf{q}_1\}^{\mathrm{T}}\int[\mathbf{N}(\omega_1)]^{\mathrm{T}}[\boldsymbol{\rho}][\mathbf{N}(\omega_2)]\,\mathrm{d}\,vol\right)\{\mathbf{q}_2\}$$
$$= \left(\{\mathbf{q}_2\}^{\mathrm{T}}[\mathbf{D}(\omega_1)]\{\mathbf{q}_1\} + \omega_1^2\{\mathbf{q}_2\}^{\mathrm{T}}\int[\mathbf{N}(\omega_2)]^{\mathrm{T}}[\boldsymbol{\rho}][\mathbf{N}(\omega_1)]\,\mathrm{d}\,vol\right)\{\mathbf{q}_1\}$$

Since $\{\mathbf{q}_1\}$ and $\{\mathbf{q}_2\}$ are not identically zero,

$$\int [\mathbf{N}(\omega_1)]^{\mathrm{T}}[\boldsymbol{\rho}][\mathbf{N}(\omega_2)]\,\mathrm{d}\,vol = \frac{[\mathbf{D}(\omega_1) - \mathbf{D}(\omega_2)]}{\omega_2^2 - \omega_1^2} \tag{2.7.4}$$

Define the mixed mass matrix,

$$[\mathbf{M}(\omega_1, \omega_2)] = \int [\mathbf{N}(\omega_1)]^{\mathrm{T}}[\boldsymbol{\rho}][\mathbf{N}(\omega_2)]\,\mathrm{d}\,vol \tag{2.7.5}$$

which is required in dealing with modal analysis. Then,

$$[\mathbf{M}(\omega)] = \lim_{\substack{\omega_1 \to \omega \\ \omega_2 \to \omega}} [\mathbf{M}(\omega_1, \omega_2)] = -\frac{\partial}{\partial \omega^2}[\mathbf{D}(\omega)] \tag{2.7.6}$$

One can verify that Eqs (2.2.6a) and (2.2.6b) are related by Eq. (2.7.6).

2.8. Simpson's Hypothesis

Two different kinds of elements have been used to model the same physical member, the finite element and the continuum element. Simpson's hypothesis states that if a sufficient number of permissible finite elements are employed, the finite element modelling and the continuum modelling are equivalent. It is permissible to use straight beam finite elements to model curved beams in linear vibration analysis but not permissible in buckling analysis. Simpson's hypothesis is very useful in transferring theorems for the finite element method to the continuum method. One application is given in the next two sections to prove the Wittrick–Williams algorithm which is analogous to Sturm's theorem for discrete systems.

2.9. Sturm's Theorem

Sturm's theorem used in solving the eigenvalue problem $\det([\mathbf{K}] - \lambda[\mathbf{M}]) = 0$ for constant matrices $[\mathbf{K}]$ and $[\mathbf{M}]$ is of interest. The theorem is extended to frequency-dependent matrices in the next section.

For a very small problem, the technique of expanding the determinant, $\det([\mathbf{K}] - \lambda[\mathbf{M}]) = 0$, into a polynomial form and then extracting the roots is very well known. However, this technique should not be used for problems having more than four unknowns because firstly, it is an expensive process to evaluate the polynomial coefficients, and secondly, the roots are always positive. The polynomial coefficients are therefore of alternate signs and of great difference in magnitudes, making the resulting polynomial equation highly ill-conditioned for most root extraction methods. In fact, for somewhat larger problems, the first few roots of lowest magnitudes may not be solved with any accuracy due to the round-off errors associated with roots of higher magnitudes. In this connection, it should be noted that an alternative method exists for dealing effectively with these problems, especially ones associated with narrowly banded matrices, from which only a few eigenvalues are needed. In this method, $\det([\mathbf{K}] - \lambda[\mathbf{M}]) = 0$ is solved directly for λ as a transcendental equation without being expanded into polynomial form.

For eigenproblems associated with large banded matrices, in which only the highest root or the lowest root is of interest, either the power iteration method or the inverse iteration method is the most suitable. Starting with an initial guess, the calculated root is refined progressively after each iteration. However, if some other roots are required, the simple iteration process cannot be applied again directly until the matrices have been deflated, after which the solutions will not converge again to the already known eigenpairs λ_i and $\{\boldsymbol{\phi}_i\}$. Unfortunately, deflation is an expensive process for large matrices, therefore it would be better to use the determinant search method mentioned earlier, in which $\det([\mathbf{K}] - \lambda[\mathbf{M}]) = 0$ is solved as if it were a transcendental equation in terms of λ. The Sturm sequence property of λ (which will be elaborated later) will prevent the possibility of missing any root within a given interval.

The subspace iteration method has been established as about the most suitable method for narrowly banded matrices from which four to 20 roots are to be extracted. The method iterates on a number of eigenpairs simultaneously, and endeavours to establish the orthogonality of the trial vectors during each iteration step. An attractive alternative is Lanczos' method which produces automatically the iterative vectors, in contrast with the assumed vectors in the subspace iteration method, and which ultimately produces a tridiagonal matrix. The solution obtained by Lanczos' method contains good approximations to the original system both in the lowest and highest spectral ranges. It is often regarded as the best method available for very large, sparse symmetrical matrices.

It is well known that Gauss elimination is in fact a congruence transformation since its standard form gives

$$[\mathbf{K}] = [\mathbf{L}][\mathbf{d}][\mathbf{L}]^{\mathrm{T}} \tag{2.9.1}$$

and therefore in spite of its wide usage in solution of static problems, it is seldom used for extracting eigenvalues directly. Owing to the fact that the matrices involved in a vibration analysis are always symmetrical, the orthogonal transformations which can preserve the symmetry of the matrix and magnitudes of eigenvalues are most widely used.

Sylvester's law of inertia states that there is a non-singular matrix $[\mathbf{S}]$ such that $[\mathbf{A}] = [\mathbf{S}][\mathbf{B}][\mathbf{S}]^{\mathrm{T}}$ if and only if the symmetrical matrices $[\mathbf{A}]$ and $[\mathbf{B}]$ have the same inertia, that is the same number of positive, negative and zero eigenvalues. In other words, the signs of the eigenvalues of a symmetric matrix do not change under congruence.

From Sylvester's law of inertia, the eigenvalues of $[\mathbf{d}]$ must all be positive since those of $[\mathbf{K}]$ are always positive for a stable conservative elastic system.

Let us now turn our attention back to the natural vibration problem with the generalized eigenvalue equation

$$[\mathbf{K}]\{\mathbf{x}\} = \lambda[\mathbf{M}]\{\mathbf{x}\}$$

which can be expressed in the form

$$([\mathbf{K}] - \lambda[\mathbf{M}])\{\mathbf{x}\} = \{\mathbf{0}\} \tag{2.9.2}$$

If $[\boldsymbol{\Phi}]$ is the matrix whose columns are modal vectors, then from the orthogonality properties,

$$\begin{aligned}[\boldsymbol{\Phi}]^{\mathrm{T}}([\mathbf{K}] - \lambda[\mathbf{M}])[\boldsymbol{\Phi}] &= [\boldsymbol{\Phi}]^{\mathrm{T}}[\mathbf{K}][\boldsymbol{\Phi}] - \lambda[\boldsymbol{\Phi}]^{\mathrm{T}}[\mathbf{M}][\boldsymbol{\Phi}] \\ &= [\boldsymbol{\Lambda}] - \lambda[\mathbf{I}]\end{aligned} \tag{2.9.3}$$

Table 2.9.1

$\bar{\lambda}$	$[\mathbf{K}]-\bar{\lambda}[\mathbf{M}]=[\mathbf{L}_{\bar{\lambda}}][\mathbf{d}_{\bar{\lambda}}][\mathbf{L}_{\bar{\lambda}}]^{\mathrm{T}}$	No. of "−ve" roots	Conclusion
1	$\begin{bmatrix} \frac{3}{2} & -1 & 0 \\ -1 & 3 & -1 \\ 0 & -1 & \frac{3}{2} \end{bmatrix} = \begin{bmatrix} 1 & 0 & 0 \\ -\frac{3}{2} & 1 & 0 \\ 0 & -\frac{3}{7} & 1 \end{bmatrix} \begin{bmatrix} \frac{3}{2} & 0 & 0 \\ 0 & \frac{3}{7} & 0 \\ 0 & 0 & \frac{15}{14} \end{bmatrix} \begin{bmatrix} 1 & -\frac{2}{3} & 0 \\ 0 & 1 & -\frac{3}{7} \\ 0 & 0 & 1 \end{bmatrix}$	0	All roots >1
1	$\begin{bmatrix} \frac{1}{2} & -1 & 0 \\ -1 & 1 & -1 \\ 0 & -1 & \frac{1}{2} \end{bmatrix} = \begin{bmatrix} 1 & 0 & 0 \\ -2 & 1 & 0 \\ 0 & 1 & 1 \end{bmatrix} \begin{bmatrix} \frac{1}{2} & 0 & 0 \\ 0 & -1 & 0 \\ 0 & 0 & \frac{3}{2} \end{bmatrix} \begin{bmatrix} 1 & -2 & 0 \\ 0 & 1 & 1 \\ 0 & 0 & 1 \end{bmatrix}$	1	1 < (one root) < 3
5	$\begin{bmatrix} -\frac{1}{2} & -1 & 0 \\ -1 & -1 & -1 \\ 0 & -1 & -\frac{1}{2} \end{bmatrix} = \begin{bmatrix} 1 & 0 & 0 \\ 2 & 1 & 0 \\ 0 & -1 & 1 \end{bmatrix} \begin{bmatrix} -\frac{1}{2} & 0 & 0 \\ 0 & 1 & 0 \\ 0 & 0 & \frac{3}{2} \end{bmatrix} \begin{bmatrix} 1 & 2 & 0 \\ 0 & 1 & -1 \\ 0 & 0 & 1 \end{bmatrix}$	2	3 < (one root) < 5
8	$\begin{bmatrix} -2 & -1 & 0 \\ -1 & -4 & -1 \\ 0 & -1 & -2 \end{bmatrix} = \begin{bmatrix} 1 & 0 & 0 \\ \frac{1}{2} & 1 & 0 \\ 0 & \frac{2}{7} & 1 \end{bmatrix} \begin{bmatrix} -2 & 0 & 0 \\ 0 & -\frac{7}{2} & 0 \\ 0 & 0 & -\frac{12}{7} \end{bmatrix} \begin{bmatrix} 1 & \frac{1}{2} & 0 \\ 0 & 1 & \frac{2}{7} \\ 0 & 0 & 1 \end{bmatrix}$	3	All roots <8

According to Eqs (4.3.6) and (4.3.7), the matrix $([\mathbf{K}]-\lambda[\mathbf{M}])$ can be decomposed as

$$([\mathbf{K}]-\lambda[\mathbf{M}])=[\mathbf{L}_\lambda][\mathbf{d}_\lambda][\mathbf{L}_\lambda]^{\mathrm{T}} \tag{2.9.4}$$

and thus,

$$[\mathbf{d}_\lambda]=[\mathbf{L}_\lambda]^{-1}([\mathbf{K}]-\lambda[\mathbf{M}])[\mathbf{L}_\lambda]^{-\mathrm{T}} \tag{2.9.5}$$

The matrices in both Eqs (2.9.3) and (2.9.5) are obtained by congruence transformations of $([\mathbf{K}]-\lambda[\mathbf{M}])$, and should therefore have the same inertia, i.e. $[\mathbf{d}_\lambda]$ and $([\mathbf{\Lambda}]-\lambda[\mathbf{I}])$ should have the same number of positive, negative and zero eigenvalues. Since the eigenvalues of the diagonal matrix $([\mathbf{\Lambda}]-\lambda[\mathbf{I}])$ are simply $\lambda_i-\lambda$, and similarly the eigenvalues of $[\mathbf{d}_\lambda]$ are simply d_{ii}, Sturm's theorem, which states that the number of negative eigenvalues of $([\mathbf{K}]-\lambda[\mathbf{M}])$ is equal to the number of negative diagonal coefficients of $[\mathbf{d}_\lambda]$, is thus established. Therefore, if a value is assumed for λ, say $\bar{\lambda}$, and $[\mathbf{d}_{\bar{\lambda}}]$ is computed according to Eq. (2.9.4), then the number of negative coefficients in $[\mathbf{d}_{\bar{\lambda}}]$ is equal to the number of eigenvalues λ_i which are smaller than the assumed value $\bar{\lambda}$.

For example, if the stiffness matrix and mass matrix are respectively,

$$[\mathbf{K}]=\begin{bmatrix} 2 & -1 & 0 \\ -1 & 4 & -1 \\ 0 & -1 & 2 \end{bmatrix} \quad \text{and} \quad [\mathbf{M}]=\begin{bmatrix} \frac{1}{2} & 0 & 0 \\ 0 & 1 & 0 \\ 0 & 0 & \frac{1}{2} \end{bmatrix}$$

by assuming various values for $\bar{\lambda}$, Table 2.9.1 can be compiled and the conclusions arrived at are self-evident.

2.10. Wittrick–Williams Algorithm

Sturm's theorem is extended to the eigensolution of $\det[\mathbf{D}(\lambda)]=\det[\mathbf{K}(\lambda)-\lambda\mathbf{M}(\lambda)]=0$ when all matrices of the conservative system are λ dependent. Use is made of Simpson's hypothesis. If the system is modelled by equivalent permissible finite elements, the following natural vibration equation is obtained:

$$[\mathbf{D}]\{\mathbf{q}\} = \begin{bmatrix} \mathbf{D}_{ss} & \mathbf{D}_{sm} \\ \mathbf{D}_{ms} & \mathbf{D}_{mm} \end{bmatrix} \begin{Bmatrix} \mathbf{q}_s \\ \mathbf{q}_m \end{Bmatrix} = \left(\begin{bmatrix} \mathbf{K}_{ss} & \mathbf{K}_{sm} \\ \mathbf{K}_{ms} & \mathbf{K}_{mm} \end{bmatrix} - \lambda \begin{bmatrix} \mathbf{M}_{ss} & \mathbf{M}_{sm} \\ \mathbf{M}_{ms} & \mathbf{M}_{mm} \end{bmatrix} \right) \begin{Bmatrix} \mathbf{q}_s \\ \mathbf{q}_m \end{Bmatrix} = \{\mathbf{0}\} \tag{2.10.1}$$

where $[\mathbf{K}_{sm}]$ and $[\mathbf{M}_{sm}]$ etc. are constant matrices. Since the finite element model has more coordinates than the continuum model, we have partitioned the equation so than $\{\mathbf{q}_m\}$ corresponds to the coordinates in the continuum model, whose equation of natural vibration is given by

$$[\mathbf{D}(\lambda)]\{\mathbf{q}_m\} = [\mathbf{K}(\lambda) - \lambda\mathbf{M}(\lambda)]\{\mathbf{q}_m\} = \{\mathbf{0}\} \tag{2.10.2}$$

For a specific λ, the Gaussian congruence of the dynamic stiffness matrix in Eq. (2.10.1) according to Eq. (2.9.5) is given by

$$\operatorname{diag}[\mathbf{d}_{ss}, \mathbf{d}_{mm}] = [\mathbf{L}]^{-1}[\mathbf{D}(\lambda)][\mathbf{L}]^{\mathrm{T}} \tag{2.10.3}$$

and the Gaussian congruence of that in Eq. (2.10.2) is given by

$$\operatorname{diag}[\mathbf{d}_m] = [\mathbf{L}_m]^{-1}[\mathbf{D}_m(\lambda)][\mathbf{L}_m]^{\mathrm{T}} \tag{2.10.4}$$

Obviously, $[\mathbf{d}_{mm}] = [\mathbf{d}_m]$. The number of negative entries in $[\mathbf{d}_m]$ is less than the number in $\operatorname{diag}[\mathbf{d}_{ss}, \mathbf{d}_{mm}]$ by an amount equal to the number of negative entries in $[\mathbf{d}_{ss}]$. Therefore, the Sturm number of Eq. (2.10.3) equals the Sturm number of Eq. (2.10.4) when $\{\mathbf{q}_m\} = \{\mathbf{0}\}$, and this is defined as the number of partial frequencies. The Wittrick–Williams algorithm uses the fact that the Sturm number of a continuum is equal to the Sturm number of its dynamic stiffness matrix plus the number of partial frequencies. The algorithm ensures that no natural frequencies are missed during a frequency search.

2.11. Derivatives of the Dynamic Stiffness

It is known that the mass matrix is equal to the negative of the first derivative of the dynamic stiffness matrix with respect to the square of the vibration frequency [8], or if

$$[\mathbf{D}(\omega)] = [\mathbf{K}(\omega) - \omega^2\mathbf{M}(\omega)] \tag{2.11.1}$$

then

$$[\mathbf{M}(\omega)] = -\frac{\partial}{\partial\omega^2}[\mathbf{D}(\omega)] \tag{2.11.2}$$

where $[\mathbf{D}]$, $[\mathbf{K}]$, $[\mathbf{M}]$ and ω are the dynamic stiffness matrix, the stiffness matrix, the mass matrix and the vibration frequency respectively. Note that $[\mathbf{K}(\omega)]$, $[\mathbf{M}(\omega)]$, $[\mathbf{D}(\omega)]$ are frequency dependent in general.

The above statement is generalized to any system parameters, such as vibration frequency, axial force and even flexural stiffness. For a given conservative structural system, if the dynamic stiffness matrix is given in the form

$$[\mathbf{D}(a_1, a_2, \ldots, a_m)] = \sum_{j=1}^{m} a_j[\mathbf{D}_j(a_1, a_2, \ldots, a_m)] \tag{2.11.3}$$

then,

$$\frac{\partial}{\partial a_j}[\mathbf{D}(a_1, a_2, \ldots, a_m)] = [\mathbf{D}_j(a_1, a_2, \ldots, a_m)] \tag{2.11.4}$$

Consider a system being described by the linear differential equation

$$\mathbf{L}(\mathbf{u}) = \sum_{j=1}^{m} a_j \mathbf{L}_j(\mathbf{u}) = \mathbf{0} \tag{2.11.5}$$

where $\mathbf{L}_j$, a_j and $\mathbf{u}$ are the differential operators, the system parameters and the dependent variables respectively. The variational statement of the corresponding problem over domain Ω is, for any arbitrary test function $\delta\mathbf{u}$,

$$\int_\Omega \delta\mathbf{u}\mathbf{L}(\mathbf{u})\,\mathrm{d}\Omega = \mathbf{0} \tag{2.11.6}$$

If the operator $\mathbf{L}$ is self-adjoint, a weaker statement on the continuity requirement of $\mathbf{u}$ can be rewritten as [9]

$$\int_\Omega \mathbf{A}(\delta\mathbf{u}, \mathbf{u})\,\mathrm{d}\Omega + \int_\Gamma \sum_{k=1}^{n} \mathbf{B}_k(\delta\mathbf{u})\mathbf{C}_k(\mathbf{u})\,\mathrm{d}\Gamma = \mathbf{0} \tag{2.11.7}$$

where Γ is the boundary of the domain Ω, $\mathbf{A}$, $\mathbf{B}$ and $\mathbf{C}$ are the functionals corresponding to the internal strain energy, essential boundary conditions and natural boundary conditions respectively. For a conservative system, $\mathbf{A}$ is symmetrical with respect to $\mathbf{u}$ and $\delta\mathbf{u}$, i.e. $\mathbf{A}(\delta\mathbf{u}, \mathbf{u}) = \mathbf{A}(\mathbf{u}, \delta\mathbf{u})$. The adjoint operations will be discussed in detail in Chap. 7.

The dynamic stiffness matrix is directly related to the functional $\mathbf{A}$. If exact solutions $\mathbf{N}_i$ can be found in terms of the system parameters alone such that

$$\mathbf{L}(\mathbf{N}_i) = \mathbf{0} \text{ in } \Omega \tag{2.11.8}$$

and

$$\mathbf{B}_k(\mathbf{N}_i) = \begin{cases} 1 \text{ on the boundary } \Gamma \text{ if } i = k \\ 0 \text{ if } i \neq k, \text{ for } i, k = 1 \text{ to } n \end{cases}$$

the general solutions for any boundary conditions $\mathbf{B}_i$ are given by

$$\mathbf{u} = \sum_{i=1}^{n} \mathbf{N}_i \mathbf{B}_i \tag{2.11.9}$$

Note that $\mathbf{N}_i(a_1, \ldots, a_m)$ need not be linear in a_j.

The stiffness matrix is given by [8]

$$\begin{aligned}[\mathbf{D}(a_1, \ldots, a_m)] &= \int_\Omega \mathbf{A}(\mathbf{N}^\mathrm{T}, \mathbf{N})\,\mathrm{d}\Omega = \int_\Omega \sum_{j=1}^{m} a_j \mathbf{A}_j(\mathbf{N}^\mathrm{T}, \mathbf{N})\,\mathrm{d}\Omega \\ &= \sum_{j=1}^{m} a_j \int_\Omega \mathbf{A}_j(\mathbf{N}^\mathrm{T}, \mathbf{N})\,\mathrm{d}\Omega = \sum_{j=1}^{m} a_j [\mathbf{D}_j(a_1, \ldots, a_m)],\end{aligned} \tag{2.11.10}$$

where $\mathbf{N} = \mathrm{row}[\mathbf{N}_i(a_1, a_2, \ldots, a_m)]$.

Differentiating Eq. (2.11.10) with respect to a_j gives

$$\frac{\partial}{\partial a_j}[\mathbf{D}(a_1, \ldots, a_m)] = [\mathbf{D}_j] + \int_\Omega \mathbf{A}(\mathring{\mathbf{N}}^\mathrm{T}, \mathbf{N})\,\mathrm{d}\Omega + \int_\Omega \mathbf{A}(\mathbf{N}^\mathrm{T}, \mathring{\mathbf{N}})\,\mathrm{d}\Omega \tag{2.11.11}$$

where a circle denotes differentiation with respect to a_j.

Since **A** is symmetrical, the last two integrals are the same and

$$\int_{\Omega} \mathbf{A}(\mathring{\mathbf{N}}^{\mathrm{T}}, \mathbf{N})\,\mathrm{d}\Omega = \int_{\Omega} \mathring{\mathbf{N}}^{\mathrm{T}}\mathbf{L}(\mathbf{N})\,\mathrm{d}\Omega - \int_{\Gamma} \sum_{k=1}^{n} \mathbf{B}_k(\mathring{\mathbf{N}}^{\mathrm{T}})\mathbf{C}_k(\mathbf{N})\,\mathrm{d}\Gamma = \mathbf{0} \tag{2.11.12}$$

since from Eq. (2.11.8), $\mathbf{L}(\mathbf{N}) = \mathbf{0}$ in Ω, and

$$\mathbf{B}_k(\mathring{\mathbf{N}}^{\mathrm{T}}) = \frac{\partial}{\partial a_j}[\mathbf{B}_k(\mathbf{N}_1(a_1, a_2, \ldots, a_m))]^{\mathrm{T}}$$

$$= (\text{either 0 or 1 on } \Gamma, \text{ independently of } a_j) = 0$$

Hence we obtain Eq. (2.11.4).

Example

Consider a Euler beam element vibrating with a frequency ω, and subject to a constant axial force P. The governing differential equation is

$$EI\frac{\mathrm{d}^4 v}{\mathrm{d}x^4} - P\frac{\mathrm{d}^2 v}{\mathrm{d}x^2} - \rho A\omega^2 v = 0$$

where P is positive when in compression. Then

$$L(\cdot) = EI\frac{\mathrm{d}^4}{\mathrm{d}x^4}(\cdot) - P\frac{\mathrm{d}^2}{\mathrm{d}x^2}(\cdot) - \rho A\omega^2(\cdot), \quad a_1 = P, \quad a_2 = \omega^2,$$

$$\mathbf{A}(\delta\mathbf{u}, \mathbf{u}) = \int_0^l EI\frac{\mathrm{d}^2}{\mathrm{d}x^2}(\delta\mathbf{u})^{\mathrm{T}}\frac{\mathrm{d}^2}{\mathrm{d}x^2}(\mathbf{u}) + P\frac{\mathrm{d}}{\mathrm{d}x}(\delta\mathbf{u})^{\mathrm{T}}\frac{\mathrm{d}}{\mathrm{d}x}(\mathbf{u}) - \omega^2\rho A\delta\mathbf{u}^{\mathrm{T}}\mathbf{u}\,\mathrm{d}x$$

The dynamic stiffness matrix $[\mathbf{D}]$, the stiffness matrix $[\mathbf{K}]$, the mass matrix $[\mathbf{M}]$ and the geometric matrix $[\mathbf{G}]$ are given by

$$[\mathbf{D}(\omega, P)] = [\mathbf{K}(\omega, P)] + P[\mathbf{G}(\omega, P)] - \omega^2[\mathbf{M}(\omega, P)]$$

$$[\mathbf{K}(\omega, P)] = \int_0^l EI[\mathbf{N}''(\omega, P, x)]^{\mathrm{T}}[\mathbf{N}''(\omega, P, x)]\,\mathrm{d}x$$

$$[\mathbf{G}(\omega, P)] = \int_0^l [\mathbf{N}'(\omega, P, x)]^{\mathrm{T}}[\mathbf{N}'(\omega, P, x)]\,\mathrm{d}x$$

$$[\mathbf{M}(\omega, P)] = \int_0^l \rho A[\mathbf{N}(\omega, P, x)]^{\mathrm{T}}[\mathbf{N}(\omega, P, x)]\,\mathrm{d}x$$

The shape function matrix $[\mathbf{N}]$ is given by [4, 5]

$$[\mathbf{N}(\omega, P, x)] = [\boldsymbol{\chi}(\xi)][\mathbf{F}], \quad \text{where } \xi = x/l$$

$$[\boldsymbol{\chi}(\xi)] = [\cos\alpha\xi \quad \sin\alpha\xi \quad \cosh\beta\xi \quad \sinh\beta\xi]$$

$$[\mathbf{F}] = \frac{1}{\alpha^2 + \beta^2}\begin{bmatrix} \beta^2 - F_4 & F_2 l & -F_3 & F_1 l \\ -F_6/\alpha & (\alpha + F_4/\alpha)/l & -F_5/\alpha & -F_3 l/\alpha \\ \alpha^2 + F_4 & -F_2 l & F_3 & -F_1 l \\ F_6/\beta & (\beta - F_4/\beta)/l & F_5/\beta & F_3 l/\beta \end{bmatrix}$$

$$F_1 = (\beta \sin\alpha - \alpha \sinh\beta)(\alpha^2 + \beta^2)/\delta$$
$$F_2 = (\alpha\cos\alpha\sinh\beta - \beta\sin\alpha\cosh\beta)(\alpha^2 + \beta^2)/\delta$$
$$F_3 = (\cos\alpha - \cos\beta)\alpha\beta(\alpha^2 + \beta^2)/\delta$$
$$F_4 = ((\beta^2 - \alpha^2)(\cos\alpha\cosh\beta - 1) + 2\,\alpha\beta\sin\alpha\sinh\beta)\alpha\beta/\delta$$
$$F_5 = (\beta\sinh\beta + \alpha\sin\alpha)(\alpha^2 + \beta^2)\alpha\beta/\delta$$
$$F_6 = -(\alpha\cosh\beta\sin\alpha + \beta\sinh\beta\cos\alpha)(\alpha^2 + \beta^2)\alpha\beta/\delta$$
$$\delta = 2\alpha\beta(\cos\alpha\cosh\beta - 1) + (\alpha^2 - \beta^2)\sin\alpha\sinh\beta$$
$$\alpha^2 = -\frac{\sigma^2}{2} + \sqrt{\frac{\sigma^4}{4} + \lambda^4}, \qquad \beta^2 = \frac{\sigma^2}{2} + \sqrt{\frac{\sigma^4}{4} + \lambda^4}$$
$$\lambda^4 = \frac{\rho A l^4 \omega^2}{EI} \quad \text{and} \quad \sigma^2 = \frac{Pl^2}{EI}$$

The dynamic stiffness is given by

$$[\mathbf{D}(\omega, P)] = \frac{EI}{l^3}\begin{bmatrix} F_6 & -F_4 l & F_5 & F_3 l \\ & F_2 l^2 & -F_3 l & F_1 l^2 \\ & & F_6 & F_4 l \\ \text{sym.} & & & F_2 l^2 \end{bmatrix}$$

The mass matrix is then found by differentiation as

$$[\mathbf{M}(\omega, P)] = \int_0^l \rho A[\mathbf{N}]^{\mathrm{T}}[\mathbf{N}]\,\mathrm{d}x = -\frac{\partial}{\partial\omega^2}[\mathbf{D}(\omega, P)]$$
$$= -\rho A l\begin{bmatrix} M_6 & -M_4 l & M_5 & M_3 l \\ & M_2 l^2 & -M_3 l & M_1 l^2 \\ & & M_6 & M_4 l \\ \text{sym.} & & & M_2 l^2 \end{bmatrix}$$

where

$$M_1 = -F_1\delta_\omega + \frac{1}{\delta}\left(\frac{2}{\alpha^2 + \beta^2}(\beta\sin\alpha - \alpha\sinh\beta) + \frac{1}{2\alpha\beta}(\alpha\sin\alpha - \beta\sinh\beta + \beta^2\cos\alpha - \alpha^2\cosh\beta)\right)$$

$$M_2 = -F_2\delta_\omega + \frac{1}{\delta}\left(\frac{2}{\alpha^2 + \beta^2}(\alpha\cos\alpha\sinh\beta - \beta\sin\alpha\cosh\beta) - \sin\alpha\sinh\beta + \frac{1}{2\alpha\beta}(\beta\cos\alpha\sinh\beta - \alpha\sin\alpha\cosh\beta) + \frac{\alpha^2 - \beta^2}{2\alpha\beta}\cos\alpha\cosh\beta\right)$$

$$M_3 = -F_3\delta_\omega + \frac{1}{\delta}\left(\frac{2\alpha\beta}{\alpha^2 + \beta^2} + \frac{\alpha^2 + \beta^2}{2\alpha\beta}\right)(\cos\alpha - \cosh\beta) - \frac{1}{2}(\beta\sin\alpha + \alpha\sinh\beta)$$

$$M_4 = -F_4\delta_\omega + \frac{1}{\delta}\left(\frac{\beta^2 - \alpha^2}{2\alpha\beta}(\cos\alpha\cosh\beta - 1) + 2\sin\alpha\sinh\beta\right.$$

$$+\frac{1}{2}\frac{\beta^2-\alpha^2}{\alpha^2+\beta^2}(-\beta\sin\alpha\cosh\beta+\alpha\cos\alpha\sinh\beta)$$

$$\left.+\frac{\alpha\beta}{\alpha^2+\beta^2}(\beta\cos\alpha\sinh\beta+\alpha\sin\alpha\cosh\beta)\right)$$

$$M_5=-F_5\delta_\omega+\frac{1}{\delta}\left(\left(\frac{2\alpha\beta}{\alpha^2+\beta^2}+\frac{\alpha^2+\beta^2}{2\alpha\beta}\right)(\alpha\sin\alpha+\beta\sinh\beta)\right.$$

$$\left.+\frac{1}{2}(\alpha\sinh\beta+\beta\sin\alpha)+\frac{\alpha\beta}{2}(\cosh\beta+\cos\alpha)\right)$$

$$M_6=-F_6\delta_\omega-\frac{1}{\delta}\left(\left(\frac{2\alpha\beta}{\alpha^2+\beta^2}+\frac{\alpha^2+\beta^2}{2\alpha\beta}\right)(\beta\cos\alpha\sinh\beta+\alpha\sin\alpha\cosh\beta)\right.$$

$$\left.+\alpha\beta\cos\alpha\cosh\beta+\frac{1}{2}(\alpha\cos\alpha\sinh\beta+\beta\sin\alpha\cosh\beta)+\frac{\alpha^2-\beta^2}{2}\sin\alpha\sinh\beta\right)$$

$$\delta_\omega=\frac{1}{\delta}\left(\frac{1}{\alpha\beta}(\cos\alpha\cosh\beta-1)+\frac{1}{\alpha^2+\beta^2}(\alpha\cos\alpha\sinh h\beta-\beta\sin\alpha\cosh h\beta)\right.$$

$$\left.+\frac{\alpha^2-\beta^2}{2\alpha\beta}\frac{1}{\alpha^2+\beta^2}(\beta\cos\alpha\sinh\beta+\alpha\sin\alpha\cosh\beta)\right)$$

The geometric matrix is also found by differentiation as

$$[G(\omega,P)]=\int_0^l[\mathbf{N}']^{\mathrm{T}}[\mathbf{N}']\,\mathrm{d}x=-\frac{\partial}{\partial P}[\mathbf{D}(\omega,P)]=\frac{1}{l}\begin{bmatrix}G_6 & -G_4l & G_5 & G_3l\\ & G_2l^2 & -G_3l & G_1l^2\\ & & G_6 & G_4l\\ \text{sym.} & & & G_2l^2\end{bmatrix}$$

where

$$G_1=-F_1\delta_P+\frac{1}{\delta}\left(\frac{\beta^2-\alpha^2}{\alpha^2+\beta^2}(\beta\sin\alpha-\alpha\sinh\beta)\right.$$

$$\left.+\frac{1}{2}(\beta\sin\alpha+\alpha\sinh\beta)-\frac{\alpha\beta}{2}(\cosh\beta+\cos\alpha)\right)$$

$$G_2=-F_2\delta_P+\frac{1}{\delta}\left(\frac{\beta^2-\alpha^2}{\alpha^2+\beta^2}(-\beta\sin\alpha-\alpha\cosh\beta+\alpha\cos\alpha\sinh\beta)\right.$$

$$\left.+\frac{\alpha^2-\beta^2}{2}\sin\alpha\sinh\beta-\frac{1}{2}(\alpha\cos\alpha\sinh\beta+\beta\sin\alpha\cosh\beta)+\alpha\beta\cos\alpha\cosh\beta\right)$$

$$G_3=-F_3\delta_P+\frac{1}{\delta}\left(\frac{\beta^2-\alpha^2}{\alpha^2+\beta^2}\alpha\beta(\cos\alpha-\cosh\beta)+\frac{\alpha\beta}{2}(\alpha\sin\alpha-\beta\sinh\beta)\right)$$

$$G_4=-F_4\delta_P+\frac{1}{\delta}\left(\alpha\beta(\cos\alpha\cosh\beta-1)+\frac{\beta^2-\alpha^2}{\alpha^2+\beta^2}\frac{\alpha\beta}{2}(\beta\cos\alpha\sinh\beta+\alpha\sin\alpha\cosh\beta)\right.$$

$$\left.-\frac{\alpha^2\beta^2}{\alpha^2+\beta^2}(\alpha\cos\alpha\sinh\beta-\beta\sin\alpha\cosh\beta)\right)$$

$$G_5 = -F_5\delta_P + \frac{1}{\delta}\left(\frac{\beta^2-\alpha^2}{\alpha^2+\beta^2}\alpha\beta(\alpha\sin\alpha+\beta\sinh\beta) - \frac{\alpha\beta}{2}(\alpha\sin\alpha-\beta\sinh\beta)\right.$$

$$\left. + \frac{\alpha\beta}{2}(\beta^2\cosh\beta-\alpha^2\cos\alpha)\right)$$

$$G_6 = -F_6\delta_P - \frac{1}{\delta}\left(\frac{\beta^2-\alpha^2}{\alpha^2+\beta^2}\alpha\beta(\beta\cos\alpha\sinh\beta+\alpha\sin\alpha\cosh\beta)\right.$$

$$+ \frac{\alpha\beta}{2}(\beta^2-\alpha^2)\cos\alpha\cosh\beta + \frac{\alpha\beta}{2}(\beta\cos\alpha\sinh\beta-\alpha\sin\alpha\cosh\beta)$$

$$\left. + \alpha^2\beta^2\sin\alpha\sinh\beta\right)$$

$$\delta_P = \frac{1}{\delta}\left(\frac{\alpha\beta}{\alpha^2+\beta^2}(\beta\cos\alpha\sinh\beta+\alpha\sin\alpha\cosh\beta) - \sin\alpha\sinh\beta\right.$$

$$\left. + \frac{1}{2}\frac{\beta^2-\alpha^2}{\alpha^2+\beta^2}(-\beta\sin\alpha\cosh\beta+\alpha\cos\alpha\sinh\beta)\right)$$

Numerically, if $EI = 1$, $\rho A = 1$, $l = 1$, $\omega = 2$ and $P = 5$,

$$[\mathbf{D}] = \begin{bmatrix} 4.463016 & 5.252110 & -6.475754 & 5.620023 \\ 5.252110 & 3.238410 & -5.620023 & 2.232281 \\ -6.475754 & -5.620023 & 4.463016 & -5.252110 \\ 5.620023 & 2.232281 & -5.252110 & 3.238410 \end{bmatrix}$$

$$[\mathbf{M}] = \begin{bmatrix} 0.376476 & 0.057635 & 0.129925 & -0.035105 \\ 0.057635 & 0.011597 & 0.035105 & -0.008995 \\ 0.129925 & 0.035105 & 0.376476 & -0.057635 \\ -0.035105 & -0.008995 & -0.057635 & 0.011597 \end{bmatrix}$$

$$[\mathbf{G}] = \begin{bmatrix} 1.217204 & 0.111908 & 1.216852 & 0.104582 \\ 0.111908 & 0.155912 & -0.104582 & -0.047934 \\ -1.216852 & -0.104582 & 1.216852 & -0.111908 \\ 0.104582 & -0.047934 & -0.111908 & 0.155912 \end{bmatrix}$$

The maximum errors when comparing the results of direct integration and the explicit formulae are 3.102E − 15 for the mass matrix and 8.500E − 15 for the geometric matrix.

Application of the derivatives of the dynamic stiffness matrix for nonlinear problems can be found in reference [10]. Also, when solving an eigenvalue problem using Newton's method, the derivatives of the dynamic stiffness matrix are required. If λ and $\{\mathbf{v}_0\}$ are approximations of an eigenvalue problem, better approximations $\lambda + \Delta\lambda$ and $\{\mathbf{v}_0 + \Delta\mathbf{v}\}$ are given by the following inverse iteration,

$$[\mathbf{D}(\lambda+\Delta\lambda)]\{\mathbf{v}_0+\Delta\mathbf{v}\} = \{\mathbf{0}\}, \text{ or}$$

$$[\mathbf{D}(\lambda)]\{\mathbf{v}_0+\Delta\mathbf{v}\} = -\Delta\lambda\frac{\partial}{\partial\lambda}[\mathbf{D}(\lambda)]\{\mathbf{v}\} + \text{neglected higher order terms}$$

Since $\{\mathbf{v}_0 + \Delta\mathbf{v}\}$ will be normalized, $\Delta\lambda$ is irrelevant and a new approximation of λ

will be obtained from $\{\mathbf{v}_0 + \Delta\mathbf{v}\}$. It can therefore be shown that for vibration problem [7],

$$[\mathbf{D}(\omega_i)]\{\mathbf{v}_{i+1}\} = -\frac{\partial}{\partial\omega^2}[\mathbf{D}(\omega, P)]\{\mathbf{v}_i\} = [\mathbf{M}(\omega_i)]\{\mathbf{v}_i\}$$

and for stability problems,

$$[\mathbf{D}(P_i)]\{\mathbf{v}_{i+1}\} = -\frac{\partial}{\partial P}[\mathbf{D}(\omega, P)]\{\mathbf{v}_i\} = -[\mathbf{G}(P_i)]\{\mathbf{v}_i\}$$

References

1. AYT Leung 1980. Dynamics of periodic structures. J Sound Vib 72, 451–467
2. JS Przemieniecki 1968. Theory of matrix structural analysis. McGraw-Hill, New York
3. NJ Fergusson, WD Pilkey 1992. Frequency dependent element mass matrices. J Appl Mech 59, 136–139
4. V Kolousek 1973. Dynamics in engineering structures. Butterworth, London
5. WP Howson, JR Benerjee, FW Williams 1983. Concise equations and program for exact eigensolutions of plane frames including member shear. Adv Engng Soft 5, 137–141
6. K Washizu 1968. Variational methods in elasticity and plasticity. Pergamon Press
7. A Simpson 1984. On the solution of $S(\omega)\,x = 0$ by a Newtonian procedure. J Sound Vib 97, 153–164
8. TH Richard, YT Leung 1977. An accurate method in structural vibration analysis, J Sound Vib 55, 363–376
9. OC Zienkiewicz 1977. The finite element method, 3rd edn. McGraw-Hill, New York
10. AYT Leung, TC Fung 1990. Nonlinear vibration of frames by the incremental dynamic stiffness method. Int J Num Meth Engng 29, 337–356

Chapter 3

Dynamic Substructures

Regardless of their simplicity, all structures have an infinite number of degrees of freedom (d.o.f.) when subjected to dynamic loading. One of the main objectives in selecting a mathematical model is to reduce the infinite d.o.f. system to a model with a limited number of d.o.f. which capture the significant physical behaviour of the system.

Finite element modelling is one of the most successful approaches for rationally establishing a relation between the displacements and forces at a finite number of discrete points in a continuous structure. These discrete points are called nodes. In a harmonic vibration analysis, the finite element method results in

$$[\bar{\mathbf{D}}]\{\mathbf{x}\} = [\bar{\mathbf{K}} - \omega^2\bar{\mathbf{M}}]\{\mathbf{x}\} = \{\mathbf{X}\} \tag{3.0.1}$$

where ω is the frequency of vibration, $[\bar{\mathbf{D}}]$, $[\bar{\mathbf{K}}]$, $[\bar{\mathbf{M}}]$ are the dynamic stiffness (or the impedance), stiffness and mass matrices respectively, and $\{\mathbf{x}\}$ and $\{\mathbf{X}\}$ are the amplitude vectors of the nodal displacements and forces. Very often, $\{\mathbf{x}\}$ has to be solved for a large number of different frequencies in order to estimate the response, either deterministically or randomly by frequency domain methods. Similar operations are required by other methods of analysis such as time step integration. Since the order of the matrices is normally large, a major portion of the computational effort is spend on this inversion process. The method of dynamic flexibility (or receptance) is designed to evaluate $\{\mathbf{x}\}$ from

$$[\bar{\mathbf{Z}}(\omega)]\{\mathbf{X}\} = \{\mathbf{x}\} \tag{3.0.2}$$

where $[\bar{\mathbf{Z}}]$ is the dynamic flexibility, the inverse of $[\bar{\mathbf{D}}]$. Explicit forms of $[\bar{\mathbf{Z}}]$ for an elastic continuum are very involved mathematically, so a simplified modal method has been developed [1]. This modal method, however, further reduces the number of d.o.f. to a limited number of modal components, so that accurate results which are comparable to the original finite element model may not be possible unless higher modes are included. To avoid the elaborate process of computing accurate higher modes, the mode-acceleration method is recommended [2].

In a large-scale structural analysis, not all the nodes are subjected to external forces, nor are all the nodal displacements of interest. The eigenvalue economization procedure [3] is developed in order to condense the matrices by eliminating the passive d.o.f. and thus minimizing the computational effort. An exact condensation method is also available [4].

Substructure methods of dynamic analysis have been developed which reduce the number of coordinates in a dynamic analysis of a complex structure, permit analysis and design of different portions of a structure to proceed independently, and permit testing of various parts of a structure to be performed individually. The terms "master" and "slave" refer to the interface coordinates and internal coordinates of a substructure, respectively. When a substructure vibrates independently with respect to the other parts of the whole structure, the modes are called component modes or partial modes.

Two categories of substructure methods are found in the literature. One is based mainly on the elimination of the slaves in the dynamic stiffness relations, and the other on defining the substructure by the partial modes. Among the former category the method of Guyan [5] and Irons [6] has been widely used [7] and is referred to as eigenvalue economization. To achieve reasonably accurate results, the masters must be chosen with care, or some of the lowest frequencies in the eigenspectrum may be lost [8]. It is further required that different sets of masters are necessary for different modes because of the redistribution of energy with respect to the coordinates. A direct extension of the method to substructures is difficult. To overcome these deficiencies, an exact formulation of such a method has been presented [4]. It is divided in two submethods. In one, fixed interface substructure modes are used, i.e. the free vibration modes of a substructure when all masters are clamped; and in the other, free interface modes are used, i.e. the free vibration modes of a substructure when all masters are unrestrained. Dynamic substructure methods [9, 10] are examples of the former; Kron's method [11] is an example of the latter.

The modal synthesis methods under the latter category may further be classified as fixed-interface methods [12–14], free-interface methods [15, 16] or hybrid methods [17, 18] depending upon whether the mode shapes used to define substructure coordinates are obtained with the master coordinates fixed, free or in combination. As noted by Benfield and Hruda [17], the fixed-interface methods produce better accuracy. One of the main disadvantages of these methods of modal synthesis in engineering practice is that modal coordinates are used in place of physical coordinates. The fact that the size of the substructure matrices depends on the number of partial modes taken and that the modal coordinates must be transformed to physical coordinates in order to assemble the substructure matrices into global matrices by means of equilibrium and compatibility conditions, makes it extremely difficult to incorporate these methods into finite element techniques. Wilson [18] attempts to improve such methods by using physical coordinates. Unfortunately, the resulting generalized matrices are of the same order as the original system and are applied to cantilever systems only.

The substructure modes are sometimes approximated by Ritz vectors [19] or Lanczos vectors [20]. However, it has been pointed out [21–23] that all the above methods are similar, and it is sufficient to concentrate on one version of the methods.

Lightly damped substructures are discussed by Hasselman [24, 25] using a modal damping method and generally damped substructures are studied by Leung [22]. Hale [26] and Hale and Bergman [27] introduce the synthesis method for non-conservative systems. Defective modes are excluded, however. Flutter of frames is studied by Leung [28] using a generally non-conservative dynamic substructure method.

A survey of the methods used up to 1981 can be found in Craig [29] and Meirovitch and Hale [30].

3.1. Exact Dynamic Condensation

The following equation results from a finite element analysis of linear systems undergoing forced harmonic oscillations

$$[\mathbf{D}]\{\mathbf{X}\} = \{\mathbf{F}\} \tag{3.1.1}$$

where

$$[\mathbf{D}(\omega)] = [\mathbf{K}] - \omega^2[\mathbf{M}]$$

$[\mathbf{D}]$ is the dynamic stiffness matrix

$[\mathbf{K}]$ and $[\mathbf{M}]$ are the stiffness and mass matrices, respectively

These matrices may be functions of frequency depending on the method of analysis.

$\{\mathbf{F}\}e^{i\omega t}$ and $\{\mathbf{X}\}e^{i\omega t}$ are the force excitation and displacement response vectors respectively.

A system may be referred to an element, a substructure or an overall system. Upon choosing a set of masters and slaves such that the slave coordinates are not subjected to driving forces, Eq. (3.1.1) is partitioned as

$$\begin{bmatrix} \mathbf{D}_{mm} & \mathbf{D}_{ms} \\ \mathbf{D}_{sm} & \mathbf{D}_{ss} \end{bmatrix} \begin{Bmatrix} \mathbf{X}_m \\ \mathbf{X}_s \end{Bmatrix} = \begin{Bmatrix} \mathbf{F}_m \\ \mathbf{0} \end{Bmatrix} \tag{3.1.2}$$

and accordingly,

$$[\mathbf{K}] = \begin{bmatrix} \mathbf{K}_{mm} & \mathbf{K}_{ms} \\ \mathbf{K}_{sm} & \mathbf{K}_{ss} \end{bmatrix} \tag{3.1.3}$$

and

$$[\mathbf{M}] = \begin{bmatrix} \mathbf{M}_{mm} & \mathbf{M}_{ms} \\ \mathbf{M}_{sm} & \mathbf{M}_{ss} \end{bmatrix} \tag{3.1.4}$$

The subscripts m and s refer to masters and slaves respectively. Eliminating $\{\mathbf{X}_s\}$ from Eq. (3.1.2), the following equations are derived:

$$\{\mathbf{X}_s\} = -[\mathbf{D}_{ss}]^{-1}[\mathbf{D}_{sm}]\{\mathbf{X}_m\} \tag{3.1.5}$$

and

$$[\mathbf{D}^*]\{\mathbf{X}_m\} = \{\mathbf{F}_m\} \tag{3.1.6}$$

where

$$[\mathbf{D}^*] = [\mathbf{D}_{mm}] - [\mathbf{D}_{ms}][\mathbf{D}_{ss}]^{-1}[\mathbf{D}_{sm}] \tag{3.1.7}$$

is the condensed dynamic stiffness matrix associated with the masters.

It has been shown [31] that the dynamic stiffness $[\mathbf{D}]$ and the mass matrix $[\mathbf{M}]$ of an elastic system are related by

$$[\mathbf{M}] = -\frac{\partial}{\partial \omega^2}[\mathbf{D}] \tag{3.1.8}$$

where $[\mathbf{M}]$ and $[\mathbf{D}]$ correspond to the same set of coordinates. $[\mathbf{M}]$ may be a function of frequency when the system is of distributed mass or when the order of the matrices is less than the actual number of d.o.f. of the system as in the case of condensed matrices.

Upon differentiating the dynamic stiffness matrix $[\mathbf{D}^*]$,

$$[\mathbf{M}^*] = [\mathbf{M}_{mm}] - [\mathbf{D}_{ms}][\mathbf{D}_{ss}]^{-1}[\mathbf{M}_{sm}] - [\mathbf{M}_{ms}][\mathbf{D}_{ss}]^{-1}[\mathbf{D}_{sm}] + [\mathbf{D}_{ms}][\mathbf{D}_{ss}]^{-1}[\mathbf{M}_{ss}][\mathbf{D}_{ss}]^{-1}[\mathbf{D}_{sm}] \tag{3.1.9}$$

where

$$[\mathbf{M}_{mm}] = -\frac{\partial}{\partial \omega^2}[\mathbf{D}_{mm}]$$

etc., are the original partitioned mass matrices as in Eq. (3.1.4).

Alternatively, $[\mathbf{M}^*]$ may be found by replacing differentials by differences. However, the author finds that the accuracy of the eigenvalues computed using the reduced eigenvalue problem depends substantially on the choice of the step length. To avoid the reduction of significant figures resulting from differencing, an optimum step length has not been found so far. This alternative method is not recommended in the numerical computation.

The reduced equation for harmonic vibration of the system is therefore,

$$[\mathbf{D}^*]\{\mathbf{X}_m\} = \{\mathbf{F}_m\}$$

or equivalently,

$$([\mathbf{K}^*] - \omega^2[\mathbf{M}^*])\{\mathbf{X}_m\} = \{\mathbf{F}_m\} \tag{3.1.10}$$

where

$$[\mathbf{K}^*] = [\mathbf{D}^*] + \omega^2[\mathbf{M}^*] \tag{3.1.11}$$

$[\mathbf{K}^*]$ and $[\mathbf{M}^*]$ being positive-definite matrices.

When the system is vibrating freely, one has

$$([\mathbf{K}^*] - \omega^2[\mathbf{M}^*])\{\mathbf{X}_m\} = \{\mathbf{0}\}$$

or

$$[\mathbf{K}^*]\{\mathbf{X}_m\} = \omega^2[\mathbf{M}^*]\{\mathbf{X}_m\} \tag{3.1.12}$$

which is the eigenvalue problem for the natural frequency ω and mode shape $\{\mathbf{X}_m\}$. The mode shape of the slaves $\{\mathbf{X}_s\}$ is determined from Eq. (3.1.5).

Apart from the approximation inherent in Eq. (3.1.1), the derivation is exact. Since the matrices involved are dependent on the vibration frequency, the method of halving the interval of frequency, i.e. bisection, in tandem with the employment of Sturm sequence properties of the dynamic stiffness matrix [32] may be used.

The previous formulation can be extended to dynamic substructuring. Suppose N substructures are to be considered and are connected to each other through the masters of the individual members only. If the superscripts denote the substructures, then the overall equation of motion is

$$\begin{bmatrix} \mathbf{D}_{ss}^{(1)} & & & & \mathbf{D}_{sm}^{(1)} \\ & \mathbf{D}_{ss}^{(2)} & & \mathbf{0} & \mathbf{D}_{sm}^{(2)} \\ & \mathbf{0} & \ddots & & \vdots \\ & & & \mathbf{D}_{ss}^{(N)} & \mathbf{D}_{sm}^{(N)} \\ \mathbf{D}_{ms}^{(1)} & \mathbf{D}_{ms}^{(2)} & \cdots & \mathbf{D}_{ms}^{(N)} & (\mathbf{D}_{mm}^{(1)} + \mathbf{D}_{mm}^{(2)} + \cdots \mathbf{D}_{mm}^{(N)}) \end{bmatrix} \begin{Bmatrix} \mathbf{X}_s^{(1)} \\ \mathbf{X}_s^{(2)} \\ \vdots \\ \mathbf{X}_s^{(N)} \\ \mathbf{X}_m \end{Bmatrix} = \begin{Bmatrix} \mathbf{0} \\ \mathbf{0} \\ \vdots \\ \mathbf{0} \\ \mathbf{F}_m \end{Bmatrix} \tag{3.1.13}$$

It can be condensed as

$$[\mathbf{D}^*]\{\mathbf{X}_m\} = \{\mathbf{F}_m\} \tag{3.1.14}$$

$$\{\mathbf{X}_s^{(i)}\} = [\mathbf{D}_{ss}^{(i)}]^{-1}[\mathbf{D}_{sm}]\{\mathbf{X}_m\} \qquad i = 1, 2, \ldots, N \tag{3.1.15}$$

where

$$[\mathbf{D}^*] = \sum_{i=1}^{N} ([\mathbf{D}_{mm}^{(i)}] - [\mathbf{D}_{ms}^{(i)}][\mathbf{D}_{ss}^{(i)}]^{-1}[\mathbf{D}_{sm}^{(i)}]) \tag{3.1.16}$$

The eigenvalue problem for natural vibration is

$$[\mathbf{K}^*]\{\mathbf{X}_m\} = \omega^2[\mathbf{M}^*]\{\mathbf{X}_m\} \tag{3.1.17}$$

where

$$[\mathbf{M}^*] = \sum_{i=1}^{N} ([\mathbf{M}_{mm}^{(i)}] - [\mathbf{D}_{ms}^{(i)}][\mathbf{D}_{ss}^{(i)}]^{-1}[\mathbf{M}_{sm}^{(i)}] - [\mathbf{M}_{ms}^{(i)}][\mathbf{D}_{ss}^{(i)}]^{-1}[\mathbf{D}_{sm}^{(i)}]$$
$$+ [\mathbf{D}_{ms}^{(i)}][\mathbf{D}_{ss}^{(i)}]^{-1}[\mathbf{M}_{ss}^{(i)}][\mathbf{D}_{ss}^{(i)}]^{-1}[\mathbf{D}_{sm}^{(i)}])$$

and

$$[\mathbf{K}^*] = [\mathbf{D}^*] + \omega^2[\mathbf{M}^*]$$

The basic computation steps for a single mode proceed as follows:

1. Choose a frequency ω_0 about which the natural frequency is of interest. If the lowest mode is required, let $\omega_0 = 0$.
2. Form the matrix $[\mathbf{D}] = [\mathbf{K}] - \omega_0^2[\mathbf{M}]$ and partition
$$[\mathbf{D}] = \begin{bmatrix} \mathbf{D}_{mm} & \mathbf{D}_{ms} \\ \mathbf{D}_{sm} & \mathbf{D}_{ss} \end{bmatrix}, \qquad [\mathbf{M}] = \begin{bmatrix} \mathbf{M}_{mm} & \mathbf{M}_{ms} \\ \mathbf{M}_{sm} & \mathbf{M}_{ss} \end{bmatrix}$$
3. Solve for the matrix $[\mathbf{Z}]$ from
$$[\mathbf{D}_{ss}][\mathbf{Z}] = [\mathbf{D}_{sm}]$$
4. Perform the following matrix operations
$$[\mathbf{A}] = [\mathbf{M}_{ms}][\mathbf{Z}]$$
$$[\mathbf{M}^*] = [\mathbf{M}_{mm}] - [\mathbf{A}]^{\mathrm{T}} - [\mathbf{A}] + [\mathbf{Z}]^{\mathrm{T}}[\mathbf{M}_{ss}][\mathbf{Z}]$$
$$[\mathbf{D}^*] = [\mathbf{D}_{mm}] - [\mathbf{D}_{ms}][\mathbf{Z}]$$
5. Solve the eigenvalue problem for the lowest absolute eigenvalue ρ and mode shape $\{\mathbf{X}_m\}$
$$[\mathbf{D}^*]\{\mathbf{X}_m\} = \rho[\mathbf{M}^*]\{\mathbf{X}_m\}$$
6. If $[\rho/\omega_0] < \varepsilon$, an acceptable error, then take the required frequency as $\sqrt{(\omega_0^2 + \rho)}$, otherwise replace ω_0 by $\sqrt{(\omega_0^2 + \rho)}$ and go to step 2.
7. Multiply
$$\{\mathbf{X}_s\} = -[\mathbf{Z}]\{\mathbf{X}_m\}$$
The required mode shape is given by
$$\{\mathbf{X}\} = \begin{Bmatrix} \mathbf{X}_m \\ \mathbf{X}_s \end{Bmatrix}$$

The condensed eigenvalue problem may be solved by inverse iteration [33]. However, it should be solved by subspace iteration [34] if the method is to be modified to generate a few modes at a time in order to avoid the difficulties of repeated natural frequencies.

Convergence to the modes of interest may be safeguarded by invoking the Sturm theorem associated with the dynamic stiffness matrix. Let $s[\mathbf{D}]$ denote the Sturm number of the matrix $[\mathbf{D}]$, that is the number of negative elements on the diagonal of the triangularized form of $[\mathbf{D}]$ obtained using Gauss elimination without interchanges, then the Sturm sequence properties of $[\mathbf{D}]$ ensure that the number of natural modes below ω^* is equal to $s[\mathbf{D}(\omega^*)]$. It can be shown that

$$s[\mathbf{D}] = s[\mathbf{D}_{ss}] + s[\mathbf{D}^*]$$

where $s[\mathbf{D}_{ss}]$ and $s[\mathbf{D}^*]$ may be obtained as in steps 3 and 5 of the algorithm when the matrices are decomposed.

The algorithm may be used for substructuring techniques with the following modification. Perform steps 2–4 to all the substructure and assemble them immediately before step 5. The Sturm number of the overall system is then given by

$$\left(\sum_{i=1}^{N} s[\mathbf{D}_{ss}^{(i)}]\right) + s[\mathbf{D}^*]$$

where $[\mathbf{D}^*]$ represents the assemblage of the condensed dynamic stiffness. The most time-consuming step in the algorithm is the solution of the equations

$$[\mathbf{D}_{ss}][\mathbf{Z}] = [\mathbf{D}_{sm}]$$

and advantage should be taken of the banded nature of the matrix $[\mathbf{D}_{ss}]$.

Experience shows that, if the method is applied to substructuring problems, the computation is best done in two phases. First, the eigenvalues are isolated in closed intervals by Sturm's theorem, then the eigenpairs are computed by the iterative process mentioned above. The subspace iteration method is recommended for solving the condensed eigenvalue problem because close eigenvalues quite often occur.

Two examples are given here. The first example illustrates some information about the possible improvement of the accuracy of the results as compared to the normal economization method, as well as the rate of convergence with respect to the full solutions. The second example demonstrates the advantage of this method when dealing with systems having repeated substructures.

Example 3.1.1

Consider the asymmetrical space frame shown in Fig. 3.1.1. The whole structure consists of 47 beams of squared cross-sections. All elements are made of aluminium and have the following properties:

Young's modulus $E = 69 \times 10^9\,\mathrm{N\,m^{-2}}$
Shear modulus $G = 26 \times 10^9\,\mathrm{N\,m^{-2}}$
Mass density $\rho = 2.7 \times 10^3\,\mathrm{kg\,m^{-2}}$
Cross-section $A = 0.4 \times 10^{-3}\,\mathrm{m^2}$
Shear factor $f = 1.2$

The length of elements 1–12 is 1.0 m, 13–26 is 1.3 m, and 27–47 is 1.5 m.

Each node has six d.o.f. and the total number of free nodes is 24. The nodes 21–24

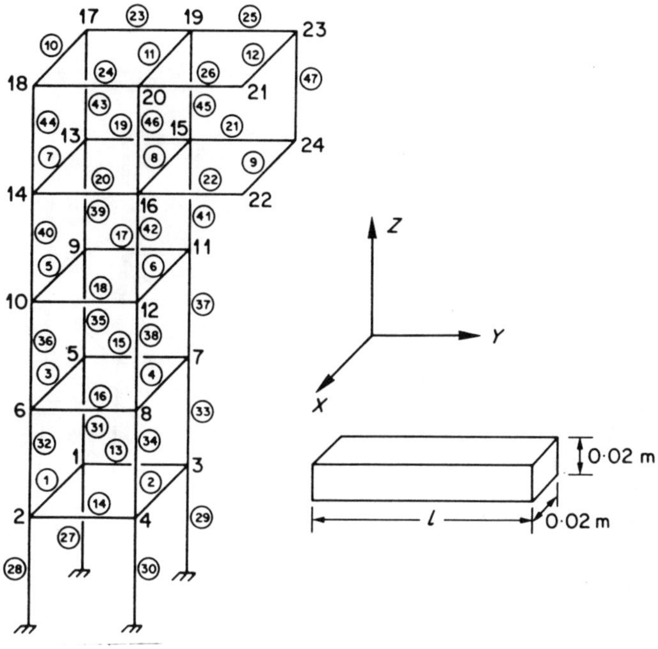

Fig. 3.1.1. A space frame

Table 3.1.1. Results of example 1

Iteration no.	Simple iteration			Iteration with Aitkin's acceleration		
	Mode 1	Mode 2	Mode 3	Mode 1	Mode 2	Mode 3
1	5.594	5.764	6.987	5.594	5.764	6.987
2	5.001	5.862	6.658	5.001	5.842	7.512
3	5.117	5.830	7.294	5.095	5.837	7.398
4	5.095	5.840	7.488	5.098	5.838	7.425
5	5.099	5.837	7.383	5.099	5.838	7.420
Exact	5.099	5.838	7.420	5.099	5.838	7.420
Economization method				5.594	6.772	10.102

constitute the masters. When convergence to a mode has been achieved, the second lowest eigenvalue is computed using Aitkin's extrapolation. The results are listed in Table 3.1.1. The exact natural frequencies are extracted from the complete matrices by subspace iteration using 12 trial vectors. The mode shapes are not sensitive to the change of the frequencies, the first four modes being depicted in Figure 3.1.2.

Example 3.1.2

Consider the structure shown in Fig. 3.1.3. It is essentially a collection composed of the three identical substructures studied in Example 3.1.1 arranged axi-symmetrically about the centre.

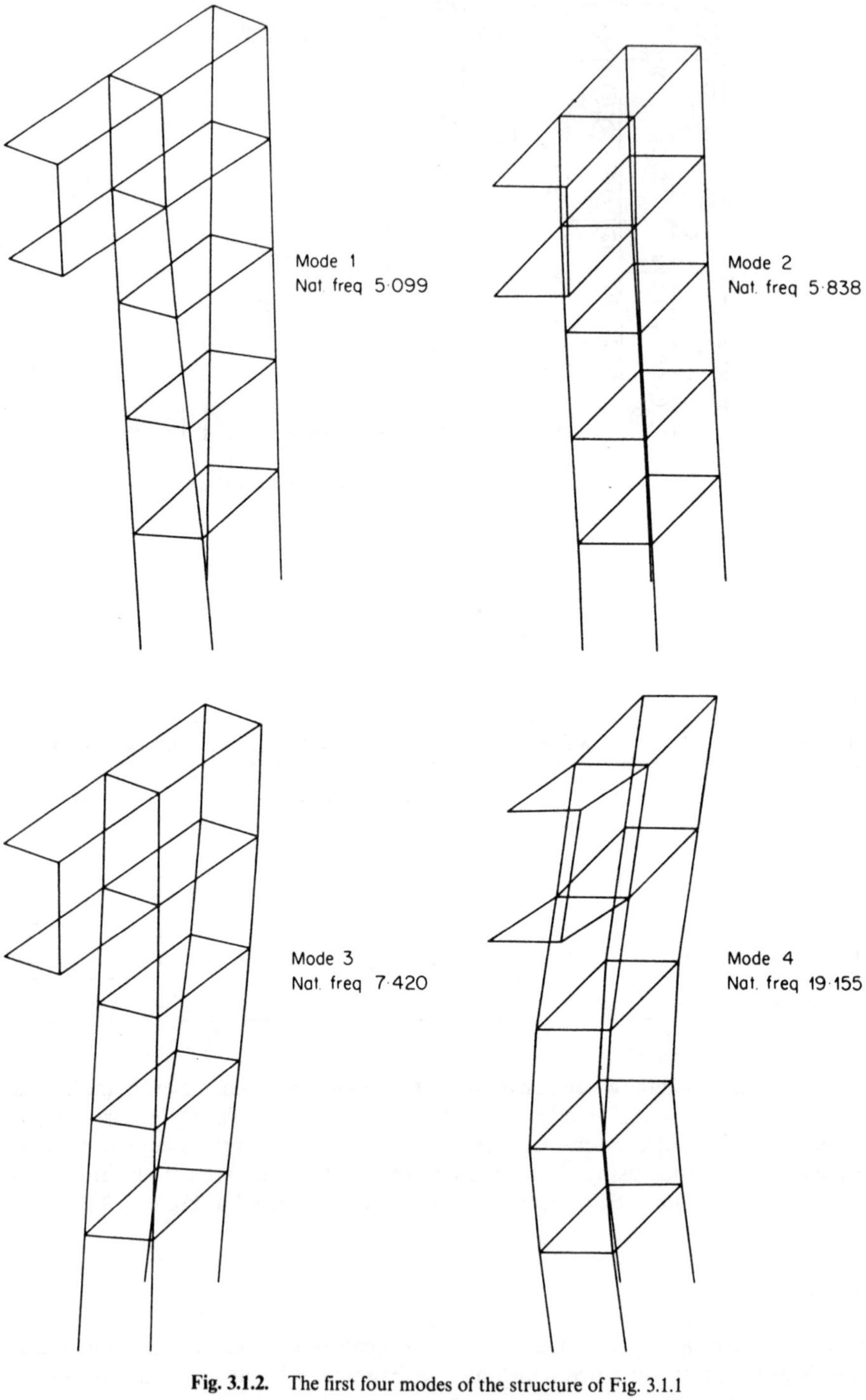

Fig. 3.1.2. The first four modes of the structure of Fig. 3.1.1

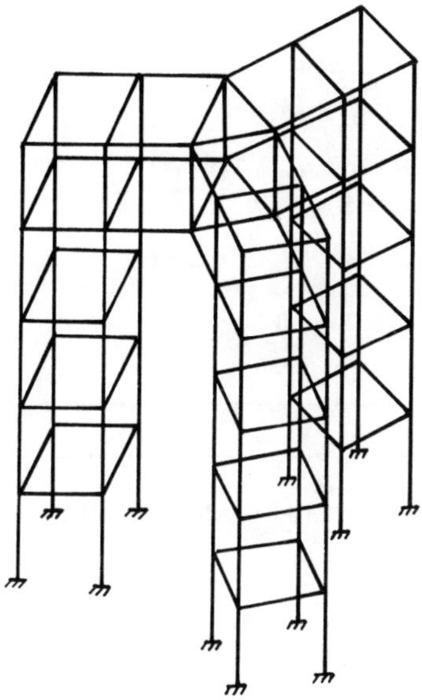

Fig. 3.1.3. A structure consisting of three substructures of Fig. 3.1.1

Table 3.1.2. Results of example 2

Frequency no.	Partial frequency rad s^{-1}	Natural frequency rad s^{-1}	No. of bisections	No. of iterations
1	15.6646	6.30355[a]	7	2
2	20.0170	7.32383	6	2
3	23.5987	17.1685[a]	3	2
4	27.4111	21.1520[a]	3	2
5	41.2982	21.5640	3	3
6	43.4355	24.1010	3	3
7	49.5433	29.7889[a]	4	2
8	59.0759	38.0273[a]	4	2
9	67.9703	39.4513	3	2
10	68.5595	42.1991[a]	3	2
11	75.3214	43.1157	3	3
12	80.1880	47.5623[a]	3	2
13		47.7948	3	3
14		52.4871[a]	3	2
Average			3.6	2.3

[a] Repeated natural frequencies of order two.

The computations are as follows. The partial frequencies for each substructure are first determined. Next, the Sturm sequence search is performed in every neighbouring pair of partial frequencies. Once the natural frequencies are isolated, they are extracted by the iterative method. The resulting natural frequencies are listed in Table 3.1.2. In this particular example, an average of 3.6 bisections and 2.3 iterations is found to be sufficient for six-digit accuracy.

The operations involved in the condensation, as well as the solution phases are

Table 3.1.3. Arithmetic operation counts

Step	Operation[a]	Arithmetic counts[b]
Matrix condensation for one iteration	$\mathbf{D}_{ss} = \mathbf{U}^T\mathbf{U}$ †	$\frac{1}{2}sb^2 + 3sb/2$
	$\mathbf{A} = \mathbf{U}^{-T}\mathbf{D}_{sm}$	$\frac{1}{2}sm(2b + 1)$
	$\mathbf{D} = \mathbf{D}_{mm} - \mathbf{A}^T\mathbf{A}$ †	$\frac{1}{2}m^2s$
	$\mathbf{B} = \mathbf{U}^{-1}\mathbf{A}$	$\frac{1}{2}sm(2b + 1)$
	$\mathbf{C} = \mathbf{M}_{ms}\mathbf{B}$ †	$\frac{1}{2}m^2s$
	$\mathbf{E} = \mathbf{M}_{mm} - \mathbf{C} - \mathbf{C}^T$ †	m^2
	$\mathbf{M}_{ss} = \mathbf{L} + \mathbf{L}^T$	s
	$\mathbf{F} = \mathbf{LB}$	bsm
	$\mathbf{H} = \mathbf{B}^T\mathbf{F}$	m^2s
	$\mathbf{M} = \mathbf{H} + \mathbf{H}^T + \mathbf{E}$ †	m^2
For p inverse iterations [7]		$4n(b + 1)p + \frac{1}{2}nb^2 + 3nb/2$

[a] m: number of masters; s: number of slaves.
[b] n: order of the overall matrices; $2b + 1$: total bandwidth of the overall matrices.
† Operation makes use of symmetry.

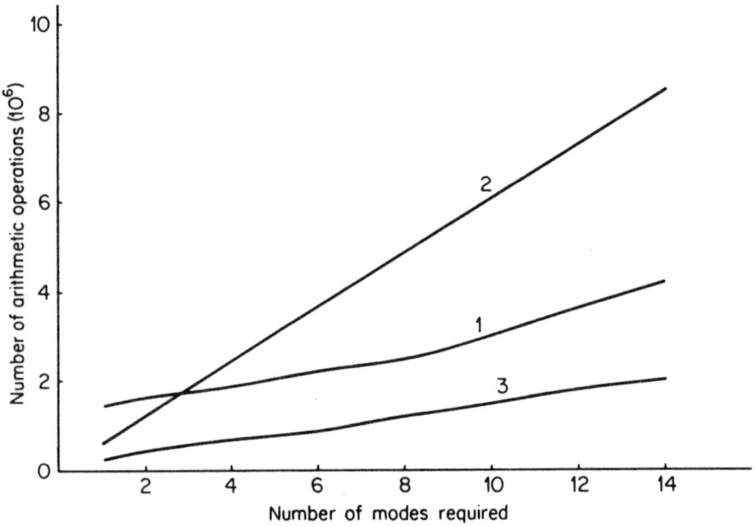

Curve 1. Unconondensed subspace iteration method
Curve 2. Condensed bisection method
Curve 3. Present method

Fig. 3.1.4. Comparison of various methods in terms of operation counts

summarized in Table 3.1.3. The computations are organized in such a way that the matrices ($\mathbf{D}_{sm}, \mathbf{A}, \mathbf{B}$) occupy the same working space assuming a working vector of length s is available. The same applies to the sets of matrices ($\mathbf{D}_{sm}, \mathbf{U}, \mathbf{C}, \mathbf{F}, \mathbf{H}$), ($\mathbf{D}, \mathbf{D}_{mm}$) and ($\mathbf{M}_{mm}, \mathbf{M}, \mathbf{E}$). If the mass matrix is assembled after the stiffness matrix is condensed, a considerable amount of work space may be saved by overwriting $\mathbf{M}_{ss}$ onto $\mathbf{D}_{ss}$ provided $\mathbf{M}_{ss}$ is large. Note that the procedures in Table 3.1.3 are slightly different from those mentioned previously to improve the programming efficiency.

The number of arithmetic operation counts is compared with that of subspace iteration on the uncondensed banded matrices and that of simple bisection with condensation given the same level of accuracy. The method of secant iteration is not used here because numerical overflow often occurs. Since the choice of method depends on the type of structure, a complete comparison is difficult. A comparison is made only for Example 3.1.2, the results being plotted in Figure 3.1.4. It should be noted that the method of simple bisection does not give the eigenvectors.

3.2. Dynamic Substructures

The exact condensation method requires the inversion of a frequency-dependent matrix for each frequency of interest. This computationally demanding process is eliminated by means of the partial modes of the substructure.

The organization of the theoretical development is as follows. First the displacement amplitudes of the substructure at vibration frequency ω are expressed in terms of the master coordinates by means of the static displacements and normal modes. Next the dynamic stiffness matrix associated with the masters is obtained for a continuum by using the reciprocal theorem. Finally the continuum formulation is modified to produce a discrete finite element model.

3.2.1. Prerequisites for the General Solution

Consider an elastic structure having m master coordinates on the boundaries and s slaves in the interior. When the structure is forced to vibrate by a boundary displacement vector $\{\mathbf{q}_m\} \sin \omega t$, where $\{\mathbf{q}_m\}$ is a collection of m master displacement coordinates, the governing differential equations of motion are represented symbolically by [16]

$$[\mathbf{L}]\{\mathbf{u}\} = \omega^2[\boldsymbol{\rho}]\{\mathbf{u}\} \quad \text{within the region} \tag{3.2.1a}$$

with prescribed boundary conditions

$$[\mathbf{B}]\{\mathbf{u}\} = \{\mathbf{q}_m\} \quad \text{on the boundaries} \tag{3.2.1b}$$

Here $[\mathbf{L}]$ is a linear self-adjoint differential operator depending on the spatial coordinates, $[\boldsymbol{\rho}]$ is a symmetrical square matrix of inertia densities, $[\mathbf{B}]$ is a differential operator depending on the spatial coordinates signifying the boundary conditions, and $\{\mathbf{u}(x, y, z)\}$ is the displacement amplitude vector to be determined. In general, the order of the arrays is six, including three translational and three rotational coodinates.

Suppose the two sets of solutions, $[\mathbf{v}(x, y, z)] = [\{\mathbf{v}_1\}, \{\mathbf{v}_2\}, \ldots, \{\mathbf{v}_m\}]$ and $[\boldsymbol{\phi}(x, y, z)] = [\{\boldsymbol{\phi}_1\}, \{\boldsymbol{\phi}_2\}, \ldots, \{\boldsymbol{\phi}_N\}]$ corresponding to the following two problems are given.

1. The static problem:

$$[\mathbf{L}][\mathbf{v}] = [\mathbf{0}] \quad \text{within the region} \tag{3.2.2a}$$

with

$$[\mathbf{B}][\mathbf{v}] = [\mathbf{I}] \quad \text{on the boundaries} \tag{3.2.2b}$$

where $[\mathbf{I}]$ is an identity matrix.

2. The fixed-interface eigenvalue problem:

$$[\mathbf{L}][\boldsymbol{\phi}] = [\boldsymbol{\rho}][\boldsymbol{\phi}][\Omega^2] \quad \text{in the region} \tag{3.2.3a}$$

where $[\Omega^2] = [\omega_k^2]$ is a diagonal matrix of natural frequencies, $k = 1, 2, \ldots, N$ with

$$[\mathbf{B}][\boldsymbol{\phi}] = [\mathbf{0}] \quad \text{on the boundaries} \tag{3.2.3b}$$

where the partial modes are normalized such that

$$\int_{vol} [\boldsymbol{\phi}]^{\mathrm{T}}[\boldsymbol{\rho}][\boldsymbol{\phi}]\,\mathrm{d}\,vol = [\mathbf{I}] \tag{3.2.4}$$

and N is the number of partial modes considered.

The aim of the following subsections is to solve the system (3.2.1) in terms of $[\mathbf{v}]$ and $[\boldsymbol{\phi}]$, and then construct the dynamic stiffness and mass matrices using these solutions.

3.2.2. General Solution for the Displacement Amplitudes

The displacement amplitudes $\{\mathbf{u}\}$ of the harmonic forced vibrations described by Eq. (3.2.1) will be solved here by means of the natural modes. All displacement boundary conditions will be satisfied automatically if the solution is expressed as

$$\{\mathbf{u}(x, y, z)\} = [\boldsymbol{\phi}]\{\boldsymbol{\alpha}\} + [\mathbf{v}]\{q_m\} \tag{3.2.5}$$

where $\{\boldsymbol{\alpha}\} = [\alpha_1, \alpha_2, \ldots, \alpha_N]^{\mathrm{T}}$ is a vector to be determined such that the differential equation (3.2.1) may also be satisfied.

Substitution of Eq. (3.2.5) into Eq. (3.2.1) and simplification according to Eqs (3.2.2) and (3.2.3), yields

$$[\boldsymbol{\rho}][\boldsymbol{\phi}][\Omega^2 - \omega^2\mathbf{I}]\{\boldsymbol{\alpha}\} = \omega^2[\boldsymbol{\rho}][\mathbf{v}]\{\mathbf{q}_m\} \tag{3.2.6}$$

Premultiplication by $[\boldsymbol{\phi}]^{\mathrm{T}}$ and integration over the whole substructure produces

$$\{\boldsymbol{\alpha}\} = [\Lambda][\mathbf{G}]^{\mathrm{T}}\{\mathbf{q}_m\} \tag{3.2.7}$$

where

$$[\Lambda] = \left[\frac{\omega^2}{\omega_k^2 - \omega^2}\right], \qquad [\mathbf{G}]^{\mathrm{T}} = \int_{vol} [\boldsymbol{\phi}]^{\mathrm{T}}[\boldsymbol{\rho}][\mathbf{v}]\,\mathrm{d}\,vol \tag{3.2.8}$$

Note that the condition of orthogonality (3.2.4) was used. Substitution of Eq. (3.2.7) into Eq. (3.2.5) gives

$$\{\mathbf{u}\} = ([\boldsymbol{\phi}][\Lambda][\mathbf{G}]^{\mathrm{T}} + [\mathbf{v}])\{\mathbf{q}_m\} \tag{3.2.9}$$

Note that the slave coordinates are not explicitly present.

When $\omega = \omega_k$, $\{\mathbf{u}\}$ tends to infinity for non-vanishing values of $\{\mathbf{q}_m\}$. In order to keep $\{\mathbf{u}\}$ finite, $\{\mathbf{q}_m\}$ must be zero, and a case of fixed interface partial mode results.

3.2.3. Dynamic Stiffness Matrix – Continuum Model

The dynamic stiffness matrix $[\mathbf{D}(\omega)]$ for a vibrating body with frequency ω is defined by

$$[\mathbf{D}(\omega)]\{\mathbf{q}\} = \{\mathbf{Q}\} \tag{3.2.10}$$

where $\{\mathbf{q}\}\sin\omega t$ and $\{\mathbf{Q}\}\sin\omega t$ are the generalized displacement vector and force vector, respectively.

Putting $\omega = 0$ in Eq. (3.2.9), gives the "static" displacement

$$\{\mathbf{u}_0(x, y, z)\} = [\mathbf{v}(x, y, z)]\{\mathbf{q}_0\} \tag{3.2.11}$$

where $\{\mathbf{q}_0\}$ corresponds to the force vector $\{\mathbf{Q}_0\}$ acting on the masters. Consider the following two equilibrium states of the substructure. One is the static state with force $\{\mathbf{Q}_0\}$ and the resulting displacement $\{\mathbf{u}_0(x, y, z)\}$ which has the values $\{\mathbf{q}_0\}$ at the master coordinates. The other is the harmonic vibrating state with force $\{\mathbf{Q}_m\}\sin\omega t$ and the response $\{\mathbf{u}(x, y, z)\}\sin\omega t$ which has the values $\{\mathbf{q}_m\}\sin\omega t$ at the masters. The work done by the force of the static state acting through the displacements of the vibrating state $\{\mathbf{Q}_0\}^{\mathrm{T}}\{q_m\}\sin\omega t$ plus the work done by the force of the vibrating state acting through the displacements of the static state is

$$\left(\{\mathbf{Q}_m\}^{\mathrm{T}}\{\mathbf{q}_m\} + \omega^2 \int_{vol} \{\mathbf{u}\}^{\mathrm{T}}[\boldsymbol{\rho}]\{\mathbf{u}_0\}\,\mathrm{d}\,vol\right)\sin\omega t$$

since the total force of the vibrating state includes the distributed inertia force $\omega^2[\boldsymbol{\rho}]\{\mathbf{u}\}\sin\omega t$ per unit volume. Applying the reciprocal theorem gives

$$\{\mathbf{Q}_0\}^{\mathrm{T}}\{\mathbf{q}_m\} = \{\mathbf{Q}_m\}^{\mathrm{T}}\{\mathbf{q}_0\} + \omega^2 \int_{vol} \{\mathbf{u}\}^{\mathrm{T}}[\boldsymbol{\rho}]\{\mathbf{u}_0\}\,\mathrm{d}\,vol \tag{3.2.12}$$

From Eqs (3.2.9)–(3.2.11)

$$\{\mathbf{q}_m\}^{\mathrm{T}}[\mathbf{D}]\{\mathbf{q}_0\} = \{\mathbf{q}_m\}^{\mathrm{T}}[\mathbf{D}_0]\{\mathbf{q}_0\} - \omega^2\{\mathbf{q}_m\}^{\mathrm{T}} \int_{vol} ([\boldsymbol{\phi}][\boldsymbol{\Lambda}][\mathbf{G}]^{\mathrm{T}} + [\mathbf{v}])^{\mathrm{T}}[\boldsymbol{\rho}][\mathbf{v}]\,\mathrm{d}\,vol\{\mathbf{q}_0\}$$

or

$$[\mathbf{D}] = [\mathbf{D}_0] - \omega^2 \int_{vol} ([\boldsymbol{\phi}][\boldsymbol{\Lambda}][\mathbf{G}]^{\mathrm{T}} + [\mathbf{v}])^{\mathrm{T}}[\boldsymbol{\rho}][\mathbf{v}]\,\mathrm{d}\,vol \tag{3.2.13}$$

since $\{\mathbf{q}_m\}$ and $\{\mathbf{q}_0\}$ are arbitrary. For the vibration frequency $\omega = 0$,

$$[\mathbf{D}_0] = [\mathbf{D}(0)] = [\mathbf{K}_0] \tag{3.2.14}$$

corresponds to the conventional stiffness matrix. Now, from the definition of $[\mathbf{G}]$,

$$\begin{aligned}\int_{vol} ([\boldsymbol{\phi}][\boldsymbol{\Lambda}][\mathbf{G}]^{\mathrm{T}} + [\mathbf{v}])^{\mathrm{T}}[\boldsymbol{\rho}][\mathbf{v}]\,\mathrm{d}\,vol &= [\mathbf{G}][\boldsymbol{\Lambda}] \int_{vol} [\boldsymbol{\phi}]^{\mathrm{T}}[\boldsymbol{\rho}][\mathbf{v}]\,\mathrm{d}\,vol \\ &\quad + \int_{vol} [\mathbf{v}]^{\mathrm{T}}[\boldsymbol{\rho}][\mathbf{v}]\,\mathrm{d}\,vol \\ &= [\mathbf{G}][\boldsymbol{\Lambda}][\mathbf{G}]^{\mathrm{T}} + [\mathbf{M}_0]\end{aligned}$$

Here, the consistent mass matrix $[\mathbf{M}_0]$ is defined as

$$[\mathbf{M}_0] = \int_{vol} [\mathbf{v}]^{\mathrm{T}}[\boldsymbol{\rho}][\mathbf{v}]\,\mathrm{d}\,vol \tag{3.2.15}$$

since $[\mathbf{v}]$ is a collection of static displacement functions as implied by Eq. (3.2.2).

Therefore, Eq. (3.2.13) becomes

$$[\mathbf{D}] = [\mathbf{K}_0] - \omega^2[\mathbf{M}_0] - \omega^2[\mathbf{G}][\boldsymbol{\Lambda}][\mathbf{G}]^T \tag{3.2.16}$$

where $[\mathbf{D}]$, $[\mathbf{K}_0]$, $[\mathbf{M}_0]$ are matrices associated with the m master coordinates. Equations (3.2.8), (3.2.9) and (3.2.16) apply directly to systems for which the functions $[\mathbf{v}(x, y, z)]$ and $[\boldsymbol{\phi}(x, y, z)]$ are available in continuous forms. Note that the formulation itself becomes exact as the number of partial modes tends to infinity, and the dimension of the dynamic stiffness matrix is independent of the number of partial modes taken. In practice, since a substructure is much stiffer than the whole structure, there is no need to take too many partial modes.

3.2.4. Dynamic Stiffness Matrix – Finite Element Model

Equations (3.2.8), (3.2.9) and (3.2.16) will now be expressed in forms convenient for incorporation with finite element techniques.

Suppose the problem described by (3.2.1) is solved by the finite element method, then

$$\{\mathbf{u}\} = [\mathbf{N}_s \quad \mathbf{N}_m]\begin{Bmatrix}\mathbf{q}_s\\ \mathbf{q}_m\end{Bmatrix} \tag{3.2.17}$$

where $[\mathbf{N}_s(x, y, z)]$ and $[\mathbf{N}_m(x, y, z)]$ are assumed frequency-independent shape functions corresponding to the slaves $\{\mathbf{q}_s\}$ and masters $\{\mathbf{q}_m\}$, respectively. The dynamic stiffness equation is

$$\begin{bmatrix}\mathbf{D}_{ss} & \mathbf{D}_{sm}\\ \mathbf{D}_{ms} & \mathbf{D}_{mm}\end{bmatrix}\begin{Bmatrix}\mathbf{q}_s\\ \mathbf{q}_m\end{Bmatrix} = \begin{Bmatrix}\mathbf{0}\\ \mathbf{Q}_m\end{Bmatrix} \tag{3.2.18}$$

This may be condensed to

$$[\mathbf{D}]\{\mathbf{q}_m\} = \{\mathbf{Q}_m\} \tag{3.2.19}$$

by the transformation

$$\{\mathbf{q}_s\} = -[\mathbf{D}_{ss}]^{-1}[\mathbf{D}_{sm}]\{\mathbf{q}_m\} \tag{3.2.20}$$

where

$$[\mathbf{D}] = [\mathbf{D}_{mm}] - [\mathbf{D}_{ms}][\mathbf{D}_{ss}]^{-1}[\mathbf{D}_{sm}] \tag{3.2.21}$$

When $[\mathbf{D}_{ss}]$ becomes singular at a certain frequency, $\{\mathbf{q}_m\}$ must be zero in order to have finite $\{\mathbf{q}_s\}$ and a partial mode results. Substituting Eq. (3.2.20) into (3.2.17) yields

$$\{\mathbf{u}\} = [\mathbf{N}_m - \mathbf{N}_s\mathbf{D}_{ss}^{-1}\mathbf{D}_{sm}]\{\mathbf{q}_m\} \tag{3.2.22}$$

Putting $\omega = 0$ gives

$$\{\mathbf{u}_0\} = [\mathbf{N}_m - \mathbf{N}_s\mathbf{K}_{ss}^{-1}\mathbf{K}_{sm}]\{\mathbf{q}_0\}$$

However, from Eq. (3.2.11)

$$\{\mathbf{u}_0\} = [\mathbf{v}]\{\mathbf{q}_0\}$$

therefore

$$[\mathbf{v}] = [\mathbf{N}_m - \mathbf{N}_s\mathbf{K}_{ss}^{-1}\mathbf{K}_{sm}] \tag{3.2.23}$$

If the same shape functions are assumed in order to calculate the partial modes $[\boldsymbol{\phi}]$, then

$$[\boldsymbol{\phi}] = [\mathbf{N}_s][\boldsymbol{\Phi}_0] \tag{3.2.24}$$

where $[\boldsymbol{\Phi}] = [\{\boldsymbol{\Phi}_1\}, \{\boldsymbol{\Phi}_2\}, \ldots, \{\boldsymbol{\Phi}_N\}]$ is a collection of the modal generalized displacements associated with the slaves, and

$$[\mathbf{K}_{ss}][\boldsymbol{\Phi}] = [\mathbf{M}_{ss}][\boldsymbol{\Phi}][\boldsymbol{\Omega}^2] \tag{3.2.25}$$

where $[\mathbf{K}_{ss}] = [\mathbf{D}_{ss}(0)]$ and $[\boldsymbol{\Omega}^2]$ is a diagonal matrix of squares of partial frequencies ω_k, $k = 1, 2, \ldots, N$.

If $[\boldsymbol{\Phi}]$ is normalized such that

$$[\boldsymbol{\Phi}]^{\mathrm{T}}[\mathbf{M}_{ss}][\boldsymbol{\Phi}] = [\mathbf{I}] \tag{3.2.26}$$

then

$$[\boldsymbol{\Phi}]^{\mathrm{T}}[\mathbf{K}_{ss}][\boldsymbol{\Phi}] = [\boldsymbol{\Omega}^2] \tag{3.2.27}$$

Pre- and post-multiplying Eq. (3.2.25) by $[\mathbf{K}_{ss}]^{-1}$ and $[\boldsymbol{\Omega}^2]^{-1}$, respectively, gives

$$[\mathbf{K}_{ss}]^{-1}[\mathbf{M}_{ss}][\boldsymbol{\Phi}] = [\boldsymbol{\Phi}][\boldsymbol{\Omega}^2] \tag{3.2.28}$$

Now, from Eqs (3.2.8), (3.2.23) and (3.2.28)

$$\begin{aligned}
[\mathbf{G}] &= \int_{vol} [\mathbf{v}]^{\mathrm{T}}[\boldsymbol{\rho}][\boldsymbol{\phi}]\,\mathrm{d}\,vol \\
&= \int_{vol} [\mathbf{N}_m - \mathbf{N}_s\mathbf{K}_{ss}^{-1}\mathbf{K}_{sm}]^{\mathrm{T}}[\boldsymbol{\rho}][\mathbf{N}_s]\,\mathrm{d}\,vol[\boldsymbol{\Phi}] \\
&= [\mathbf{M}_{ms}][\boldsymbol{\Phi}] - [\mathbf{K}_{ms}][\mathbf{K}_{ss}]^{-1}[\mathbf{M}_{ss}][\boldsymbol{\Phi}] \\
&= [\mathbf{M}_{ms}][\boldsymbol{\Phi}] - [\mathbf{K}_{ms}][\boldsymbol{\Phi}][\boldsymbol{\Omega}^2]^{-1}
\end{aligned} \tag{3.2.29}$$

Substituting into Eq. (3.2.16) results in

$$[\mathbf{D}] = [\mathbf{K}_0] - \omega^2[\mathbf{M}_0] - \omega^2[\mathbf{G}][\boldsymbol{\Lambda}][\mathbf{G}]^{\mathrm{T}} \tag{3.2.30}$$

where

$$[\mathbf{K}_0] = [\mathbf{D}(0)] = [\mathbf{K}_{mm}] - [\mathbf{K}_{ms}][\mathbf{K}_{ss}]^{-1}[\mathbf{K}_{sm}] \tag{3.2.31}$$

$$\begin{aligned}
[\mathbf{M}_0] &= \int_{vol} [\mathbf{v}]^{\mathrm{T}}[\boldsymbol{\rho}][\mathbf{v}]\,\mathrm{d}\,vol \\
&= [\mathbf{M}_{mm}] - [\mathbf{M}_{ms}][\mathbf{K}_{ss}]^{-1}[\mathbf{K}_{sm}]^{-1} - [\mathbf{K}_{ms}][\mathbf{K}_{ss}]^{-1}[\mathbf{M}_{sm}]^{-1} \\
&\quad + [\mathbf{K}_{ms}][\mathbf{K}_{ss}]^{-1}[\mathbf{M}_{ss}][\mathbf{K}_{ss}]^{-1}[\mathbf{K}_{sm}]
\end{aligned} \tag{3.2.32}$$

and

$$[\mathbf{G}] = [\mathbf{M}_{ms}][\boldsymbol{\Phi}] - [\mathbf{K}_{ms}][\boldsymbol{\Phi}][\boldsymbol{\Omega}^2]^{-1}$$

Therefore, the dynamic properties of the substructure are preserved if the partial frequencies and the matrices $[\mathbf{K}_0]$, $[\mathbf{M}_0]$ and $[\mathbf{G}]$ are retained.

Note that the matrices $[\mathbf{K}_0]$ and $[\mathbf{M}_0]$ have the same forms as those obtained by the eigenvalue economization process [3]. The third term on the right-hand side of Eq. (3.2.30) has been neglected by Guyan and Irons and many other authors. It will be shown later that this term is not negligible if accuracy is to be maintained and completeness is to be guaranteed.

3.2.5. Computational Methods

The computational procedures required to generate information necessary to identify a substructure are summarized in Table 3.2.1. Note that the inversion of the positive definite band matrix $\mathbf{K}_{ss}$ is performed implicitly by symmetrical decomposition, i.e. $\mathbf{K}_{ss} = \mathbf{U}^T\mathbf{U}$, where $\mathbf{U}$ is an upper triangular matrix, in order to minimize computer operations and storage. The organization is in such a way that $(\mathbf{K}_{sm}, \mathbf{A}, \mathbf{B})$ may occupy the same space if a working vector of length s is available. The same applies to the sets $(\mathbf{K}_{ss}, \mathbf{U}, \mathbf{C}, \mathbf{F}, \mathbf{H})$, $(\mathbf{K}_0, \mathbf{K}_{mm})$, $(\mathbf{M}_0, \mathbf{M}_{mm}, \mathbf{E})$, and $(\mathbf{M}_{ss}, \mathbf{L})$ where $\mathbf{L}$ is a lower triangular matrix. If the mass matrix is assembled after the stiffness matrix is condensed, a considerable amount of working space may be saved by overwriting $\mathbf{M}_{ss}$ onto $\mathbf{K}_{ss}$. Then, at a particular frequency, the dynamic stiffness matrix is obtained from Eq. (3.2.30). These matrices are assembled over all the substructures according to a standard finite element procedure and the following matrix equation of motion is obtained:

$$\overline{\mathbf{D}}(\omega)\mathbf{q} = \mathbf{Q}$$

where $\mathbf{q}\sin\omega t$ and $\mathbf{Q}\sin\omega t$ are the generalized displacement vector and force vector, respectively. For free vibration, this becomes

$$\overline{\mathbf{D}}(\omega)\mathbf{q} = \mathbf{0}$$

or

$$\mathbf{K}(\omega)\mathbf{q} = \omega^2\mathbf{M}(\omega)\mathbf{q}$$

An efficient method for estimating the natural frequencies with certainty is the bisection method of Wittrick and Williams [35]. The technique has been extended to ill-conditioned matrices. Once an approximate natural frequency ω_0 is located,

Table 3.2.1. Procedures required to generate information to identify a substructure

Step	Description	Formula[a]	No. of arithmetic operations[b]
1	Subspace iteration for q modes	$\mathbf{K}_{ss}\boldsymbol{\phi} = \mathbf{M}_{ss}\boldsymbol{\phi}\boldsymbol{\Omega}^2$	$sb^2 + 4sb - 16sq(2b + q + \frac{3}{2})$
2	Multiplication	$\mathbf{G} = \mathbf{M}_{ss}\boldsymbol{\phi} - \mathbf{K}_{ms}\boldsymbol{\phi}\boldsymbol{\Omega}^{-2}$	$2msq + sq$
3	Matrix condensation	$\mathbf{K}_{ss} = \mathbf{U}^T\mathbf{U}$	done in step 1
		$\mathbf{A} = \mathbf{U}^{-T}\mathbf{K}_{sm}$	$\frac{1}{2}sm(2b + 1)$
		$\mathbf{K}_0 = \mathbf{K}_{mm} - \mathbf{A}^T\mathbf{A}$ †	$\frac{1}{2}m^2s$
		$\mathbf{B} = \mathbf{U}^{-1}\mathbf{A}$	$\frac{1}{2}sm(2b + 1)$
		$\mathbf{C} = \mathbf{M}_{ms}\mathbf{B}$ †	$\frac{1}{2}m^2s$
		$\mathbf{E} = \mathbf{M}_{mm} - \mathbf{C} - \mathbf{C}^T$ †	m^2
		$\mathbf{M}_{ss} = \mathbf{L} + \mathbf{L}^T$	s
		$\mathbf{F} = \mathbf{L}\mathbf{B}$	bsm
		$\mathbf{H} = \mathbf{B}^T\mathbf{F}$ †	$\frac{1}{2}m^2s$
		$\mathbf{M}_0 = \mathbf{H} + \mathbf{H}^T$ †	$\frac{1}{2}m^2$

[a] m: number of masters; s: number of slaves.
[b] $(2b + 1)$: total bandwidth.
† Formula makes use of symmetry.

Table 3.2.2. Operation counts for p iterations

	Banded matrices	Full matrices
Bisection	$\frac{1}{2}(nb^2 + \frac{3}{2}nb)p$	$\frac{1}{6}n^2(n+1)p$
Inverse iteration	$4n(b+1)p + \frac{1}{2}nb^2 + \frac{3}{2}nb$	$2n(n+1)p + \frac{1}{6}n^2(n+1)$

the corresponding modal vector may be extracted accurately by the method of inverse iteration similar to that for the linear eigenvalue problem

$$\overline{\mathbf{D}}(\omega_0)\mathbf{q} = p\overline{\mathbf{M}}(\omega_0)\mathbf{q}$$

When $\overline{\mathbf{M}}$ is positive-definite, the inverse iteration has been confirmed to be numerically stable, even if $\overline{\mathbf{D}}$ is ill-conditioned, by Wilkinson. When the eigenvector $\mathbf{q}$ is accurately determined, the natural frequency is improved using Rayleigh's quotient, i.e.

$$\omega^2 = \omega_0^2 + \frac{\mathbf{q}^{\mathrm{T}}\overline{\mathbf{D}}\mathbf{q}}{\mathbf{q}^{\mathrm{T}}\overline{\mathbf{M}}\mathbf{q}}$$

Normally, three bisections and three iterations are sufficient to give a natural frequency to six-digit accuracy. The operation counts are listed in Table 3.2.2.

A few numerical examples are considered here and comparisons with the other methods are presented. In order to illustrate the efficiency of the present method without going into computational complications, for the first example, we consider the lateral free vibration of a cantilevered beam using both the continuous model and the discrete model.

Consider a cantilever beam element having unit structural dimensions. The stiffness and mass matrices corresponding to the static state are given by

$$[\mathbf{K}_0] = \begin{bmatrix} 12 & -6 \\ -6 & 4 \end{bmatrix}$$

$$[\mathbf{M}_0] = \frac{1}{420}\begin{bmatrix} 156 & -22 \\ -22 & 4 \end{bmatrix}$$

and

$$[\mathbf{v}(x)] = [\mathbf{v}_1 \quad \mathbf{v}_2] = [(3x^2 - 2x^3)(-x^2 + x^3)]$$

$$\phi_k(x) = \cosh\lambda_k x - \cos\lambda_k x - \sigma_k(\sinh\lambda_k x - \sin\lambda_k x)$$

$$G_{ik} = \int_0^1 v_i(x)\phi_k(x)\,\mathrm{d}x \qquad i = 1, 2;\ k = 1, 2, \ldots, N$$

The values λ_k and σ_k can be found from reference [1], while the G_{ik} are listed in Table 3.2.3. The dynamic stiffness matrix has the form

$$[\mathbf{D}] = [\mathbf{K}_0] - \omega^2[\mathbf{M}_0] - \omega^2[\mathbf{G}][\mathbf{\Lambda}][\mathbf{G}]^{\mathrm{T}}$$

The first six modes are calculated by using different values of N for a single element. Note that the order of matrices $[\mathbf{D}]$, $[\mathbf{K}_0]$, $[\mathbf{M}_0]$ is unaffected by N and is always equal to 2. The percentage errors are listed in Table 3.2.4. It is seen that the error decreases rapidly with the increasing number of terms taken. The ability to extract the higher modes accurately makes the present method more attractive than the others.

Table 3.2.3. G_{ik} for a uniform beam member

k	λ_k	σ_k	G_{1k}	G_{2k}
1	4.73004	0.9825022	0.415431	−0.0893923
2	7.85320	1.0007773	−0.254871	0.0324292
3	10.9956	0.9999665	0.181885	−0.0165421
4	14.1372	1.0000015	−0.141471	0.0100070
5	17.2788	0.9999999	0.115749	−0.0066989
6	20.4204	1.0	−0.097942	0.0047963
7	23.5619	1.0	0.084883	−0.0036025
8	26.7035	1.0	−0.074896	0.0028047
9	29.8451	1.0	0.067013	−0.0022453
10	32.9867	1.0	−0.060630	0.0018380

Table 3.2.4. The comparison of natural frequency parameters by taking different numbers of terms. Figures in parentheses are the ratios to the exact values

Terms N	Mode					
	1	2	3	4	5	6
0	1.8796 (1.0024)					
1	1.8754 (1.0002)	4.7039 (1.0021)				
2	1.8751 (1.0000)	4.6965 (1.0005)	7.8693 (1.0018)			
3	1.8751 (1.0000)	4.6949 (1.0002)	7.8613 (1.0008)	11.012 (1.0015)		
4	1.8751 (1.0000)	4.6944 (1.0001)	7.8584 (1.0005)	11.005 (1.0008)	14.154 (1.0012)	
5	1.8751 (1.0000)	4.6942 (1.0000)	7.8561 (1.0002)	11.000 (1.0004)	14.148 (1.0008)	17.297 (1.0010)
6	1.8751 (1.0000)	4.6942 (1.0000)	7.8554 (1.0001)	10.998 (1.0002)	14.144 (1.0005)	17.291 (1.0007)
7	1.8751 (1.0000)	4.6941 (1.0000)	7.8551 (1.0000)	10.997 (1.0001)	14.141 (1.0003)	17.287 (1.0005)
8	1.8751 (1.0000)	4.6941 (1.0000)	7.8550 (1.0000)	10.997 (1.0001)	14.140 (1.0002)	17.285 (1.0003)
9	1.8751 (1.0000)	4.6941 (1.0000)	7.8549 (1.0000)	10.996 (1.0000)	14.139 (1.0001)	17.283 (1.0002)
10	1.8751 (1.0000)	4.6941 (1.0000)	7.8549 (1.0000)	10.996 (1.0000)	14.138 (1.0001)	17.282 (1.0002)
Exact	1.8751	4.6941	7.8548	10.996	14.137	17.279

Consider, as the second example, the finite element representation of the same cantilever using two equal elements. The conventional dynamic stiffness is given by

$$[\mathbf{D}] = \begin{bmatrix} 192 & 0 & -96 & 24 \\ 0 & 16 & -24 & 4 \\ -96 & -24 & 96 & -24 \\ 24 & 4 & -24 & 8 \end{bmatrix} - \frac{\omega^2}{420}\begin{bmatrix} 156 & 0 & 27 & -3.25 \\ 0 & 1 & 3.25 & -0.375 \\ 27 & 3.25 & 78 & -5.5 \\ -3.25 & -0.375 & -5.5 & 0.5 \end{bmatrix}$$

If we take the last two coordinates as masters, the condensed stiffness and mass matrices according to the method of eigenvalue economization, Eqs (3.2.31) and (3.2.32), are

$$[\mathbf{K}_0] = \begin{bmatrix} 12 & -6 \\ -6 & 4 \end{bmatrix} \quad \text{and} \quad [\mathbf{M}_0] = \frac{1}{420}\begin{bmatrix} 156 & -22 \\ -22 & 4 \end{bmatrix}$$

It is interesting to note that the same matrices are obtained as when using one element. Therefore, by subdividing the structure while keeping the same master coordinates, no increase of accuracy is achieved. If we subdivide the cantilever into ten elements, calculate the natural modes, when the masters are fixed, we obtain the matrix [G] for the first six modes as follows:

$$[\mathbf{G}] = \begin{bmatrix} 0.415431 & -0.254870 & 0.181874 & -0.141514 & 0.114687 & -0.104685 \\ -0.0893423 & 0.0324292 & -0.165419 & 0.0100094 & -0.00666613 & 0.00504667 \end{bmatrix}$$

The rate of convergence is similar to that of Table 3.2.4.

Consider the system in Fig. 3.1.3, consisting of three substructures (Fig. 3.1.1), as the third example. The fixed-interface modes are computed using subspace iteration and listed in Table 3.2.5. The natural frequencies of the overall structure with an increasing number of terms are listed in Table 3.2.6 (superscript denotes double natural frequencies). In spite of the complexity of the structure, the convergent rate is similar to that of the cantilever beam studied above. It is seen that the convergence is slightly slower for multiple roots than for simple ones.

As a fourth example, we consider a square plate with $l = E = \rho = h = 1$, Fig. 3.2.1a, as a substructure whose edges are subjected to rotational displacements only. It consists of 36 identical elements. Conforming 16-d.o.f. elements with $\nu = 0.3$ are used. The normal rotations of the nodes along the edges are taken as masters. With all the 20 masters clamped, the first 12 partial modes are computed using subspace iterations with 16 trial vectors. The superstructures shown in Fig. 3.2.1b, c are analysed by the present method. The resulting frequencies are tabulated in Table 3.2.5. A full finite element technique was carried out to investigate the structure of Fig. 3.2.1b, the results agreeing up to the fourth digit. Since the structure in Fig. 3.2.1c is

Table 3.2.5. Computed results for Examples 3 and 4

Mode No.	Example 3					Example 4		
	Partial frequency (rad s^{-1})	Guyan [5] and Irons [6] (rad s^{-1})	Present method			Partial frequency (rad s^{-1})	Fig. 3.2.2b	Fig. 3.2.2c
			(rad s^{-1})	No. of bisections	No. of iterations			
1	15.66	6.32	6.30	7	2	36.0	30.9	23.6
2	20.02	7.37	7.32	6	2	73.5	33.1	24.3
3	23.60	18.49	17.17	3	2	73.5	61.5	26.2
4	27.41	24.91	21.15	3	2	108.5	66.7	28.9
5	41.30	46.90	21.56	3	3	132.5	71.3	31.0
6	43.44	47.40	24.10	3	3	133.1	72.3	33.6
7	49.54	60.51	29.79	4	2	166.0	101.1	53.3
8	59.08	100.71	38.03	4	2	166.0	104.4	54.1
9	67.97	102.91	39.45	3	2	214.0	110.7	55.3
10	68.56	125.72	42.20	3	2	214.7	119.7	58.0
11	75.32	141.91	43.12	3	3	222.2	131.3	59.5
12	80.19	184.46	47.56	3	2	246.0	132.9	61.9
13		186.98	47.79	3	3		151.8	62.6
14		192.95	52.42	3	2		158.3	65.7
Average				3.6	2.3			

Table 3.2.6. The convergence of the natural frequencies of Example 3 (rad s^{-1})

Terms N	Mode 1	2	3	4	5	6
2	6.3187	7.3693	17.2198			
	(1.0024)	(1.0062)	(1.0030)			
4	6.3059	7.3243	17.1791	21.1758	21.5751	24.1167
	(1.0004)	(1.0001)	(1.0006)	(1.0011)	(1.0005)	(1.0007)
6	6.3042	7.3242	17.1705	21.1615	21.5650	24.1022
	(1.0001)	(1.0000)	(1.0001)	(1.0004)	(1.0000)	(1.0000)
8	6.3040	7.3238	17.1700	21.1588	21.5641	24.1011
	(1.0001)	(1.0000)	(1.0001)	(1.0003)	(1.0000)	(1.0000)
10	6.3136	7.3238	17.1686	21.1532	21.5640	24.1011
	(1.0000)	(1.0000)	(1.0000)	(1.0001)	(1.0000)	(1.0000)
Exact	6.3036[a]	7.3238	17.1685[a]	21.1520[a]	21.5640	24.1010

[a] Double natural frequencies.

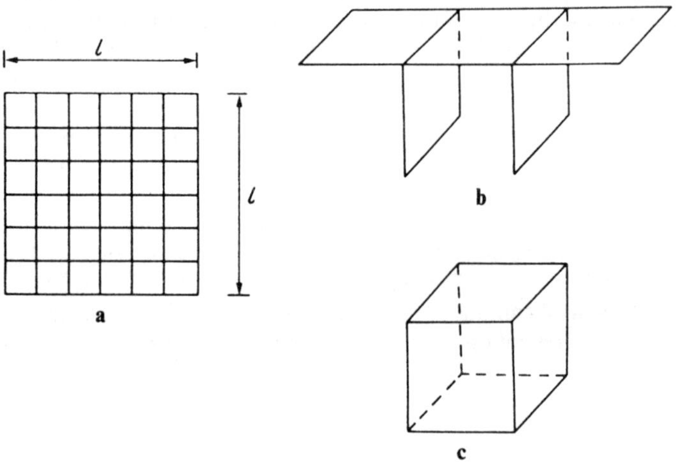

Fig. 3.2.1. Structures of fourth example: **a** square-plate substructure; **b** plate system with all edges clamped; **c** closed box with all edges supported

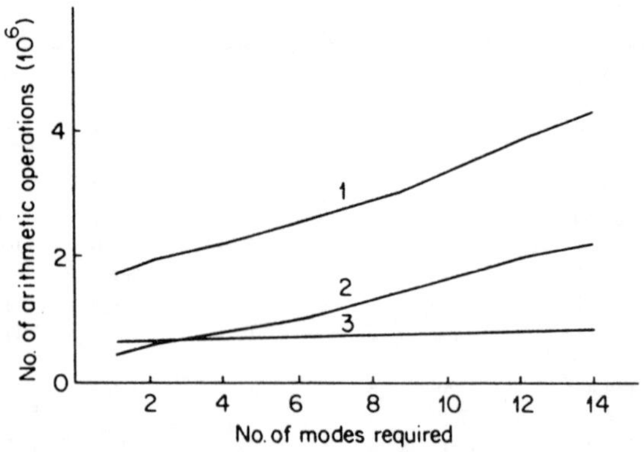

Fig. 3.2.2. Comparison of methods in terms of operation counts: (1) finite element method with band matrices; (2) exact substructure method; (3) simplified substructure method

of six-fold symmetry, all the natural frequencies shown are repeated. It should be noted that natural frequencies which are very close to partial frequencies are excluded from Table 3.2.5. The theoretical operation counts for the solution of the third example are plotted in Fig. 3.2.2 against the number of modes required, showing that the present method is favourable.

3.3. Dynamic Flexibility

The dynamic stiffness and flexibility formulation are defined by Eqs (3.0.1) and (3.0.2) respectively. We shall find a computationally efficient form of the dynamic flexibility method by means of the partial modes. Passive coordinates are referred to the nodes which are not subjected to external forces and whose displacements are not of immediate interest. The undamped harmonic vibration equation (3.0.1) may be partitioned according to the active and passive coordinates as follows:

$$\begin{bmatrix} \mathbf{D}_{11} & \mathbf{D}_{12} \\ \mathbf{D}_{21} & \mathbf{D}_{22} \end{bmatrix} \begin{Bmatrix} \mathbf{x}_1 \\ \mathbf{x}_2 \end{Bmatrix} = \begin{Bmatrix} \mathbf{X}_1 \\ \mathbf{0} \end{Bmatrix} \tag{3.3.1}$$

where the subscripts 1 and 2 denote the active and passive components respectively and $[\mathbf{D}_{ij}] = [\mathbf{K}_{ij}] - \omega^2[\mathbf{M}_{ij}]$, where $[\mathbf{D}_{ij}]$, $[\mathbf{K}_{ij}]$ and $[\mathbf{M}_{ij}]$, $i, j = 1, 2$ are submatrices of $[\mathbf{D}]$, $[\mathbf{K}]$ and $[\mathbf{M}]$ respectively. When $\{\mathbf{x}_2\}$ is eliminated, Eq. (3.3.1) gives

$$[\mathbf{D}]\{\mathbf{x}_1\} = \{\mathbf{X}_1\} \tag{3.3.2}$$

where

$$[\mathbf{D}(\omega)] = [\mathbf{D}_{11} - \mathbf{D}_{12}\mathbf{D}_{22}^{-1}\mathbf{D}_{21}] = [\mathbf{K}(\omega) - \omega^2\mathbf{M}(\omega)] \tag{3.3.3}$$

The matrices $[\mathbf{M}]$ and $[\mathbf{K}]$ are positive definite and associated with the kinetic and strain energies respectively. They may be obtained from

$$\begin{aligned} [\mathbf{M}(\omega)] &= -\frac{\partial}{\partial\omega^2}[\mathbf{D}(\omega)] \\ &= [\mathbf{M}_{11} - \mathbf{M}_{12}\mathbf{D}_{22}^{-1}\mathbf{D}_{21}^{-1} - \mathbf{D}_{12}\mathbf{D}_{22}^{-1}\mathbf{M}_{21}^{-1} + \mathbf{D}_{12}\mathbf{D}_{22}^{-1}\mathbf{M}_{22}\mathbf{D}_{22}^{-1}\mathbf{D}_{21}] \end{aligned} \tag{3.3.4}$$

$$[\mathbf{K}(\omega)] = [\mathbf{D}(\omega) + \omega^2\mathbf{M}(\omega)] \tag{3.3.5}$$

Let

$$[\mathbf{D}^{-1}(\omega)] = [\mathbf{A}_0 + \omega^2\mathbf{A}_2 + \omega^4\mathbf{A}_4 + \cdots] \tag{3.3.6}$$

where $[\mathbf{A}_0]$, $[\mathbf{A}_2]$, $[\mathbf{A}_4]$ are frequency-independent matrices to be determined. Since

$$\begin{aligned} [\mathbf{D}^{-1}(0)] &= [\mathbf{A}_0] \\ \frac{\partial}{\partial\omega^2}[\mathbf{D}^{-1}(0)] &= [\mathbf{A}_2] \\ \left(\frac{\partial}{\partial\omega^2}\right)^2[\mathbf{D}^{-1}(0)] &= [\mathbf{A}_4] \text{ etc.} \end{aligned} \tag{3.3.7}$$

it is found that

$$[\mathbf{A}_0] = [\mathbf{D}^{-1}(0)], \qquad [\mathbf{A}_2] = [\mathbf{D}_0^{-1}\mathbf{M}_0\mathbf{D}_0^{-1}]$$

$$[\mathbf{A}_4] = [\mathbf{D}_0^{-1}\mathbf{M}_0\mathbf{D}_0^{-1}\mathbf{M}_0\mathbf{D}_0^{-1}] + \frac{1}{2}\left[\mathbf{D}_0^{-1}\left(\frac{\partial}{\partial\omega^2}\mathbf{M}_0\mathbf{D}_0^{-1}\right)\right] \text{ etc.} \tag{3.3.8}$$

where $[\mathbf{D}_0] = [\mathbf{D}(0)]$ and $[\mathbf{M}_0] = [\mathbf{M}(0)]$.

Now $[\mathbf{D}_0] = [\mathbf{K}(0)] = [\mathbf{K}_0]$, so Eq. (3.3.6) becomes

$$\begin{aligned}[\mathbf{D}^{-1}(\omega)] &= [\mathbf{K}_0^{-1}] + \omega^2[\mathbf{K}_0^{-1}\mathbf{M}_0\mathbf{K}_0^{-1}] \\ &\quad + \omega^4\left[\mathbf{K}_0^{-1}(\mathbf{M}_0\mathbf{K}_0^{-1})^2 + \frac{1}{2}\mathbf{K}_0^{-1}\left(\frac{\partial}{\partial\omega^2}\mathbf{M}_0\right)\mathbf{K}_0^{-1}\right] \\ &\quad + \omega^6[\mathbf{R}_6(\omega)]\end{aligned} \tag{3.3.9}$$

Alternatively, $[\mathbf{D}^{-1}(\omega)]$ for the first n modes may be formulated as follows. The original natural vibration problem is solved for the first n modes $[\boldsymbol{\Omega}^2] = [\omega_i^2]$ and $[\boldsymbol{\Phi}] = [\boldsymbol{\phi}_1, \boldsymbol{\phi}_2, \ldots, \boldsymbol{\phi}_n]$ so that

$$[\bar{\mathbf{K}}][\boldsymbol{\Phi}] = [\bar{\mathbf{M}}][\boldsymbol{\Phi}][\boldsymbol{\Omega}^2] \tag{3.3.10}$$

where $[\boldsymbol{\Phi}]$ is normalized according to

$$[\boldsymbol{\Phi}^{\mathrm{T}}\bar{\mathbf{K}}\boldsymbol{\Phi}] = [\boldsymbol{\Omega}^2] \qquad \text{and} \qquad [\boldsymbol{\Phi}^{\mathrm{T}}\bar{\mathbf{M}}\boldsymbol{\Phi}] = [\mathbf{I}] \tag{3.3.11}$$

Therefore

$$[\boldsymbol{\Phi}^{\mathrm{T}}\bar{\mathbf{D}}\boldsymbol{\Phi}] = [\boldsymbol{\Omega}^2 - \omega^2\mathbf{I}] \tag{3.3.12}$$

Suppose n equals the order of the original matrices N, the inversion of Eq. (3.3.12) is

$$[\boldsymbol{\Phi}^{-1}\bar{\mathbf{D}}^{-1}\boldsymbol{\Phi}^{-\mathrm{T}}] = [\boldsymbol{\Omega}^2 - \omega^2\mathbf{I}]^{-1}$$

or

$$[\bar{\mathbf{D}}^{-1}] = [\boldsymbol{\Phi}][\boldsymbol{\Omega}^2 - \omega^2\mathbf{I}]^{-1}[\boldsymbol{\Phi}]^{\mathrm{T}} \tag{3.3.13}$$

Let the matrices be partitioned according to the active and the passive coordinates

$$[\bar{\mathbf{D}}^{-1}] = \begin{pmatrix}\mathbf{E}_{11} & \mathbf{E}_{12}\\ \mathbf{E}_{21} & \mathbf{E}_{22}\end{pmatrix} = \begin{Bmatrix}\boldsymbol{\Phi}_1\\ \boldsymbol{\Phi}_2\end{Bmatrix}[\boldsymbol{\Omega}^2 - \omega^2\mathbf{I}]^{-1}[\boldsymbol{\Phi}_1^{\mathrm{T}} \quad \boldsymbol{\Phi}_2^{\mathrm{T}}] \tag{3.3.14}$$

where 1 and 2 denote the active and the passive components as before. However, from Eq. (3.3.1)

$$\begin{Bmatrix}\mathbf{x}_1\\ \mathbf{x}_2\end{Bmatrix} = \begin{pmatrix}\mathbf{E}_{11} & \mathbf{E}_{12}\\ \mathbf{E}_{21} & \mathbf{E}_{22}\end{pmatrix}\begin{Bmatrix}\mathbf{X}_1\\ \mathbf{0}\end{Bmatrix}$$

or

$$\{\mathbf{x}_1\} = [\mathbf{E}_{11}]\{\mathbf{X}_1\} \qquad \text{and} \qquad \{\mathbf{x}_2\} = [\mathbf{E}_{21}]\{\mathbf{X}_1\} \tag{3.3.15}$$

Compared with Eq. (3.3.14), $[\mathbf{E}_{11}] = [\mathbf{D}]^{-1}$, and therefore, from Eq. (3.3.14)

$$[\mathbf{D}^{-1}] = [\boldsymbol{\Phi}_1][\boldsymbol{\Omega}^2 - \omega^2\mathbf{I}]^{-1}[\boldsymbol{\Phi}_1]^{\mathrm{T}} \tag{3.3.16}$$

The convergence rate of Eqs (3.3.13) and (3.3.16) is very slow if ω_i, $i = 1, 2, \ldots,$ are not increasing rapidly. This may be accelerated by making use of the fact that

$$(\omega_i^2 - \omega^2)^{-1} = \omega_i^{-2} + \omega^2\omega_i^{-4} + \cdots + \omega^{2j-2}\omega_i^{-2j} + \omega^{2j}\omega_i^{-2j}(\omega_i^2 - \omega^2)^{-1}$$

Thus, Eq. (3.3.6) becomes

$$[\mathbf{D}^{-1}(\omega)] = [\mathbf{\Phi}_1\mathbf{\Omega}^{-2}\mathbf{\Phi}_1^{\mathrm{T}}] + \omega^2[\mathbf{\Phi}_1\mathbf{\Omega}^{-4}\mathbf{\Phi}_1^{\mathrm{T}}] + \omega^4[\mathbf{\Phi}_1]\left[\mathbf{\Omega}^{-6} + \frac{\omega^2}{\omega_i^6(\omega_i^2-\omega^2)}\right][\mathbf{\Phi}_1]^{\mathrm{T}} \quad (3.3.17)$$

Comparing Eqs (3.3.9) and (3.3.17) according to like powers of ω,

$$[\mathbf{K}_0^{-1}] = [\mathbf{\Phi}_1\mathbf{\Omega}^{-2}\mathbf{\Phi}_1^{\mathrm{T}}]$$

$$[\mathbf{K}_0^{-1}\mathbf{M}_0\mathbf{K}_0^{-1}] = [\mathbf{\Phi}_1\mathbf{\Omega}^{-4}\mathbf{\Phi}_1^{\mathrm{T}}]$$

$$[\mathbf{K}_0^{-1}(\mathbf{M}_0\mathbf{K}_0^{-1})^2] + \frac{1}{2}\left[\mathbf{K}_0^{-1}\left(\frac{\partial}{\partial\omega^2}\mathbf{M}_0\right)\mathbf{K}_0^{-1}\right] + \omega^2[\mathbf{R}_6(\omega)] = [\mathbf{\Phi}_1][\omega_i^{-4}(\omega_i^2-\omega^2)^{-1}][\mathbf{\Phi}_1]^{\mathrm{T}} \quad (3.3.18)$$

For uncondensed matrices, when $[\mathbf{K}]$ and $[\mathbf{M}]$ are frequency independent, the above relations may alternatively be obtained by simpler approaches. The present approach, however, is believed to be more appropriate for frequency-dependent matrices.

Finally, from Eqs (3.3.17) and (3.3.18):

$$[\mathbf{Z}(\omega)] = [\mathbf{D}^{-1}(\omega)] = [\mathbf{K}_0^{-1}] + \omega^2[\mathbf{K}_0^{-1}\mathbf{M}_0\mathbf{K}_0^{-1}] + \omega^4[\mathbf{\Phi}_1][\omega_i^{-4}(\omega_i^2-\omega^2)^{-1}][\mathbf{\Phi}_1]^{\mathrm{T}} \quad (3.3.19)$$

This is the condensed dynamic flexibility required. It is noted that in this exact form all matrices are of reduced size and the convergent rate of the last term is very much faster than that of Eq. (3.3.13). Further accelerated convergence is limited by the uncertain physical meaning of $\partial[\mathbf{M}]/\partial\omega^2$. It should also be noted that only the active components $[\mathbf{\Phi}_1]$ of the eigenvectors $[\mathbf{\Phi}]$ are required, and therefore, in place of the original eigenvalue problem (3.3.10), the reduced equation (3.3.2) may be solved directly by letting $\{\mathbf{X}_1\} = \{\mathbf{0}\}$, thus minimizing the computational effort.

Very often, the normal modes of a freely vibrating system subjected to a particular set of boundary conditions are much easier to obtain than the other sets. Therefore, it is desirable to be able to calculate the natural modes for different sets of boundary conditions when the natural modes for a particular set of boundary conditions are known. Here, the method of dynamic flexibility is applied to estimate the normal modes of the structure in the presence of additional constraints. The method is first applied to some beam systems and then to plate systems.

Some simple examples concerning the natural vibrations of a simply supported beam are considered. If E, I, l, ρ, A denote Young's modulus, second moment of area, length, density and cross-sectional area respectively, then the normal modes are

$$\phi_i(x) = \sqrt{\frac{2}{m}}\sin\frac{i\pi x}{l}, \qquad m = \rho A l, \quad i = 1, 2, \ldots$$

and

$$\omega_i^2 = \frac{(i\pi)^4}{\lambda}, \qquad \lambda = \frac{\rho A l^4}{EI}$$

If the end rotations are chosen as active coordinates, the normal modes are represented by

$$[\mathbf{\Phi}_1] = \sqrt{\frac{2}{m}\frac{\pi}{l}}\begin{bmatrix} 1 & 2 & \dots & i & \dots \\ -1 & 2 & \dots & (-1)^i i & \dots \end{bmatrix}$$

For a beam idealized by two finite elements, then the following stiffness matrix is obtained:

$$[\bar{\mathbf{K}}] = \frac{8EI}{l^3}\begin{bmatrix} l^2 & & & \\ -3l & 24 & & \text{sym.} \\ \frac{1}{2}l^2 & 0 & 2l^2 & \\ 0 & 3l & \frac{1}{2}l^2 & l^2 \end{bmatrix}$$

With the first and the last coordinates as actives, $[\bar{\mathbf{K}}]$ is condensed to $[\mathbf{K}_0]$ according to Eq. (3.3.3)

$$[\mathbf{K}_0] = \frac{EI}{l}\begin{bmatrix} 4 & 2 \\ 2 & 4 \end{bmatrix}$$

which is the same as the stiffness matrix of a beam when it is idealized by one finite element. Therefore, $[\mathbf{K}_0]$ remains the same regardless of the number of finite elements if the same active coordinates are used. Similarly, the consistent mass matrix is

$$[\mathbf{M}_0] = \frac{\rho A l^3}{420}\begin{bmatrix} 4 & -3 \\ -3 & 4 \end{bmatrix}$$

The dynamic flexibility matrices, according to Eqs (3.3.18), (3.3.9) and (3.3.19) respectively, are

$$\begin{aligned}[\mathbf{Z}(\omega)] = \frac{l}{EI}\left(\frac{1}{6}\begin{bmatrix} 2 & -1 \\ -1 & 2 \end{bmatrix} + \frac{\omega^2\lambda}{15\,120}\begin{bmatrix} 32 & -31 \\ -31 & 32 \end{bmatrix}\right. \\ \left. + \frac{\omega^4\lambda^2}{23\,950\,080}\begin{bmatrix} 512 & -511 \\ -511 & 512 \end{bmatrix}\right) \\ + \omega^6[\mathbf{\Phi}_1][\omega_i^{-6}(\omega_i^2 - \omega^2)^{-1}][\mathbf{\Phi}_1]^{\mathrm{T}}\end{aligned} \quad (3.3.20)$$

$$\begin{aligned}[\mathbf{Z}(\omega)] = \frac{l}{EI}\left(\frac{1}{6}\begin{bmatrix} 2 & -1 \\ -1 & 2 \end{bmatrix} + \frac{\omega^2\lambda}{15\,120}\begin{bmatrix} 32 & -31 \\ -31 & 32 \end{bmatrix}\right. \\ \left. + \frac{\omega^4\lambda^2}{38\,102\,400}\begin{bmatrix} 662 & -661 \\ -661 & 662 \end{bmatrix}\right) \\ + \frac{\omega^4}{2}\left[\mathbf{K}_0^{-1}\left(\frac{\partial}{\partial\omega^2}\mathbf{M}_0\right)\mathbf{K}_0^{-1}\right] + \omega^6[\mathbf{R}_6(\omega)]\omega^6\end{aligned} \quad (3.3.21)$$

$$\begin{aligned}[\mathbf{Z}(\omega)] = \frac{l}{EI}\left(\frac{1}{6}\begin{bmatrix} 2 & -1 \\ -1 & 2 \end{bmatrix} + \frac{\omega^2\lambda}{15\,120}\begin{bmatrix} 32 & -31 \\ -31 & 32 \end{bmatrix}\right. \\ + \omega^4[\mathbf{\Phi}_1][\omega_i^{-4}(\omega_i^2 - \omega^2)^{-1}][\mathbf{\Phi}_1]^{\mathrm{T}}\end{aligned} \quad (3.3.22)$$

where the following closed form summation formulae have been used in deriving Eq. (3.3.20) for the purpose of comparison:

$$\sum_{n=1}^{\infty} n^{-2} = \frac{\pi^2}{6}, \qquad \sum_{n=1}^{\infty} n^{-6} = \frac{\pi^6}{945}, \qquad \sum_{n=1}^{\infty} n^{-10} = \frac{\pi^{10}}{93\,555}$$
$$\sum_{n=1}^{\infty} (-1)^n n^{-2i} = (1 - 2^{1-2i}) \sum_{n=1}^{\infty} n^{-2i} \tag{3.3.23}$$

It is noted in these exact presentations that in general, the form (3.3.20) is not always possible due to the irregular distribution of normal modes which makes the application of Eq. (3.3.23) impossible, and that the form (3.3.21) is incomplete as $\partial[\mathbf{M}_0]/\partial\omega^2$ has not been defined physically.

Now, the natural frequencies of a clamped–clamped beam and a clamped–hinged beam are calculated by letting $\det[\mathbf{Z}(\omega)] = 0$, and by equating the first element of $[\mathbf{Z}(\omega)]$ to zero respectively. Using the same flexibility, the symmetrical frame shown in Fig. 3.3.1 is also analysed for the symmetrical modes. The convergence of each example for the first few modes is tabulated in Table 3.3.1, where $\lambda_i^4 = \omega_i^2 \rho A l^2/EI$ and numbers in brackets are the percentage errors. It is shown that the convergence is very rapid and that even higher modes may be determined accurately.

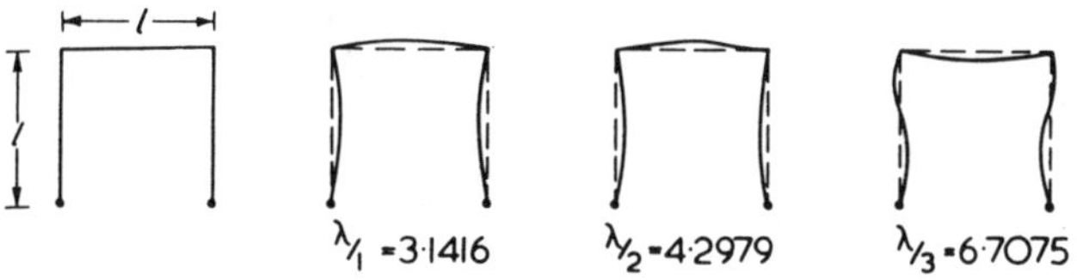

Fig. 3.3.1. Symmetrical modes of a symmetrical frame

Table 3.3.1. Convergence to the natural frequency parameter λ_i

Mode	No. of terms	Clamped–hinged beam	Clamped–clamped beam	Symmetrical frame
λ_1	1	3.9321 (0.14)	4.7334 (0.08)	3.1416[a] (0.00)
	2	3.9267 (0.00)	4.7318 (0.04)	
	3	3.9266 (0.00)	4.7300 (0.00)	
λ_2	2	7.0934 (0.35)	7.8652 (0.14)	4.3073 (0.22)
	3	7.0697 (0.02)	7.8584 (0.06)	4.2981 (0.00)
	4	7.0686 (0.00)	7.8540 (0.00)	4.2979 (0.00)
λ_3	3	10.261 (0.50)	11.026 (0.27)	6.7211 (0.20)
	4	10.214 (0.04)	10.997 (0.00)	6.7079 (0.00)
	5	10.210 (0.00)	10.996 (0.00)	6.7075 (0.00)

[a] Partial frequency.

Consider next a simply supported rectangular plate with sides a and b. The natural frequencies ω_{mn} and the corresponding modes $\phi_{mn}(x, y)$ are given by

$$\omega_{mn}^2 = D\pi^4 \frac{(m^2/a^2) + (n^2/b^2)}{\rho h} \qquad m, n = 1, 2, \ldots$$

$$\phi_{mn} = \frac{2\sin(m\pi x/a)\sin(n\pi y/b)}{\sqrt{\rho abh}}$$

where D, ρ, h are flexural rigidity, density and thickness of the plate. If a concentrated vertical harmonic force $Q \sin \omega t$ is acting on the centre of the plate, then the amplitude of the centre deflection, q is given as

$$q = Z(\omega)Q$$

where the dynamic flexibility $Z(\omega)$ according to the conventional method is

$$Z(\omega) = \frac{4}{\rho abh}\sum_m\sum_n \frac{1}{(\omega_{mn}^2 - \omega^2)} \qquad m, n = 1, 3, 5, \ldots \tag{3.3.24}$$

The asymptotic convergence rate

$$\int_R^\infty \int_0^{\pi/2} \frac{r d\theta\, dr}{r^4} \propto \frac{1}{R^2} = \frac{1}{m^2 + n^2}$$

When $\omega_{mn}^2 \gg \omega^2$, a 1% accuracy requires $m, n > 10$. That is, for every frequency ω, 55 terms are to be summed for two-digit accuracy. However, the convergence of the following equation is very much more rapid if Eq. (3.3.19) is used, i.e.

$$Z(\omega) = K^{-1} + \omega^2 K^{-2} M + \frac{4\omega^4}{\rho abh}\sum_m\sum_n \frac{1}{[\omega_{mn}^4(\omega_{mn}^2 - \omega^2)]} \qquad m, n = 1, 3, 5, \ldots \tag{3.3.25}$$

where K and M are the condensed stiffness and consistent mass respectively. When $b = 2a$, a finite element analysis using four conforming 16 d.o.f. elements for a quarter of the plate gives $K = 60.518D/a^2$ and $M = 0.323605\rho ha^2$ and therefore, Eq. (3.3.25) becomes

$$Z(\omega) = \left[0.016524 + 8.8358 \times 10^{-5}\lambda + 2\lambda^2 \sum_m\sum_n \lambda_{mn}^{-2}(\lambda_{mn} - \lambda)^{-1}\right] a^2/D \tag{3.3.26}$$

where $\lambda_{mn} = \pi^4(m^2 + n^2/4)^2$ and $\lambda = \omega^2 \rho h a^4/D$. The dynamic flexibility is plotted against λ in Fig. 3.3.2.

The natural frequencies of the plate when the centre is simply supported are given by the interceptions of $Z(\lambda)$ with the λ axis. The first natural frequency is given by

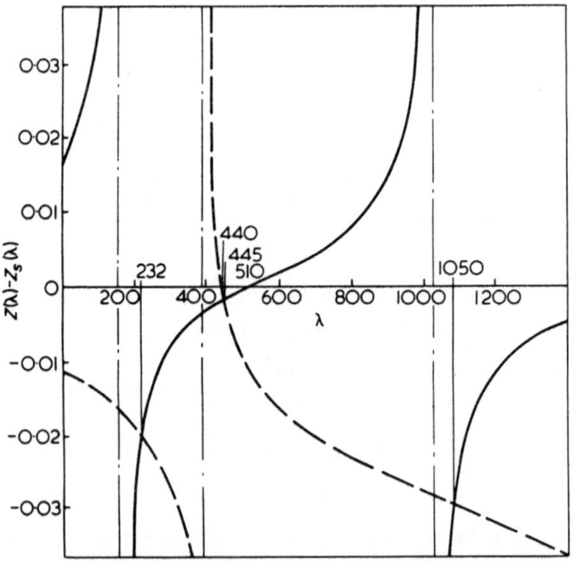

Fig. 3.3.2. Dynamic flexibility against λ

$\lambda = 510$. If the centre support is replaced by a vertical spring with stiffness k, then the horizontal line with distance l/k below the λ axis will intercept $Z(\lambda)$ at natural frequencies which approach those of the original system. A mass attached to the centre of the plate may be analysed similarly.

The dynamic flexibility of a square plate of side length a and with the same physical properties as the above mentioned rectangular plate is

$$Z_s(\lambda) = \left[0.011599 + 2.6438 \times 10^{-5}\lambda + \lambda^2 \sum_m \sum_n \nu_{mn}^{-2}(\nu_{mn} - \lambda)^{-1}\right]\frac{a^2}{D} \quad (3.3.27)$$

where $\nu_{mn} = \pi^4(m^2 + n^2)^2$ and $\lambda = \omega^2 \rho h a^4/D$. If the centres of those two plates are connected by a vertical light rigid rod, then the natural frequencies of the whole system may be solved from

$$Z(\lambda) + Z_s(\lambda) = 0 \quad \text{or} \quad Z(\lambda) = -Z_s(\lambda)$$

When $-Z_s(\lambda)$ is plotted as shown in Fig. 3.3.2 by dashed curves, the roots are the intersections of these two sets of curves. This example illustrates the application of dynamic flexibility to substructures.

3.4. Dynamic Transformation

It will be shown that most substructure methods are equivalent to matrix transformation. By means of the flexibility method given in the previous section, we establish a new dynamic substructure method called the flexibility dynamic substructure method. All the previously discussed methods are compared. First a review is given of some important formulae which will later be used extensively.

In an undamped harmonic finite element analysis with frequency ω, the response $\{\mathbf{u}\}e^{i\omega t}$ is related to the excitation $\{\mathbf{f}\}e^{i\omega t}$ by

$$[\mathbf{D}]\{\mathbf{u}\} = [\mathbf{K} - \omega^2\mathbf{M}]\{\mathbf{u}\} = \{\mathbf{f}\} \quad (3.4.1)$$

where $[\mathbf{D}]$, $[\mathbf{K}]$ and $[\mathbf{M}]$ are the substructure dynamic stiffness, stiffness and mass matrices respectively. If the coordinates are partitioned according to the slave coordinates (subscript s) and the master coordinates (subscript m), then

$$\begin{bmatrix} \mathbf{D}_{ss} & \mathbf{D}_{sm} \\ \mathbf{D}_{ms} & \mathbf{D}_{mm} \end{bmatrix} \begin{Bmatrix} \mathbf{u}_s \\ \mathbf{u}_m \end{Bmatrix} = \begin{Bmatrix} \mathbf{0} \\ \mathbf{f}_m \end{Bmatrix} \quad (3.4.2)$$

Eliminating the slave coordinates $\{\mathbf{u}_s\}$, one has the condensed dynamic stiffness equation in exact form,

$$[\mathbf{D}_m^*]\{\mathbf{u}_m\} = [\mathbf{D}_{mm} - \mathbf{D}_{ms}\mathbf{D}_{ss}^{-1}\mathbf{D}_{sm}]\{\mathbf{u}_m\} = \{\mathbf{f}_m\} \quad (3.4.3)$$

The inversion of the frequency-dependent matrix $[\mathbf{D}_{ss}]$ in Eq. (3.4.3) is given by

$$[\mathbf{D}_{ss}^{-1}] = [\mathbf{K}_{ss} - \omega^2\mathbf{M}_{ss}]^{-1} = [\mathbf{\Phi}][\mathbf{\Omega}^2 - \omega^2\mathbf{I}]^{-1}[\mathbf{\Phi}]^T \quad (3.4.4)$$

where $[\mathbf{\Phi}]$ is the fixed interface modal matrix defined such that

$$[\mathbf{\Phi}^T\mathbf{M}_{ss}\mathbf{\Phi}] = [\mathbf{I}] \quad \text{and} \quad [\mathbf{\Phi}^T\mathbf{K}_{ss}\mathbf{\Phi}] = [\mathbf{\Omega}^2] \quad (3.4.5)$$

in which $[\mathbf{\Omega}^2] = [\Omega_1^2, \Omega_2^2, \ldots, \Omega_n^2]$ is a collection of the natural frequencies. Recognizing the following identities,

$$\frac{1}{\Omega_i^2 - \omega^2} = \frac{1}{\Omega_i^2} + \frac{\omega^2}{\Omega_i^4} + \frac{\omega^4}{\Omega_i^4(\Omega_i^2 - \omega^2)}$$
$$[\mathbf{\Phi}\mathbf{\Omega}^{-2}\mathbf{\Phi}^{\mathrm{T}}] = [\mathbf{K}_{ss}^{-1}] \tag{3.4.6}$$
$$[\mathbf{\Phi}\mathbf{\Omega}^{-4}\mathbf{\Phi}^{\mathrm{T}}] = [\mathbf{K}_{ss}^{-1}\mathbf{M}_{ss}\mathbf{K}_{ss}^{-1}]$$

one can rewrite Eq. (3.4.4) as

$$[\mathbf{D}_{ss}^{-1}] = [\mathbf{K}_{ss}^{-1} - \omega^2\mathbf{K}_{ss}^{-1}\mathbf{M}_{ss}\mathbf{K}_{ss}^{-1} + \omega^4\mathbf{\Phi}\,\mathrm{diag}[\Omega_i^{-4}(\Omega_i^2 - \omega^2)^{-1}]\mathbf{\Phi}^{\mathrm{T}}] \tag{3.4.7}$$

The computation of $\mathbf{K}_{ss}^{-1}$ (or its equivalent) is required by all of the dynamic substructure methods considered (including the simplest eigenvalue economizer). In practice, the computation of $\mathbf{K}_{ss}^{-1}\mathbf{K}_{sm}$ instead of the explicit inverse $\mathbf{K}_{ss}^{-1}$, is performed, and this may be accomplished by any equation solver which takes advantage of the sparsity of $\mathbf{K}_{ss}$. When substituting Eq. (3.4.7) into Eq. (3.4.3), one notes that even $\mathbf{K}_{ss}^{-1}\mathbf{M}_{ss}\mathbf{K}_{ss}^{-1}$ is not required explicitly, but rather $(\mathbf{K}_{ss}^{-1}\mathbf{K}_{sm})^{\mathrm{T}}\mathbf{M}_{ss}\mathbf{K}_{ss}^{-1}\mathbf{K}_{sm}$. Therefore, the expansion of Eq. (3.4.7) does not introduce a large amount of computation because $\mathbf{K}_{ss}^{-1}\mathbf{K}_{sm}$ and $\mathbf{\Phi}$ are needed anyway.

3.4.1. Methods of Guyan [5] and Irons [6]

If the modal contribution given in the last term of Eq. (3.4.7) is ignored, equation (3.4.3) is identical to the result of Guyan and Irons. The method has been widely used and is referred to as an "eigenvalue economizer" [3]. To achieve reasonably accurate results the "masters" must be chosen with care, or some of the lowest frequencies in the eigenspectrum may be lost [8]. It is further required that different sets of masters are necessary for different modes due to the redistribution of energy in various modes; hence a direct extension of the method to substructures is difficult. To overcome these deficiencies an exact application of Eq. (3.4.3) is given in Ref. [4]. The flexibility method also uses Eq. (3.4.3) but evaluates the inversion of the dynamic stiffness matrix in terms of the component modes. The modal contribution in Eq. (3.4.7) is significant, as may be seen in the examples.

3.4.2. Ritz Vector Method [19]

The primary objective of the Ritz vector method is to obtain the dynamic response. However, it also produces natural modes if the loading patterns assumed are close to the natural modes. In contrast to the transformation inherent in the first block of Eq. (3.4.2), which is

$$\begin{Bmatrix} \mathbf{u}_s \\ \mathbf{u}_m \end{Bmatrix} = \begin{bmatrix} -\mathbf{D}_{ss}^{-1}\mathbf{D}_{mm} \\ \mathbf{I} \end{bmatrix} \mathbf{u}_m \tag{3.4.8}$$

the Ritz vector method uses

$$\begin{Bmatrix} \mathbf{u}_s \\ \mathbf{u}_m \end{Bmatrix} = \begin{bmatrix} \mathbf{X} & -\mathbf{K}_{ss}^{-1}\mathbf{K}_{sm} \\ \mathbf{0} & \mathbf{I} \end{bmatrix} \begin{Bmatrix} \mathbf{y} \\ \mathbf{u}_m \end{Bmatrix} \tag{3.4.9}$$

where $\mathbf{X}$ and $\mathbf{y}$ are the collections of Ritz vectors and Ritz coordinates respectively.

Equation (3.4.8) is exact while Eq. (3.4.9) is approximate. The advantages of Eq. (3.4.9) include the frequency-independent characteristics and the relative ease in generating the Ritz vectors rather than the substructure modes. However, the unknowns retained in Eq. (3.4.9) are $\mathbf{y}$ and $\mathbf{u}_m$, compared with $\mathbf{u}_m$ alone in Eq. (3.4.8). Owing to the redistribution of energy in various global modes, the Ritz vectors have to be regenerated if the energy distribution patterns of current interest have not been previously considered. In any case, the number of global modes predicted will not exceed the number of unknowns retained. The flexibility method is able to predict more global modes than the number of unknowns retained due to the frequency-dependent transformation (3.4.8). On the other hand, Eq. (3.4.9) requires solving the standard eigenvalue problem only, while Eq. (3.4.8) requires solving a frequency-dependent eigenvalue problem. It is shown in the examples that by retaining only two unknowns of a cantilever the present method can predict ten modes accurately. The "static" components in the Ritz vector method are indeed considered in the first two terms of Eq. (3.4.7).

3.4.3. Substructure Synthesis and Dynamic Substructures

Sotiropoulos [21] noted that a version of the substructure synthesis [36] is in fact identical to the dynamic substructure methods in formulation. This version is a great improvement over the method originally proposed by Hurty [37]. The difference between substructure synthesis and the dynamic substructure method is in the solution sequences. Both methods employ the following transformation, transforming the slave coordinates $\{\mathbf{u}_s\}$ to modal coordinates $\{\mathbf{u}_p\}$

$$\begin{Bmatrix} \mathbf{u}_s \\ \mathbf{u}_m \end{Bmatrix} = \begin{bmatrix} \mathbf{\Phi} & -\mathbf{K}_{ss}^{-1}\mathbf{K}_{sm} \\ \mathbf{0} & \mathbf{I} \end{bmatrix} \begin{Bmatrix} \mathbf{u}_p \\ \mathbf{u}_m \end{Bmatrix} \tag{3.4.10}$$

Substituting Eqs (3.4.5) and (3.4.10) into (3.4.2), we have

Table 3.4.1. Comparison of the methods

Method	Transformation formulae $\{\mathbf{u}_s, \mathbf{u}_m\} =$
Guyan [5] and Irons [6]	$\begin{bmatrix} -\mathbf{K}_{ss}^{-1}\mathbf{K}_{sm} \\ \mathbf{I} \end{bmatrix} \mathbf{u}_m$
Ritz vector	$\begin{bmatrix} \mathbf{X} & -\mathbf{K}_{ss}^{-1}\mathbf{K}_{sm} \\ \mathbf{0} & \mathbf{I} \end{bmatrix} \begin{Bmatrix} \mathbf{y} \\ \mathbf{u}_m \end{Bmatrix}$
Substructure synthesis	$\begin{bmatrix} \mathbf{\Phi} & -\mathbf{K}_{ss}^{-1}\mathbf{K}_{sm} \\ \mathbf{0} & \mathbf{I} \end{bmatrix} \begin{Bmatrix} \mathbf{u}_p \\ \mathbf{u}_m \end{Bmatrix}$
Dynamic substructure	$\begin{bmatrix} \mathbf{\Phi} & -\mathbf{K}_{ss}^{-1}\mathbf{K}_{sm} \\ \mathbf{0} & \mathbf{I} \end{bmatrix} \begin{Bmatrix} \mathbf{u}_p \\ \mathbf{u}_m \end{Bmatrix}$ and $\mathbf{u}_p = \omega^2(\mathbf{\Omega}^2 - \omega^2\mathbf{I})^{-1}\mathbf{G}^{\mathrm{T}}\mathbf{u}_m$
Exact dynamic condensation	$\begin{bmatrix} -\mathbf{D}_{ss}^{-1}\mathbf{D}_{mm} \\ \mathbf{I} \end{bmatrix} \mathbf{u}_m$
Present method	$\begin{bmatrix} -\mathbf{D}_{ss}^{-1}\mathbf{D}_{mm} \\ \mathbf{I} \end{bmatrix} \mathbf{u}_m$ and evaluate $\mathbf{D}_{ss}^{-1}$ from Eq. (3.4.7)

$$\left(\begin{bmatrix} \mathbf{K}_0 & \mathbf{0} \\ \mathbf{0} & \mathbf{\Omega}^2 \end{bmatrix} - \omega^2 \begin{bmatrix} \mathbf{M}_0 & \mathbf{G} \\ \mathbf{G} & \mathbf{I} \end{bmatrix}\right) \begin{Bmatrix} \mathbf{u}_s \\ \mathbf{u}_m \end{Bmatrix} = \begin{Bmatrix} \mathbf{f}_m \\ \mathbf{0} \end{Bmatrix} \tag{3.4.11}$$

where $\mathbf{K}_0 = \mathbf{D}^*(0) = \mathbf{K}_{mm} - \mathbf{K}_{ms}\mathbf{K}_{ss}^{-1}\mathbf{K}_{sm}$

$$-\mathbf{M}_0 = \frac{\partial}{\partial \omega^2}\mathbf{D}^*(\mathbf{0}) = \begin{bmatrix} -\mathbf{K}_{ss}^{-1}\mathbf{K}_{sm} \\ \mathbf{I} \end{bmatrix}^{\mathrm{T}} \begin{bmatrix} \mathbf{M}_{ss} & \mathbf{M}_{sm} \\ \mathbf{M}_{ms} & \mathbf{M}_{mm} \end{bmatrix} \begin{bmatrix} -\mathbf{K}_{ss}^{-1}\mathbf{K}_{sm} \\ \mathbf{I} \end{bmatrix}$$

and

$$\mathbf{G} = \mathbf{M}_{ms}\mathbf{\Phi} - \mathbf{K}_{ms}\mathbf{\Phi}\mathbf{\Omega}^{-2} \tag{3.4.12}$$

The substructure synthesis method retains $\mathbf{u}_m$ and $\mathbf{u}_p$ in Eq. (3.4.11) whereas the dynamic substructure method retains only $\mathbf{u}_m$. It is obvious from Eq. (3.4.12) that the convergence with respect to the substructure modes is proportional to Ω_i^{-2}, similar to Eq. (3.4.4). The convergence of the present new method is proportional to $\Omega_i^{-4}(\Omega_i^2 - \omega^2)^{-1}$, which is faster. The comparison of the methods is summarized in Table 3.4.1.

Numerical Example

Consider the vibration of a cantilever beam having unit properties. The finite element representation by two equal elements gives

$$\mathbf{D} = \begin{bmatrix} 192 & & & \text{sym.} \\ 0 & 16 & & \\ -96 & -24 & 96 & \\ 24 & 4 & -24 & 8 \end{bmatrix} - \frac{\omega^2}{3360} \begin{bmatrix} 1248 & & & \text{sym.} \\ 0 & 8 & & \\ 216 & 26 & 624 & \\ -26 & -3 & -44 & 4 \end{bmatrix}$$

If one takes the last two coordinates as masters, the method of Guyan and Irons gives

$$\mathbf{K}_0 = \begin{bmatrix} 12 & -6 \\ -6 & 4 \end{bmatrix} \quad \text{and} \quad \mathbf{M}_0 = \frac{1}{420}\begin{bmatrix} 156 & -22 \\ -22 & 4 \end{bmatrix}$$

It is interesting to note that the same matrices are obtained using just one element with one elimination. Therefore, by subdividing the structure while keeping the same masters coordinates, no increase of accuracy is achieved in this case.

If the cantilever is divided into ten finite elements, the first ten natural frequencies obtained by solving the resulting 20×20 eigenvalue problem are 3.5216, 22.035, 61.713, 121.017, 200.363, 300.167, 421.149, 564.223, 729.534, 906.16 rad s^{-1}. If the ten-element system is taken as one substructure in which the last two degrees of freedom, i.e. the displacement and rotation at the free end, are retained as masters, the dynamic substructure method (or the equivalent substructure synthesis) gives the natural frequencies in Table 3.4.2 by solving the resulting 2×2 matrix with different numbers of terms k, in which k represents the number of partial modes when the masters are clamped. The corresponding results as computed by the flexibility method are tabulated in Table 3.4.3. It is evident that the flexibility method gives better results for the same number of partial modes. The flexibility method is even more effective in the higher frequency range, predicting more modes than the number of the retained unknowns, as shown.

Table 3.4.2.

No. of terms k	Dynamic substructure method (% error in parentheses)									
	Mode 1	Mode 2	Mode 3	Mode 4	Mode 5	Mode 6	Mode 7	Mode 8	Mode 9	Mode 10
1	3.51724	22.1270								
	(0.0347)	(0.4166)								
2	3.51622	22.0590	61.9376							
	(0.0057)	(0.1080)	(0.3641)							
3	3.51607	22.0429	61.8068	121.359						
	(0.0014)	(0.0349)	(0.1522)	(0.2826)						
4	3.51604	22.0381	61.7540	121.199	200.797					
	(0.0006)	(0.0132)	(0.0666)	(0.1504)	(0.2166)					
5	3.51603	22.0365	61.7321	121.111	200.626	300.648				
	(0.0003)	(0.0059)	(0.0311)	(0.0777)	(0.1313)	(0.1602)				
6	3.51602	22.0358	61.7222	121.066	200.513	300.481	421.616			
	(0.0000)	(0.0027)	(0.0151)	(0.0405)	(0.0749)	(0.1046)	(0.1109)			
7	3.51602	22.0355	61.7175	121.043	200.446	300.354	421.470	564.614		
	(0.0000)	(0.0014)	(0.0075)	(0.0215)	(0.0414)	(0.0623)	(0.0762)	(0.0693)		
8	3.51602	22.0353	61.7152	121.030	200.407	300.272	421.345	564.506	729.807	
	(0.0000)	(0.0005)	(0.0037)	(0.0107)	(0.0220)	(0.0350)	(0.0465)	(0.0502)	(0.0374)	
9	3.51602	22.0353	61.7141	121.024	200.389	300.229	421.272	564.418	729.764	907.055
	(0.0000)	(0.0005)	(0.0019)	(0.0058)	(0.0130)	(0.0207)	(0.0292)	(0.0346)	(0.0315)	(0.0153)
10	3.51602	22.0352	61.7137	121.022	200.379	300.208	421.231	564.358	729.704	907.038
	(0.0000)	(0.0000)	(0.0013)	(0.0041)	(0.0080)	(0.0137)	(0.0195)	(0.0239)	(0.0233)	(0.0135)
FEM	3.5160	22.0352	61.7129	121.017	200.363	300.167	421.149	564.223	729.534	906.916

Table 3.4.3.

No. of terms k	Flexibility method (% error in parentheses)									
	Mode 1	Mode 2	Mode 3	Mode 4	Mode 5	Mode 6	Mode 7	Mode 8	Mode 9	Mode 10
1	3.51718 (0.0330)	22.11946 (0.3824)								
2	3.51619 (0.0048)	22.0555 (0.0921)	61.8885 (0.2845)							
3	3.51606 (0.0011)	22.041 (0.0263)	61.7786 (0.1065)	121.225 (0.1719)						
4	3.51603 (0.0003)	22.0371 (0.0086)	61.7376 (0.0400)	121.114 (0.0802)	200.556 (0.0963)					
5	3.51602 (0.0000)	22.0359 (0.0032)	61.7225 (0.0156)	121.059 (0.0347)	200.465 (0.0509)	300.316 (0.0496)				
6	3.51602 (0.0000)	22.0355 (0.0014)	61.7167 (0.0062)	121.035 (0.0149)	200.411 (0.0240)	300.251 (0.0280)	421.245 (0.0228)			
7	3.51602 (0.0000)	22.0353 (0.0005)	61.7144 (0.0024)	121.024 (0.0058)	200.384 (0.0105)	300.207 (0.0133)	421.206 (0.0135)	564.270 (0.0083)		
8	3.51602 (0.0000)	22.0353 (0.0005)	61.7134 (0.0008)	121.02 (0.0025)	200.371 (0.0040)	300.184 (0.0057)	421.176 (0.0064)	564.253 (0.0053)	729.548 (0.0019)	
9	3.51602 (0.0000)	22.0353 (0.0005)	61.7131 (0.0003)	121.018 (0.0008)	200.367 (0.0020)	300.174 (0.0023)	421.161 (0.0028)	564.240 (0.0030)	729.547 (0.0018)	906.919 (0.0003)
10	3.51602 (0.0000)	22.0353 (0.0005)	61.7131 (0.0003)	121.018 (0.0005)	200.365 (0.0010)	300.171 (0.0013)	421.155 (0.0014)	564.233 (0.0018)	729.543 (0.0012)	906.918 (0.0002)
FEM	3.51602	22.0352	61.7129	121.017	200.363	300.167	421.149	564.223	729.534	906.916

3.5. Damped Substructures

The equations of motion obtained by finite element discretization for a generally damped substructure have the following form:

$$[\mathbf{M}]\{\ddot{\mathbf{v}}\} + [\mathbf{C}]\{\dot{\mathbf{v}}\} + [\mathbf{K}]\{\mathbf{v}\} = \{\mathbf{F}\} \tag{3.5.1}$$

where [**M**], [**C**] and [**K**] are the substructure mass, damping and stiffness matrices respectively. {**F**} and {**v**} are the force and displacement vectors respectively. The coordinates considered include both slaves and masters. For natural vibration analysis, we assume

$$\{\mathbf{v}(t)\} = e^{\lambda t}\{\mathbf{u}\} \quad \text{and} \quad \{\mathbf{F}(t)\} = e^{\lambda t}\{\mathbf{f}\} \tag{3.5.2}$$

where λ is a complex frequency parameter to be determined. $\{\mathbf{F}(t)\}$ and $\{\mathbf{f}\}$ vanish everywhere except at master coordinates and represent the interactions between substructures. Substituting Eq. (3.5.2) into (3.5.1) gives

$$[\mathbf{D}(\lambda)]\{\mathbf{u}\} = [\lambda^2\mathbf{M} + \lambda\mathbf{C} + \mathbf{K}]\{\mathbf{u}\} = \{\mathbf{f}\} \tag{3.5.3}$$

where $[\mathbf{D}(\lambda)]$ is the substructure dynamic stiffness matrix. [**C**] is positive-definite and corresponds to external damping. Internal damping is represented by complex elastic moduli to give a complex stiffness matrix [**K**].

Partitioning Eq. (3.5.3) according to the slave coordinates $\{\mathbf{u}_s\}$ and master coordinates $\{\mathbf{u}_m\}$, we have

$$\begin{bmatrix} \mathbf{D}_{ss} & \mathbf{D}_{sm} \\ \mathbf{D}_{ms} & \mathbf{D}_{mm} \end{bmatrix} \begin{Bmatrix} \mathbf{u}_s \\ \mathbf{u}_m \end{Bmatrix} = \begin{Bmatrix} \mathbf{0} \\ \mathbf{f}_m \end{Bmatrix} \tag{3.5.4}$$

Using the first partitioned equation, the slave coordinates can be eliminated by

$$\{\mathbf{u}_s\} = [\mathbf{D}_{ss}^{-1}\mathbf{D}_{sm}]\{\mathbf{u}_m\} \tag{3.5.5}$$

to give the condensed dynamic stiffness relation

$$[\mathbf{D}^*(\lambda)]\{\mathbf{u}_m\} = [\mathbf{D}_{mm} - \mathbf{D}_{ms}\mathbf{D}_{ss}^{-1}\mathbf{D}_{sm}]\{\mathbf{u}_m\} = \{\mathbf{f}_m\} \tag{3.5.6}$$

where $[\mathbf{D}^*(\lambda)]$ is the condensed dynamic stiffness matrix. Then, the condensed dynamic stiffness matrices of individual substructures are assembled by means of the usual finite element method to give the global system dynamic stiffness matrix. No additional approximation has been introduced so far.

The difficulty in Eq. (3.5.6) is in the inversion of the λ-dependent dynamic stiffness matrix associated with the slave coordinates,

$$[\mathbf{D}_{ss}(\lambda)]^{-1} = [\lambda^2\mathbf{M}_{ss} + \lambda\mathbf{C}_{ss} + \mathbf{K}_{ss}]^{-1} \tag{3.5.7}$$

However, if the fixed interface modes λ_r, $\{\mathbf{\Phi}_r\}$, $r = 1, 2, \ldots, 2n$, where n is the number of slave coordinates, of the following eigenvalue problem are known,

$$[\lambda_r^2\mathbf{M}_{ss} + \lambda_r\mathbf{C}_{ss} + \mathbf{K}_{ss}]\{\mathbf{\Phi}_r\} = \{\mathbf{0}\} \tag{3.5.8}$$

and the complex $\{\mathbf{\Phi}_r\}$ are normalized so that

$$\{\mathbf{\Phi}_r\}^{\mathrm{T}}[2\lambda_r\mathbf{M}_{ss} + \mathbf{C}_{ss}]\{\mathbf{\Phi}_r\} = 1 \tag{3.5.9}$$

then one can show that

$$[\mathbf{D}_{ss}(\lambda)]^{-1} = \sum_{r=1}^{2n} (\lambda - \lambda_r)^{-1}\{\mathbf{\Phi}_r\}\{\mathbf{\Phi}_r\}^{\mathrm{T}} \tag{3.5.10}$$

Therefore, when substituting into Eq. (3.5.6), the evaluation of the condensed dynamic stiffness $[\mathbf{D}^*(\lambda)]$ may not involve inversion at all.

The free interface model does not partition the governing equations of a substructure according to masters and slaves, as in Eq. (3.5.4). The unconnected system equation in free vibration with frequency parameter λ is written as

$$[\mathbf{D}(\lambda)]\{\mathbf{u}\} = \{\mathbf{F}\} \tag{3.5.11}$$

where $[\mathbf{D}(\lambda)] = \mathrm{diag}[\mathbf{D}_i(\lambda)]$, $\{\mathbf{u}\} = \mathrm{col}\{\mathbf{u}_i\}$ and, for free vibration, $\{\mathbf{F}\}$ is zero everywhere except at masters (the tear coordinates) where the connection forces are non-vanishing. $[\mathbf{D}_i(\lambda)]$ is the dynamic stiffness matrix and $\{\mathbf{u}_i\}$ is the displacement vector of the ith substructure. The master coordinates appear in multiples associated with all substructures concerned and these multiple master coordinates are denoted as $\{\mathbf{u}_i\}$, the tear coordinates. Since the master coordinates are single-valued, the following constraint equation is necessary,

$$[\mathbf{R}]\{\mathbf{u}_t\} = \{\mathbf{0}\} \tag{3.5.12}$$

where $[\mathbf{R}]$ is a $p \times q$ matrix of elements 1, 0, or -1, p being the number of system master coordinates and q the total number of tear coordinates. Also, $\{\mathbf{u}_t\}$ is a subset of $\{\mathbf{u}\}$,

$$\{\mathbf{u}_t\} = [\mathbf{T}]\{\mathbf{u}\} \tag{3.5.13}$$

where $[\mathbf{T}]$ is a Boolean matrix. By the introduction of a Lagrange multiplier $\{\boldsymbol{\beta}\}$, the interconnecting forces necessary to make Eq. (3.5.12) possible are

$$\{\mathbf{F}\} = [\mathbf{RT}]^{\mathrm{T}}\{\boldsymbol{\beta}\} \tag{3.5.14}$$

Therefore, Eq. (3.5.11) requires that

$$\begin{bmatrix} \mathbf{D}(\lambda) & -(\mathbf{RT})^{\mathrm{T}} \\ -\mathbf{RT} & \mathbf{0} \end{bmatrix} \begin{Bmatrix} \mathbf{u} \\ \boldsymbol{\beta} \end{Bmatrix} = \{\mathbf{0}\} \tag{3.5.15}$$

The natural frequencies are the non-trivial roots of the equation

$$[(\mathbf{RT})\mathbf{D}^{-1}(\lambda)(\mathbf{RT})^{\mathrm{T}}]\{\boldsymbol{\beta}\} = \{\mathbf{0}\} \tag{3.5.16}$$

However, recalling that $[\mathbf{D}(\lambda)]$ is a diagonal block matrix consisting of the individual substructure dynamic stiffness matrices, we have

$$[\mathbf{D}(\lambda)]^{-1} = \mathrm{diag}[\mathbf{D}_i(\lambda)]^{-1} \tag{3.5.17}$$

We may also express $[\mathbf{D}_i(\lambda)]^{-1}$ in terms of the free interface modes λ_s and $\{\boldsymbol{\Phi}_s\}$ of the substructure as suggested by Simpson,

$$[\mathbf{D}_i(\lambda)]^{-1} = \sum_{s=1}^{2m} (\lambda - \lambda_s)^{-1}\{\boldsymbol{\Phi}_s\}\{\boldsymbol{\Phi}_s\}^{\mathrm{T}} \tag{3.5.18}$$

where m is the number of total degrees of freedom of the substructure concerned. In comparing Eqs (3.5.10) and (3.5.18), it is clear that the free interface approach requires one to solve for the natural modes with more degrees of freedom than the fixed interface modal model for each substructure. The numbers of degrees of freedom, however, in the final systems given by equations (3.5.6) and (3.5.16) are the same.

Because of the similar forms of Eqs (3.5.6), (3.5.16), (3.5.10) and (3.5.18), further study of the fixed interface model only is given below.

Very often, it is impractical to compute all the $2n$ damped natural modes. If there are less than $2n$ available modes, truncation errors are introduced. Reference [18] suggests a method to minimize the truncation errors contributed by higher natural modes of an undamped structure and reference [19] extends the method to generally damped structures. This method makes use of the Taylor series expansion of the dynamic flexibility matrix $[\mathbf{Z}(\lambda)] = [\mathbf{D}_{ss}(\lambda)]^{-1}$ about $\lambda = 0$,

$$\begin{aligned}[\mathbf{D}_{ss}(\lambda)]^{-1} &= [\mathbf{Z}(\lambda)] \\ &= [\mathbf{Z}_0] + \lambda[\mathbf{Z}_0'] + \tfrac{1}{2}\lambda^2[\mathbf{Z}_0''] + \tfrac{1}{6}\lambda^3[\mathbf{Z}_0'''] \\ &\quad + \lambda^4 \sum_{s=1}^{2n} \lambda_r^{-4}(\lambda - \lambda_r)^{-1}\{\mathbf{\Phi}_r\}\{\mathbf{\Phi}_r\}^{\mathrm{T}}\end{aligned} \tag{3.5.19}$$

where primes denote derivatives with respect to λ, and a zero subscript denotes evaluation at $\lambda = 0$.

The derivatives of $[\mathbf{Z}(\lambda)]$ can be obtained by successive differentiation of the following identity:

$$[\mathbf{DZ}] = [\mathbf{I}]$$

i.e.

$$\begin{aligned}[\mathbf{Z}'\mathbf{D} + \mathbf{ZD}'] &= [\mathbf{0}] \\ [\mathbf{Z}''\mathbf{D} + 2\mathbf{Z}'\mathbf{D}' + \mathbf{ZD}''] &= [\mathbf{0}] \\ [\mathbf{Z}'''\mathbf{D} + 3\mathbf{Z}''\mathbf{D}' + 3\mathbf{Z}'\mathbf{D}'' + \mathbf{ZD}'''] &= [\mathbf{0}]\end{aligned}$$

When these are evaluated at $\lambda = 0$,

$$\begin{aligned}[\mathbf{Z}_0] = [\mathbf{K}]^{-1}, \qquad [\mathbf{Z}_0'] &= -[\mathbf{K}^{-1}\mathbf{CK}^{-1}] \\ [\mathbf{Z}_0''] &= 2[\mathbf{K}^{-1}(\mathbf{CK}^{-1})^2 - \mathbf{K}^{-1}\mathbf{MK}^{-1}] \\ [\mathbf{Z}_0'''] &= -6[\mathbf{K}^{-1}(\mathbf{CK}^{-1})^3 - \mathbf{K}^{-1}\mathbf{CK}^{-1}\mathbf{MK}^{-1} - \mathbf{K}^{-1}\mathbf{MK}^{-1}\mathbf{CK}^{-1}]\end{aligned} \tag{3.5.20}$$

where the subscripts ss are omitted for clarity. The last term in Eq. (3.5.19) represents the modal contribution, the contribution of the rth mode being proportional to $\lambda_r^{-4}(\lambda - \lambda_r)^{-1}$, which is much smaller than the contribution of the rth mode in Eq. (3.5.10) when λ is small. Therefore, the truncation errors in Eq. (3.5.19) are smaller than those in Eq. (3.5.10) according to consideration of the rth mode.

One may also differentiate Eq. (3.5.10) with respect to λ to obtain Eq. (3.5.20), thus:

$$\begin{aligned}[\mathbf{Z}_0] &= -\sum_{s=1}^{2n} \lambda_r^{-1}\{\mathbf{\Phi}_r\}\{\mathbf{\Phi}_r\}^{\mathrm{T}} \\ [\mathbf{Z}_0'] &= -\sum_{s=1}^{2n} \lambda_r^{-2}\{\mathbf{\Phi}_r\}\{\mathbf{\Phi}_r\}^{\mathrm{T}} \\ [\mathbf{Z}_0''] &= -2\sum_{s=1}^{2n} \lambda_r^{-3}\{\mathbf{\Phi}_r\}\{\mathbf{\Phi}_r\}^{\mathrm{T}} \\ [\mathbf{Z}_0'''] &= -6\sum_{s=1}^{2n} \lambda_r^{-4}\{\mathbf{\Phi}_r\}\{\mathbf{\Phi}_r\}^{\mathrm{T}}\end{aligned} \tag{3.5.21}$$

Since, from Eq. (3.5.20), if external damping only is concerned, $\sum \lambda_r^{-m}\{\mathbf{\Phi}_r\}\{\mathbf{\Phi}_r\}^{\mathrm{T}}$ are always real, even though $\{\mathbf{\Phi}_r\}$ and λ_r are complex for any integer m.

Consider the heavily damped mass–spring–dashpot system shown in Fig. 3.5.1. The system matrices are

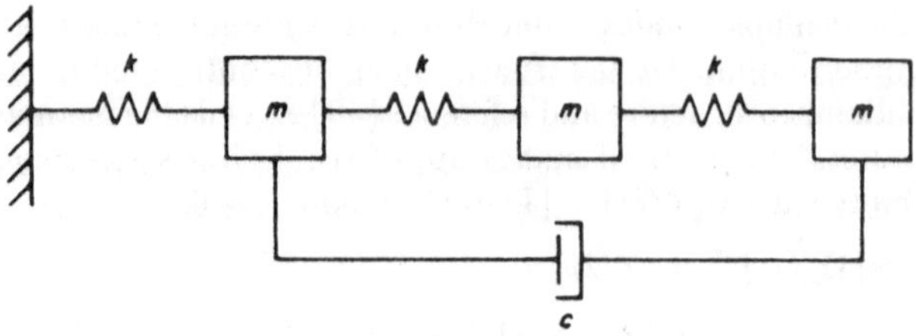

Fig. 3.5.1. A damped mass–spring–dashpot system

Table 3.5.1.

	Mode		
	1	2	3
Eigenvalue	−0.006732 + 0.48912i	−0.92601 + 0.65230i	−0.00667 + 1.78811i
Eigenvector	−0.41873 + 0.64137i	−0.65242 + 0.42145i	−0.14313 + 0.18027i
	−0.47637 + 0.47720i	−0.10056 − 0.06901i	0.28156 − 0.28954i
	−0.39077 + 0.23240i	0.32449 − 0.46781i	−0.20089 + 0.15967i

$$[\mathbf{K}] = \begin{bmatrix} 1 & -1 & 0 \\ -1 & 2 & -1 \\ 0 & -1 & 2 \end{bmatrix}, \quad [\mathbf{C}] = \begin{bmatrix} 1 & 0 & -1 \\ 0 & 0 & 0 \\ -1 & 0 & 1 \end{bmatrix}, \quad [\mathbf{M}] = \begin{bmatrix} 1 & 0 & 0 \\ 0 & 1 & 0 \\ 0 & 0 & 1 \end{bmatrix}$$

and the natural modes are shown in Table 3.5.1. It can be easily checked by either Eq. (3.5.20) or (3.5.21) that

$$\begin{aligned} [\mathbf{D}(\lambda)]^{-1} &= [\mathbf{Z}(\lambda)] \\ &= [\mathbf{D}(\lambda)]^{-1} \\ &= \lambda^4 \sum_{s=1}^{2n} \lambda_r^{-4}(\lambda - \lambda_r)^{-1}\{\mathbf{\Phi}_r\}\{\mathbf{\Phi}_r\}^{\mathrm{T}} \\ &\quad + \begin{bmatrix} 3 & 2 & 1 \\ 2 & 2 & 1 \\ 1 & 1 & 1 \end{bmatrix} - \lambda \begin{bmatrix} 4 & 2 & 0 \\ 2 & 1 & 0 \\ 0 & 0 & 0 \end{bmatrix} - \lambda^2 \begin{bmatrix} 6 & 7 & 6 \\ 7 & 7 & 5 \\ 6 & 5 & 3 \end{bmatrix} + \lambda^3 \begin{bmatrix} 16 & 12 & 6 \\ 12 & 8 & 3 \\ 6 & 3 & 0 \end{bmatrix} \end{aligned}$$

After all the substructures are assembled, the damped natural modes of the system are determined by the eigenvalue problem, which is highly non-linear in λ,

$$[\mathbf{D}(\lambda)]\{\mathbf{\Phi}\} = \{\mathbf{0}\} \tag{3.5.22}$$

The non-trivial solution for λ is obtained by solving the equivalent non-linear algebraic equation

$$\det[\mathbf{D}(\lambda)] = 0 \tag{3.5.23}$$

Because of non-linearity in λ, there are more roots than p, the order of the matrix $[\mathbf{D}(\lambda)]$. For undamped systems, Wittrick and Williams [32] proposed a Strum sequence method to locate the roots with certainty. In the following, we shall develop a method which takes the undamped modes as initial approximations and determines the damped modes with arbitrary damping.

One may solve for the $p + 1$ unknowns $\{\mathbf{\Phi}\}$ and λ of the eigenvalue problem by solving the following p algebraic equations $\{\mathbf{f}(\lambda)\} = [\mathbf{D}(\lambda)]\{\boldsymbol{\phi}\} = \{\mathbf{0}\}$ plus the normalization condition

$$\{\boldsymbol{\phi}\}^{\mathrm{T}}[\mathbf{D}'(\lambda)]\{\boldsymbol{\phi}\} = 1 \tag{3.5.24}$$

where a prime denotes derivatives with respect to the argument. The degree of dependency of $[\mathbf{D}(\lambda)]$ on λ is not restricted here, as long as $[\mathbf{D}(\lambda)]$ is symmetrical.

If an approximate solution λ^0, $\{\boldsymbol{\phi}^0\}$ is available, Newtonian iteration requires the expansion of $\{\mathbf{f}\}$ about the initial approximation by the first-order Taylor series,

$$\{\mathbf{f}(\lambda^0 + \mathrm{d}\lambda)\} = [\mathbf{D}(\lambda^0) + \mathbf{D}'(\lambda^0)\,\mathrm{d}\lambda]\{\boldsymbol{\phi}^0 + \mathrm{d}\boldsymbol{\phi}\} = \{\mathbf{0}\} \tag{3.5.25}$$

After neglecting the diminishing higher order terms,

$$[\mathbf{D}(\lambda^0)]\{\boldsymbol{\phi}^0 + \mathrm{d}\boldsymbol{\phi}\} = -(\mathrm{d}\lambda)[\mathbf{D}'(\lambda^0)]\{\boldsymbol{\phi}^0\} \tag{3.5.26}$$

from which the improved eigenvector $\{\boldsymbol{\phi}^0 + \mathrm{d}\boldsymbol{\phi}\}$ can be determined. Since the normalization condition (3.5.24) was chosen, the multiplying factor $-(\mathrm{d}\lambda)$ is immaterial.

If λ_r and $\{\boldsymbol{\phi}_r\}$ are one pair of solutions, one can prove that λ_r is stationary with respect to a small change in $\{\boldsymbol{\phi}_r\}$ by taking differentials of the identify,

$$\{\boldsymbol{\phi}_r\}^{\mathrm{T}}[\mathbf{D}(\lambda_r)]\{\boldsymbol{\phi}_r\} = 0 \tag{3.5.27}$$

resulting in

$$\{\boldsymbol{\phi}_r\}^{\mathrm{T}}[\mathbf{D}(\lambda_r)]\{\mathrm{d}\boldsymbol{\phi}_r\} + (\mathrm{d}\lambda_r)\{\boldsymbol{\phi}_r\}^{\mathrm{T}}[\mathbf{D}'(\lambda_r)]\{\boldsymbol{\phi}_r\} = 0$$

In view of the fact that $[\mathbf{D}(\lambda_r)]\{\boldsymbol{\phi}_r\} = \{\mathbf{0}\}$, and $\{\boldsymbol{\phi}_r\}^{\mathrm{T}}[\mathbf{D}'(\lambda_r)]\{\boldsymbol{\phi}_r\} \neq 0$, $\mathrm{d}\lambda_r$ must be equal to zero subject to a small change in $\{\boldsymbol{\phi}_r\}$.

When an approximated eigenpair λ^0, $\{\boldsymbol{\phi}^0\}$ is available, an improved eigenvalue can be obtained by the Newtonian algorithm for the following equation:

$$g(\lambda) = \{\boldsymbol{\phi}^0\}^{\mathrm{T}}[\mathbf{D}(\lambda^0)]\{\boldsymbol{\phi}^0\} = 0 \tag{3.5.28}$$

which is

$$\lambda = \lambda^0 - \frac{g(\lambda^0)}{g'(\lambda^0)} = \lambda^0 - \frac{\{\boldsymbol{\phi}^0\}^{\mathrm{T}}[\mathbf{D}(\lambda^0)]\{\boldsymbol{\phi}^0\}}{\{\boldsymbol{\phi}^0\}^{\mathrm{T}}[\mathbf{D}'(\lambda^0)]\{\boldsymbol{\phi}^0\}} \tag{3.5.29}$$

the generalized form of Rayleigh's quotient.

Usually, the eigenvector is improved by inverse iteration first. When the eigenvector converges, the improved eigenvalue is obtained by Rayleigh's quotient. If the inverse iteration is repeated by using the improved eigenvalue, the process is sometimes called Rayleigh's iteration.

When the damping is very heavy, the iteration may converge to other undesired modes. The following continuation parameter method is designed for very heavy damping. The dynamic stiffness is written as

$$[\mathbf{D}(\lambda, \alpha)] = [\mathbf{D}_{\mathrm{u}}(\lambda)] + \alpha[\mathbf{D}_{\mathrm{d}}(\lambda)] \tag{3.5.30}$$

where $[\mathbf{D}_{\mathrm{u}}(\lambda)]$ is the undamped dynamic stiffness and $[\mathbf{D}_{\mathrm{d}}(\lambda)]$ includes all the damping effects (external or internal). The continuation parameter α varies from zero (undamped case) to one (the desired case) in steps according to the degree of damping. Tables 3.5.2 and 3.5.3 show the numerical results for the system in Fig. 3.5.1 with and without continuation respectively. The damping is very heavy in this test case.

For systems having internal damping characterized by complex elastic moduli, e.g. $E(1 + \mathrm{i}\gamma)$, the damping parameter γ can serve the purpose of the continuation parameter. Therefore, heavy internal damping can be considered similarly.

Table 3.5.2. Inverse iteration for damped system

Mode	Alpha	Iteration eigenvalue	Eigenvector		
1 0.00	0	0.00000+0.44504i	−0.73698+0.00000i	−0.59101+0.00000i	−0.32799+0.00000i
1 1.00	1	−0.07601+0.48484i	0.52702−0.63272i	0.54715−0.44943i	0.41582−0.19719i
1 1.00	1	−0.06724+0.48904i	0.41288−0.64844i	0.47289−0.48430i	0.38993−0.23764i
1 1.00	1	−0.06732+0.48912i	−0.41883+0.64140i	−0.47646+0.47721i	−0.39083+0.23240i
1 1.00	1	−0.06732+0.48912i	−0.41873+0.64137i	−0.47637+0.47720i	−0.39077+0.23240i
1 1.00	1	−0.06732+0.48912i	−0.41873+0.64137i	−0.47637+0.47720i	−0.39077+0.23240i
2 0.00	0	0.00000+1.24698i	0.59101+0.00000i	−0.32799+0.00000i	−0.73698+0.00000i
2 1.00	1	−0.45714+0.79854i	−0.24833−0.04625i	0.17230−0.12249i	0.39483−0.30821i
2 1.00	1	−1.00149+0.24115i	1.03222+0.32443i	0.21226+0.53527i	−0.46751+0.54675i
2 1.00	1	−0.41932+0.21581i	0.08723+0.46723i	−0.07735+0.43689i	−0.14553+0.30487i
2 1.00	1	−0.06151+0.26315i	−0.27629+0.86645i	−0.21314+0.64099i	−0.10117+0.30027i
2 1.00	1	0.03256+0.70532i	0.69202 *******i	0.83104−0.86874i	0.71294−0.29918i
3 0.00	0	0.00000+1.80194i	0.32799+0.00000i	−0.73698+0.00000i	0.59101+0.00000i
3 1.00	1	−0.00663+1.78815i	−0.14241+0.17861i	0.28089−0.28762i	−0.20088+0.15975i
3 1.00	1	−0.00667+1.78811i	0.14313−0.18027i	−0.28156+0.28953i	0.20089−0.15967i
3 1.00	1	−0.00667+1.78811i	−0.14313+0.18027i	0.28156−0.28954i	−0.20089+0.15967i
3 1.00	1	−0.00667+1.78811i	0.14313−0.18027i	−0.28156+0.28954i	0.20089−0.15967i
3 1.00	1	−0.00667+1.78811i	−0.14313+0.18027i	0.28156−0.28954i	−0.20089+0.15967i

******* Overflow.

Table 3.5.3. Successive inverse iteration for damped system (Mode 2 only)

Mode	Alpha	Iteration eigenvalue	Eigenvector		
2 0.00	0	0.00000+1.24698i	0.59101+0.00000i	−0.32799+0.00000i	−0.73698+0.00000i
2 0.50	1	−0.37747+1.07335i	−0.32841+0.07864i	0.16077−0.14311i	0.35646−0.34369i
2 0.50	1	−0.45377+1.13135i	0.43158−0.20737i	−0.04440+0.14339i	−0.31119+0.40677i
2 0.50	1	−0.45084+1.13213i	0.39583−0.21366i	−0.05576+0.14543i	−0.29881+0.40472i
2 0.50	1	−0.45084+1.13213i	−0.39589+0.21320i	0.05594−0.14522i	0.29920−0.40413i
2 0.50	1	−0.45084+1.13213i	−0.39589+0.21320i	0.05594−0.14522i	0.29920−0.40413i
2 1.00	2	−0.75419+0.71199i	0.50624−0.06139i	0.04030+0.18756i	−0.33364+0.36205i
2 1.00	2	−0.90122+0.65055i	−0.69695+0.36908i	−0.14297−0.05383i	0.29669−0.40118i
2 1.00	2	−0.92599+0.65181i	0.65613−0.40487i	0.09580+0.07110i	−0.33549+0.45873i
2 1.00	2	−0.92601+0.65230i	−0.65272+0.42157i	−0.10059−0.06905i	0.32466−0.46797i
2 1.00	2	−0.92601+0.65230i	−0.65242+0.42145i	−0.10056−0.06901i	0.32449−0.46781i

Consider the natural vibration of the space frame shown in Fig. 3.5.2 as a first example. Complex elastic moduli $E(1 + i\alpha\gamma)$ and $G(1 + i\alpha\gamma)$, $E = 210\,\mathrm{kN\,mm^{-2}}$, $G = 89\,\mathrm{kN\,mm^{-2}}$, $\gamma = 0.1$, are used to simulate the internal damping. There are two dashpots of magnitude $1000\alpha\,\mathrm{N\,s\,m^{-1}}$ at each node in the x and y directions respectively and the mass density is $7800\,\mathrm{kg\,m^{-3}}$. All members are circular, the vertical members being 0.3 m in diameter and other members 0.2 m in diameter.

When the tip node, node 9, is taken as a master coordinate, the eigenvalues of the first 12 natural modes, when node 9 is fixed, are tabulated in Table 3.5.4. Five Rayleigh iterations are performed in each damping case for $\alpha = 0.2(0.2)1$, although it is found that three iterations are sufficient.

When node 9 is released, the damped natural modes may be computed either using the undamped fixed interface modes with a continuation parameter to handle the damping, or using the damped fixed interface modes directly. Both approaches give the same results. However, the former requires less data storage and less matrix

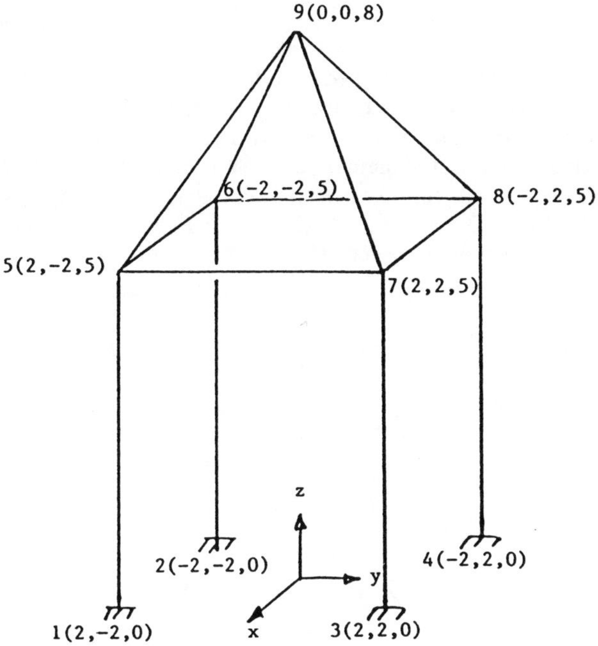

Fig. 3.5.2. A substructure

Table 3.5.4. Eigenvalues when the masters are fixed (partial modes)

Mode	Eigenvalue
1	−4.31 + 80.93i
2	−14.64 + 293.46i
3	−15.67 + 311.88i
4	−15.67 + 311.88i
5	−18.89 + 378.69i
6	−19.33 + 387.55i
7	−19.44 + 389.85i
8	−19.44 + 389.85i
9	−21.62 + 432.70i
10	−21.65 + 434.01i
11	−21.65 + 434.01i
12	−24.29 + 486.73i

Table 3.5.5. The natural modes when the masters are released (partial modes are excluded)

Mode	Eigenvalue	Eigenvector
1	−2.58 + 47.38i	−0.1626 − 0.1128i, 0.6915 − 0.7353i, 0 + 0i, 0.1010 − 0.1066i, 0.0236 + 0.0165i, 0 + 0i
2	−2.58 + 47.38i	−0.6022 + 0.6158i, −0.3230 + 0.3694i, 0 + 0i, −0.0472 + 0.0536i, 0.0879 − 0.0893i, 0 + 0i
3	−3.75 + 70.03i	0 + 0i, 0 + 0i, 0 + 0i, 0 + 0i, 0 + 0i, −0.3758 + 0.3944i
4	−15.33 + 306.67i	0.0532 − 0.0539i, 0.0610 − 0.0656i, 0 + 0i, −0.7559 + 0.8146i, 0.6593 − 0.6699i, 0 + 0i
5	−15.33 − 306.67i	−0.0643 + 0.0371i, 0.0556 − 0.0807i, 0 + 0i, −0.6888 + 1.0016i, −0.7969 + 0.4611i, 0 + 0i
6	−15.68 + 314.06i	0 + 0i, 0 + 0i, 0 + 0i, 0 + 0i, 0 + 0i, 1.2723 − 1.3376i

condensation, and is therefore adopted here for presentation purposes. The mode shapes for the six master coordinates are given in Table 3.5.5.

Figure 3.5.3 shows the same space frame with an additional column of diameter 0.2 m of the same material, the similar results are tabulated in Table 3.5.6. In all cases, $\alpha = 0.2(0.2)1$, and the results corresponding to $\alpha = 1$ only are listed.

It is shown in Tables 3.5.5 and 3.5.6 that the method predicts very close eigenvalues and repeated eigenvalues successfully. Reorthogonalization of the eigenvec-

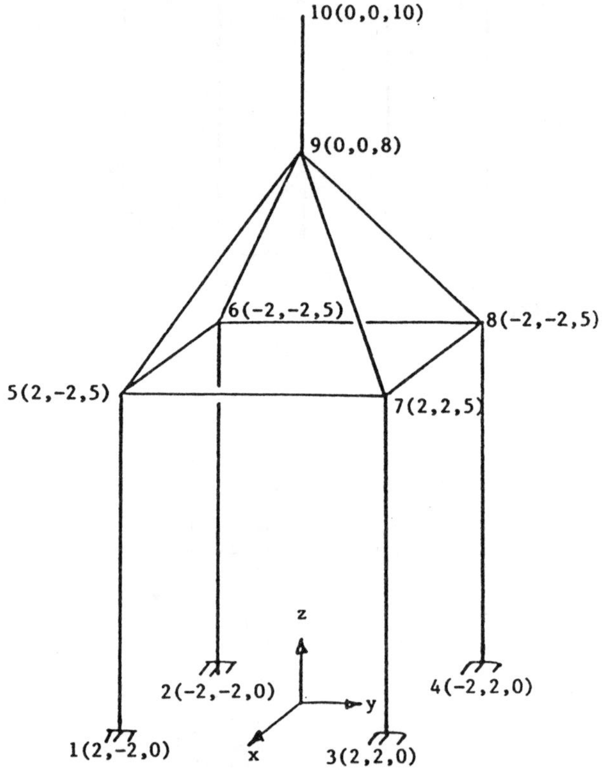

Fig. 3.5.3. A system

Table 3.5.6. Damped natural modes of Fig. 3.5.3 at the tip (partial modes are excluded)

Mode	Eigenvalue	Eigenvector
1	$-2.53 + 46.62i$	$0.2214 - 0.0631i$, $-0.5146 + 0.5842i$, $0 + 0i$, $-0.0563 + 0.0638i$, $-0.0242 + 0.0069i$, $0 + 0i$
2	$-2.53 + 46.62i$	$-0.4769 + 0.3677i$, $0.3008 - 0.4628i$, $0 + 0i$, $0.0329 - 0.0506i$, $0.0521 - 0.0402i$, $0 + 0i$
3	$-3.75 + 70.03i$	$0 + 0i$, $0 + 0i$, $0 + 0i$, $0 + 0i$, $0 + 0i$, $0.3762 + 0.3948i$
4	$-8.14 + 163.24i$	$-2.4588 + 2.7003i$, $-1.2722 + 1.0825i$, $0 + 0i$, $0.7563 - 0.6435i$, $-1.4617 + 1.6053i$, $0 + 0i$
5	$-8.14 + 163.24i$	$0.8105 - 1.6019i$, $-2.7011 + 2.4872i$, $0 + 0i$, $1.6057 - 1.4786i$, $0.4818 - 0.9523i$, $0 + 0i$
6	$-15.64 + 313.29i$	$0 + 0i$, $0 + 0i$, $0 + 0i$, $0 + 0i$, $0 + 0i$, $-1.2948 + 1.3613i$

tors is completely eliminated. We have not tried the critical damping nor asymmetrical damping cases.

3.6. Multilevel Substructures

The dynamic substructure method is extended to multilevel (recursive) substructures. The obvious distinction of the two approaches is that the stiffness and mass matrices before condensation are no longer frequency independent. The dynamic stiffness matrix at any substructure level is proved to be a function of the vibrating frequency in terms of some constant matrices which are derivable from the dynamic stiffness matrix at one lower substructure level. The method can accurately predict more modes than the number of degrees of freedom retained. The computational procedure, the generalized inverse iteration, the stationary principle of the system natural frequency and the generalized Rayleigh's quotient are derived for the frequency-dependent matrices. Numerical examples are given to illustrate some engineering applications. A transcendental dynamic stiffness matrix can be transformed to a more convenient algebraic form.

If the dynamic stiffness equation is obtained by previous condensation, Eq. (3.0.1) still holds, except that [**K**] and [**M**] are no longer constant owing to condensation. At substructure level r,

$$[\mathbf{D}_r(\lambda)]\{\mathbf{u}_r\} = [\mathbf{K}_r(\lambda) - \lambda\mathbf{M}_r(\lambda)]\{\mathbf{u}_r\} = \{\mathbf{f}_r\} \tag{3.6.1}$$

where $[\mathbf{K}_r(\lambda)]$ an $[\mathbf{M}_r(\lambda)]$ can be obtained from $[\mathbf{D}_r(\lambda)]$ by the theorem

$$\frac{\mathrm{d}}{\mathrm{d}\lambda}[\mathbf{D}_r(\lambda)] = [\mathbf{D}_r'(\lambda)] = -[\mathbf{M}_r(\lambda)] \tag{3.6.2}$$

and

$$[\mathbf{K}_r(\lambda)] = [\mathbf{D}_r(\lambda)] + \lambda[\mathbf{M}_r(\lambda)] \tag{3.6.3}$$

When Eq. (3.6.1) is partitioned according to slaves sr and masters mr at substructure level r,

$$[\mathbf{D}_r(\lambda)]\{\mathbf{u}_r\} = \begin{bmatrix} \mathbf{D}_{sr} & \mathbf{D}_{cr}^{\mathrm{T}} \\ \mathbf{D}_{cr} & \mathbf{D}_{mr} \end{bmatrix} \begin{Bmatrix} \mathbf{u}_{sr} \\ \mathbf{u}_{mr} \end{Bmatrix} = \begin{Bmatrix} \mathbf{0} \\ \mathbf{f}_{mr} \end{Bmatrix} \tag{3.6.4}$$

After eliminating $\{\mathbf{u}_{sr}\}$ and performing the necessary assembling process, one has the condensed dynamic stiffness equation at substructure level $r + 1$,

$$[\mathbf{D}_{r+1}(\lambda)]\{\mathbf{u}_{r+1}\} = [\overline{\mathbf{D}}_{mr}(\lambda)]\{\mathbf{u}_{mr}\} = \{\mathbf{f}_{mr}\} \tag{3.6.5}$$

That is to say, the dynamic stiffness equation at substructure level $r + 1$ is exactly the condensed dynamic stiffness equation at level r. From Eq. (3.1.7)

$$[\mathbf{D}_{r+1}(\lambda)] = [\mathbf{D}_{mr} - \mathbf{D}_{cr}\mathbf{D}_{sr}^{-1}\mathbf{D}_{rr}^{\mathrm{T}}] \tag{3.6.6}$$

and from Eq. (3.2.30)

$$\begin{aligned} [\mathbf{D}_{r+1}(\lambda)] &= [\overline{\mathbf{K}}_{mr} - \lambda\overline{\mathbf{M}}_{mr}] - \lambda^2 \sum_i (\lambda - \lambda_{ir})^{-1}\{\mathbf{G}_{ir}\}\{\mathbf{G}_{ir}\}^{\mathrm{T}} \\ \{\mathbf{G}_{ir}\} &= -\lambda_{ir}^{-1}[\mathbf{D}_{cr}(\lambda_{ir})]\{\mathbf{\Phi}_{ir}\} \end{aligned} \tag{3.6.7}$$

where $[\overline{\mathbf{K}}_{mr}]$ and $[\overline{\mathbf{M}}_{mr}]$ are constant matrices calculated according to Eqs (3.2.31) and (3.2.32), and λ_{ir} and $\{\mathbf{\Phi}_{ir}\}$ are the ith natural fixed interface modes at substructure level r. Equation (3.6.7) is more advantageous than Eq. (3.6.6) because all matrices are constant (λ independent). $[\mathbf{D}_{cr}(\lambda_{ir})]$ is also a constant matrix evaluated at λ_{ir}.

Suppose $[\mathbf{D}_r(\lambda)]$ in Eq. (3.6.4) is known. One can then transform Eq. (3.6.6) to Eq. (3.6.7) in the following manner.

1. $[\overline{\mathbf{K}}_{mr}] = [\mathbf{D}_{r+1}(0)]$ is obtained by substituting $\lambda = 0$ on the right-hand side of Eq. (3.6.6).
2. $[\overline{\mathbf{M}}_{mr}] = -[\mathbf{D}'_{r+1}(0)]$ is obtained either by explicit differentiation, or Romberg's algorithm (discussed below).
3. λ_{ir} and $\{\mathbf{\Phi}_{ir}\}$ are obtained by solving the eigenvalue problem at substructure level r.
4. $\{\mathbf{G}_{ir}\} = -\lambda_{ir}^{-1}[\mathbf{D}_{cr}(\lambda_{ir})]\{\mathbf{\Phi}_{ir}\}$ is the product of the cross-dynamic stiffness at level r evaluated at $\lambda = \lambda_{ir}$ and the ith fixed interface mode resulting from step 3.

Therefore, the substructure at level $r + 1$ is completely defined by the quantities $[\overline{\mathbf{K}}_{mr}]$, $[\overline{\mathbf{M}}_{mr}]$, λ_{ir} and $\{\mathbf{G}_{ir}\}$, associated with the masters $\{\mathbf{u}_{mr}\}$. The evaluation of $[\overline{\mathbf{K}}_{mr}]$ according to Eq. (3.2.31) is simply

$$[\overline{\mathbf{K}}_{mr}] = [\mathbf{T}_r]^{\mathrm{T}}[\mathbf{K}_r][\mathbf{T}_r] \tag{3.6.8}$$

where $[\mathbf{T}_r] = \mathrm{col}[-\mathbf{K}_{sr}^{-1}\mathbf{K}_{cr}^{\mathrm{T}}, \mathbf{I}]$, and explicitly,

$$[\overline{\mathbf{K}}_{mr}] = [\mathbf{K}_{mr} - \mathbf{K}_{cr}\mathbf{K}_{sr}^{-1}\mathbf{K}_{cr}^{\mathrm{T}}] \tag{3.6.9}$$

Also, expressing Eq. (3.6.7) in Taylor's series in λ, one can prove that

$$\begin{aligned}[\overline{\mathbf{M}}_{mr}] &= -[\mathbf{D}'_{r+1}(0)] \\ &= [\mathbf{M}_{mr} - \mathbf{M}_{cr}\mathbf{K}_{sr}^{-1}\mathbf{K}_{cr}^{\mathrm{T}} - \mathbf{K}_{cr}\mathbf{K}_{sr}^{-1}\mathbf{M}_{cr}^{\mathrm{T}} + \mathbf{K}_{cr}\mathbf{K}_{sr}^{-1}\mathbf{M}_{sr}\mathbf{K}_{sr}^{-1}\mathbf{K}_{cr}^{\mathrm{T}}] \\ &= [\mathbf{T}_r]^{\mathrm{T}}[\mathbf{M}_r][\mathbf{T}_r]\end{aligned} \tag{3.6.10}$$

Since $[\mathbf{M}_r]$ is determined in the previous level of substructures, $[\overline{\mathbf{M}}_{mr}]$ can be evaluated explicitly according to Eq. (3.6.10).

Very often, the dynamic stiffness equation is derived directly from solution of the differential equation of motion, and the form (3.2.30) is not readily available. The evaluation of $[\overline{\mathbf{M}}_{mr}]$ can conveniently be performed by numerical differentiation of Eq. (3.6.6) according to Romberg's algorithm [38]. In essence, one chooses a small step length h for λ, evaluates $[\mathbf{D}^{0p}] = [\mathbf{D}'_{r+1}(2^{-p}h)]$ for $p = 0, 1, 2, \ldots$ and forms the sequences

$$[\mathbf{D}^{1p}] = [\mathbf{D}^{0p}] + (2^{-2} - 1)[\mathbf{D}^{0p} - \mathbf{D}^{0,p-1}], \qquad p > 1$$

$$[\mathbf{D}^{2p}] = [\mathbf{D}^{1p}] + (2^{-4} - 1)[\mathbf{D}^{1p} - \mathbf{D}^{1,p-1}], \qquad p > 2$$

$$[\mathbf{D}^{np}] = [\mathbf{D}^{n-1,p}] + (2^{-2n} - 1)[\mathbf{D}^{n-1,p} - \mathbf{D}^{n-1,p-1}], \qquad p > n$$

The orders of error of the above sequences approaching $[\mathbf{D}'_{r+1}(0)]$ are $(2^{-p}h)^2$, $(2^{-p}h)^4$ and $(2^{-p}h)^{2n}$ respectively. In practice, if one takes h equal to 1% of the lowest eigenvalue of the partial modes at level r, and $p = 0, 1, 2$, the order of error is $(2^{-2} \times 0.01)^4 = 4 \times 10^{-11}$, which is very acceptable for engineering applications. In fact, three evaluations of the dynamic stiffness $[\mathbf{D}_{r+1}(\lambda)]$ at different λ are often more computationally efficient than one explicit condensation according to Eq. (3.6.10). A series of numerical tests has shown that the natural modes computed using the explicit and implicit methods for evaluating $[\overline{\mathbf{M}}_{mr}]$ are indistinguishable.

The computation of the fixed interface substructure modes λ_{ir}, $\{\mathbf{\Phi}_{ir}\}$ at level r constitutes a non-linear eigenvalue problem

$$[\mathbf{D}_{sr}(\lambda)]\{\mathbf{\Phi}_{\rho}\} = \{\mathbf{0}\}$$

or, equivalently, the solution of the non-linear equations for $\{\mathbf{\Phi}_i\}$ and λ_i,

$$\{\mathbf{f}(\lambda_i)\} = [\mathbf{D}(\lambda_i)]\{\mathbf{\Phi}_i\} = \{\mathbf{0}\} \tag{3.6.11}$$

and

$$\{\mathbf{\Phi}_i\}^{\mathrm{T}}[\mathbf{M}(\lambda_i)]\{\mathbf{\Phi}_i\} = 1$$

where the subscripts r are dropped for clearer presentation, and $[\mathbf{M}(\lambda_i)] = -[\mathbf{D}'(\lambda_i)]$. Suppose initial approximations λ_i^0 and $\{\mathbf{\Phi}_i^0\}$ are available, the Newtonian algorithm to improve the solution requires the expansion of Eq. (3.6.11) in a Taylor series, $\lambda_i = \lambda_i^0 + \mathrm{d}\lambda_i$, $\{\mathbf{\Phi}_i\} = \{\mathbf{\Phi}_i^0\} + \mathrm{d}\{\mathbf{\Phi}_i\}$,

$$\{\mathbf{f}(\lambda_i^0 + \mathrm{d}\lambda_i)\} = [\mathbf{D}(\lambda_i^0) + \mathbf{D}'(\lambda_i)\,\mathrm{d}\lambda_i]\{\mathbf{\Phi}_i^0 + \mathrm{d}\mathbf{\Phi}_i\} = \{\mathbf{0}\} \tag{3.6.12}$$

Neglecting the diminishing higher order terms, we have

$$[\mathbf{D}(\lambda_i^0)]\{\mathbf{\Phi}_i^0 + \mathrm{d}\mathbf{\Phi}_i\} = [\mathbf{M}(\lambda_i)]\{\mathbf{\Phi}_i^0\}\,\mathrm{d}\lambda_i \tag{3.6.13}$$

from which the improved eigenvector $\{\mathbf{\Phi}_i^0 + \mathrm{d}\mathbf{\Phi}_i\}$ can be determined up to a constant multiplier. If the eigenvectors are to be normalized according to Eq. (3.6.11), the constant multiplier is immaterial. Premultiplying Eq. (3.6.10) by $\{\mathbf{\Phi}_i\}^{\mathrm{T}}$,

$$g(\lambda_i) = \{\mathbf{\Phi}_i\}^{\mathrm{T}}[\mathbf{D}(\lambda_i)]\{\mathbf{\Phi}_i\} = 0 \tag{3.6.14}$$

and taking differentials on both sides, one can prove that $\mathrm{d}\lambda_i = 0$ subject to a small (non-zero) change of $\{\mathbf{\Phi}_i\}$, which is the stationary principle of the eigenvalue with respect to the eigenvector for general non-linear $[\mathbf{D}(\lambda)]$.

The improved eigenvalue can be obtained by the following generalized Rayleigh quotient. With the readily available mode $\{\mathbf{\Phi}_i\}$, an improved eigenvalue λ_i according to the Newtonian algorithm is given by

$$\lambda_i = \lambda_i^0 - \frac{g(\lambda_i^0)}{g'(\lambda_i^0)} = \lambda_i^0 - \frac{\{\mathbf{\Phi}_i\}^{\mathrm{T}}[\mathbf{D}(\lambda_i^0)]\{\mathbf{\Phi}_i\}}{\{\{\mathbf{\Phi}_i\}^{\mathrm{T}}[\mathbf{M}(\lambda_i^0)]\{\mathbf{\Phi}_i\}} \tag{3.6.15}$$

The remaining question of locating λ_i^0 approximately for general λ-dependent matrices in structural dynamics can be solved by the well known Wittrick and Williams algorithm [32] and will not be repeated here.

Example 3.6.1. Construction of **G** *Matrices for Continuous Models*

Very often, the dynamic stiffness is formed by direct solution of the governing differential equations of motion. The dynamic stiffness is then expressed in terms of elementary functions (trigonometric and hyperbolic) or orthogonal functions (Bessel and Henkel) in the frequency parameter λ. When λ approaches zero, the indefinite form 0/0 results. Reference [39] expands the dynamic stiffness matrix in Taylor's series in λ to avoid numerical problems. When λ is large, the λ functions involve differences of large numbers and accuracy is reduced. Therefore, special treatment is also required for large λ.

The above-mentioned numerical problems do not exist if the dynamic stiffness is expressed in the form of Eq. (3.2.30). Methods involving the integrations of the natural modes to find the **G** matrix for such models have been recommended. However, the application of Eq. (3.2.29) does not involve integration at all. The following example demonstrates the construction of the dynamic stiffness matrix in algebraic form when the transcendental form is available.

Consider a uniform Euler beam whose dynamic stiffness matrix is

$$[\mathbf{D}(\lambda)] = \frac{EI}{l^3}\begin{bmatrix} F_6 & & & \text{sym.} \\ -F_4 l & F_2 l^2 & & \\ F_5 & -F_3 l & F_6 & \\ F_3 l & F_1 l^2 & F_4 l & F_2 l^2 \end{bmatrix}$$

and whose rth fixed interface modes are

$$\phi_r(\xi) = \cosh \lambda_r \xi - \cos \lambda_r \xi - \sigma_r(\sinh \lambda_r \xi - \sin \lambda_r \xi)$$

where

$$\begin{aligned} \sigma_r &= \cosh \lambda_r - \cos \lambda_r)/(\sinh \lambda_r - \sin \lambda_r) \\ \lambda_r^4 &= \rho A l^4 \omega_r^2 / EI \\ \lambda_r &= 4.73004075,\ 7.85320462,\ 10.99560784,\ 14.13716549,\ 17.27875966 \\ &= (r + 0.5)\pi, \quad r > 5 \end{aligned}$$

and F_i, $i = 1, 2, \ldots, 6$ are Kolousek functions. If the beam is treated as a substructure of two beam elements of length $0.5l$, then

$$[\mathbf{D}(\lambda)] = \frac{8EI}{l^3}\begin{bmatrix} 2F_6 & & & & & \\ 0 & 0.5F_2 l^2 & & & & \text{sym.} \\ F_5 & 0.5F_3 l & F_6 & & & \\ -0.5F_3 l & 0.25F_1 l^2 & -0.5F_4 l & 0.25F_2 l^2 & & \\ F_5 & -0.5F_3 l & 0 & 0 & F_6 & \\ 0.5F_2 l & 0.25F_1 l^2 & 0 & 0 & 0.5F_4 l & 0.25F_2 l^2 \end{bmatrix}$$

$$= \begin{bmatrix} \mathbf{D}_s & \mathbf{D}_c^{\mathrm{T}} \\ \mathbf{D}_c & \mathbf{D}_m \end{bmatrix}$$

Therefore, since the sign of **G** is insignificant,

$$\begin{aligned} \{\mathbf{G}_r\} &= [\mathbf{D}_c(\lambda_r)]\{\mathbf{\Phi}_r\}/\lambda_r^4 = \text{col}\{G_{1r} \quad G_{2r} \quad G_{2r} \quad G_{2r}\} \\ \{\mathbf{\Phi}_r\} &= \text{col}\{\phi_r(\tfrac{1}{2}), \phi_r'(\tfrac{1}{2})\} \end{aligned}$$

which are evaluated for the first 20 modes, and G_{1r} and G_{2r} are listed in the second and third columns of Table 3.6.1. $G_{3r} = (-1)^{r+1}G_{1r}$, $G_{4r} = (-1)^{r+1}G_{2r}$.

It is demonstrated that the matrix [**G**] can be evaluated by means of [$\mathbf{D}_c$] without integration at all. Evaluation of the index J_0 in the Wittrick–Williams algorithm is straightforward because λ_r is readily available.

Example 3.6.2. Multilevel Beam Substructure

The dynamic stiffness matrix of a uniform beam of length l has been constructed in algebraic form rather than transcendental form in the previous example. One may

Table 3.6.1. *G* matrices in multilevel beam substructures

Mode	Level 1		Level 2		Level 3		Level 4	
	G_1	G_2	G_1	G_2	G_1	G_2	G_1	G_2
1	0.415431	0.0893923	0.587508	0.2528397	0.830862	0.7151386	1.17502	2.0227174
2	0.254871	0.0324292	0.360442	0.0917236	0.509742	0.2594336	0.720884	0.7337889
3	0.181885	0.0165421	0.257224	0.0467882	0.363769	0.1323371	0.514448	0.3743057
4	0.141471	0.0100070	0.200071	0.0283042	0.282943	0.0800563	0.400141	0.2264333
5	0.115749	0.0066989	0.163694	0.0189474	0.231498	0.0535914	0.327388	0.1515793
6	0.097942	0.0047963	0.138510	0.0135659	0.195883	0.0383702	0.277020	0.1085272
7	0.084883	0.0036025	0.120042	0.0101895	0.169765	0.0288202	0.240084	0.0815160
8	0.074896	0.0028047	0.105920	0.0079330	0.149793	0.0224379	0.211839	0.0634640
9	0.067013	0.0022453	0.094770	0.0063508	0.134025	0.0179628	0.189540	0.0508064
10	0.060630	0.0018380	0.085744	0.0051987	0.121261	0.0147042	0.171489	0.0415898
11	0.055358	0.0015323	0.078288	0.0043339	0.110716	0.0122581	0.156577	0.0346713
12	0.050930	0.0012969	0.072025	0.0036682	0.101859	0.0103753	0.144051	0.0293457
13	0.047157	0.0011119	0.066690	0.0031449	0.094314	0.0088951	0.133380	0.0251593
14	0.043905	0.0009638	0.062091	0.0027261	0.087810	0.0077105	0.124182	0.0218087
15	0.041072	0.0008435	0.058085	0.0023857	0.082144	0.0067477	0.116170	0.0190854
16	0.038583	0.0007443	0.054565	0.0021053	0.077166	0.0059546	0.109129	0.0168421
17	0.036378	0.0006617	0.051447	0.0018715	0.072757	0.0052935	0.102893	0.0149723
18	0.034412	0.0005921	0.048666	0.0016747	0.068824	0.0047367	0.097331	0.0133974
19	0.032647	0.0005329	0.046170	0.0015073	0.065294	0.0042633	0.092340	0.0120586
20	0.031055	0.0004822	0.043918	0.0013639	0.062109	0.0038576	0.087836	0.0109108

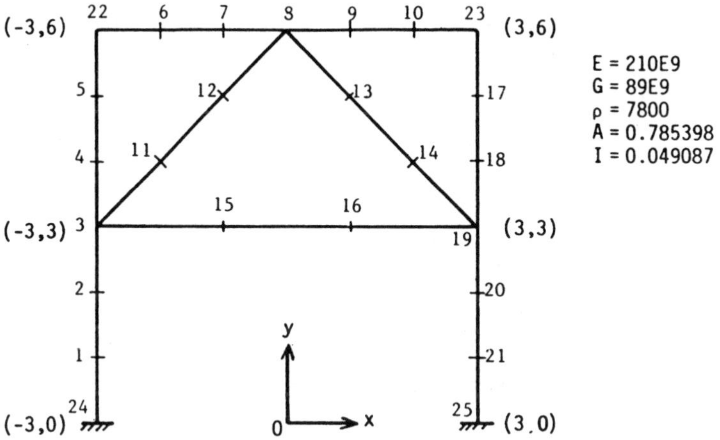

Fig. 3.6.1. A substructure

apply the present multilevel dynamic substructure method to construct dynamic stiffness matrices for beam elements of lengths $2l, 4l, \ldots, 2^{n-1}l$. Although the example is trivial, the numerical results are useful in checking the algorithm as well as the program. The **G** matrices are tabulated in Table 3.6.1 for various substructure levels.

Example 3.6.3. Multilevel Frame Substructure

An economical method to obtain the natural modes of uniform multistorey building frames is introduced here. The natural modes of an idealized single storey frame

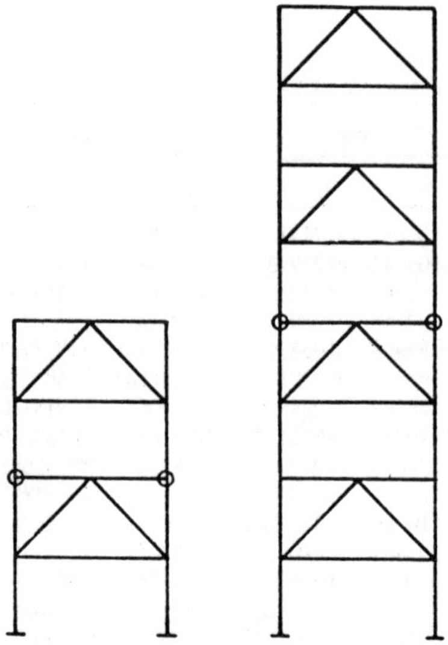

2 storey model 4 storey model etc. **Fig. 3.6.2.** Elimination of slave coordinates (o)

Table 3.6.2. Natural frequencies when the four corner nodes are fixed (rad s^{-1})

Mode number	Number of storeys						
	1	2	4	8	16	32	64
1	655.68	225.08	74.19	25.67	7.219	1.841	0.458
2	676.61	605.05	127.21	55.51	18.160	4.940	1.254
3	957.29	642.56	176.39	88.89	32.200	9.366	2.436
4	1216.32	652.83	219.97	115.33	47.861	14.878	3.981
5	1358.35	818.35	328.18	130.19	64.622	21.271	5.867
6	1682.59	917.40	380.11	150.08	67.839	28.326	8.069
7	2032.54	936.43	503.37	174.65	82.064	34.781	10.561
8	2177.00	1141.70	618.41	189.40	98.485	35.956	13.316
9	2363.45	1245.20	692.89	205.55	115.340	43.976	16.384
10	2603.43	1272.82	772.33	257.85	121.746	52.200	17.502
11	2649.71	1341.36	804.89	305.70	133.019	60.724	20.588
12	2661.43	1410.72	850.33	342.84	145.832	67.157	29.007

Table 3.6.3. Natural frequencies when the lower two corner nodes only are fixed (rad s^{-1})

Mode number	Number of storeys						
	1	2	4	8	16	32	64
1	182.07	94.52	17.62	4.56	1.159	0.286	0.0712
2	583.06	249.44	68.05	24.20	6.842	1.768	0.445
3	768.76	376.22	132.21	56.09	17.836	4.848	1.239
4	838.37	523.33	139.72	69.63	31.590	9.221	2.408
5	1136.10	644.21	184.91	91.90	35.800	14.701	3.940
6	1234.24	732.93	280.41	121.35	47.980	17.623	5.813
7	1494.04	839.03	333.53	151.31	65.095	21.151	7.800
8	1538.09	961.71	373.10	165.90	81.788	28.224	8.795
9	1667.15	1147.01	534.63	183.43	98.768	36.361	10.500
10	1912.69	1151.81	611.13	191.28	99.466	46.968	13.543
11	2114.21	1200.82	686.47	233.57	117.288	53.841	18.508
12	2321.22	1293.99	772.96	280.28	132.099	68.257	26.150

shown in Fig. 3.6.1 and the **K**, **M** and **G** matrices associated with nodes 22, 23, 24 and 25 are first evaluated. When two identical substructures are assembled by means of the present method, the **K**, **M**, and **G** matrices of the resulting substructure are generated. The substructure matrices **K**, **M**, **G** are then generated recurrently for the 4-, 8-, 16-, 32- and 64-storey models. Note that only two nodes are involved in each elimination cycle (Fig. 3.6.2). To obtain the substructure matrices for the 64-storey model, 6 × 2 nodes are eliminated in total. The natural frequencies when the masters are fixed are given in Table 3.6.2 and those when the base only are fixed are given in Table 3.6.3.

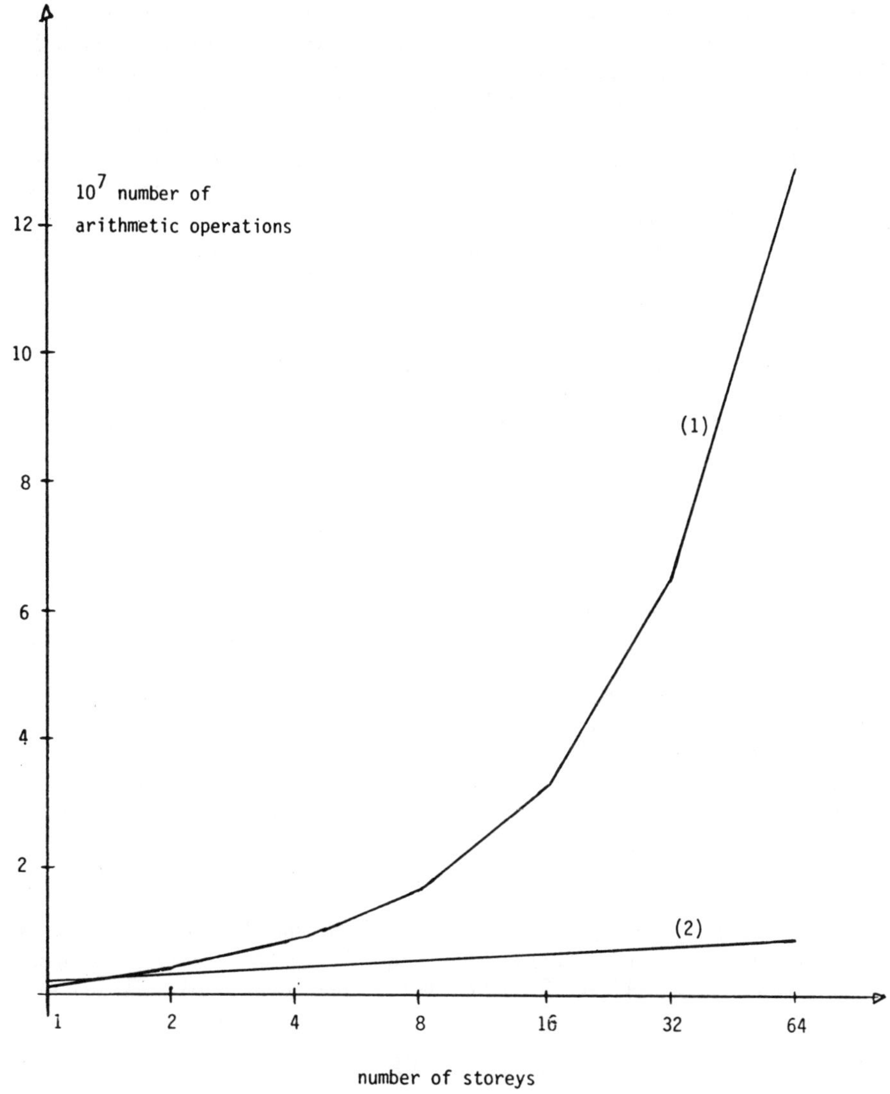

Fig. 3.6.3. Number of arithmetic operations required for 12 modes: (1) subspace iteration with band matrices; (2) multilevel dynamic substructure method

The arithmetic operation counts for the 64-storey model are compared with the subspace iteration in Fig. 3.6.3. It is evident that the present method is superior for structures having many identical substructures. The resulting substructure can be regarded as a superelement and is ready to be included in any finite element analysis.

3.7. Non-conservative Substructures

The dynamic substructure method is extended to non-conservative systems including defective modes. The condensed dynamic stiffness is expressed in terms of the fixed-interface dynamic flexibility. Then the dynamic flexibility is decomposed into the partial modes and the principal vectors (in the case of defective modes). To accelerate the convergence of the dynamic flexibility with respect to the modes, static effects are introduced to account for the contributions of the higher modes. A solution algorithm is introduced to solve the resulting system matrix. If defective modes are specifically required, the system parameters which make the matrix defective are determined by a generalized Newtonian algorithm. A numerical example including the Beck column as a substructure is given for illustration.

Suppose that a linear substructure is discretized by the finite-element method. The dynamic-stiffness equation in terms of the frequency parameter λ is

$$[\mathbf{D}(\lambda)]\{\mathbf{u}\} = \{\mathbf{f}\} \tag{3.7.1}$$

where $\{\mathbf{u}\}$ is the generalized nodal displacement vector and $\{\mathbf{f}\}$ is the force vector. The dynamic stiffness matrix $[\mathbf{D}(\lambda)]$ is not necessarily symmetrical. If (3.7.1) is partitioned according to the slave coordinates $\{\mathbf{u}_s\}$ and the master coordinates $\{\mathbf{u}_m\}$, where $\{\mathbf{u}\} = [\mathbf{u}_s^T\mathbf{u}_m^T]^T$, then

$$\begin{bmatrix} \mathbf{D}_{ss} & \mathbf{D}_{sm} \\ \mathbf{D}_{ms} & \mathbf{D}_{mm} \end{bmatrix} \begin{Bmatrix} \mathbf{u}_s \\ \mathbf{u}_m \end{Bmatrix} = \begin{Bmatrix} \mathbf{0} \\ \mathbf{f}_m \end{Bmatrix} \tag{3.7.2}$$

where, without loss of generality, $\{\mathbf{f}_s\} = \{\mathbf{0}\}$; otherwise $\{\mathbf{f}_m\}$ will be modified to $\{\mathbf{f}_m - \mathbf{D}_{ms}\mathbf{D}_{ss}^{-1}\mathbf{f}_s\}$ in the subsequent analysis. Eliminating $\{\mathbf{u}_s\}$, one has

$$[\mathbf{D}^*(\lambda)]\{\mathbf{u}_m\} = [\mathbf{D}_{mm} - \mathbf{D}_{ms}\mathbf{D}_{ss}^{-1}\mathbf{D}_{sm}]\{\mathbf{u}_m\} = \{\mathbf{f}_m\} \tag{3.7.3}$$

where $[\mathbf{D}^*(\lambda)]$ is the required condensed substructure dynamic stiffness matrix. The difficulty in evaluating $[\mathbf{D}^*(\lambda)]$ by Eq. (3.7.3) is in the inversion of $[\mathbf{D}_{ss}]$ for every value of λ required. This will be overcome in the following.

The reciprocal of the fixed-interface dynamic stiffness matrix $[\mathbf{D}_{ss}(\lambda)]$, that is, the dynamic flexibility matrix, will be expressed in terms of the fixed-interface modes. Since the problem is originally formulated by finite elements,

$$[\mathbf{D}_{ss}(\lambda)] = [\mathbf{A} - \lambda\mathbf{B}] \tag{3.7.4}$$

where $[\mathbf{A}]$ and $[\mathbf{B}]$ are constant matrices and $[\mathbf{B}]$ is assumed to be non-singular as the unnecessary coordinates can easily be removed. From the theory of the linear algebraic eigenvalue problem [40], there exists a non-singular matrix $[\mathbf{P}]$, such that the matrix

$$[\mathbf{P}]^{-1}[\mathbf{B}^{-1}\mathbf{A}][\mathbf{P}] = [\mathbf{J}] \tag{3.7.5}$$

is in a canonical form:

$$[\mathbf{J}] = \operatorname{diag}[\mathbf{J}_0, \mathbf{J}_1, \ldots, \mathbf{J}_p] \tag{3.7.6}$$

where $[\mathbf{J}_0] = \mathrm{diag}[\lambda_0, \lambda_1, \ldots, \lambda_q]$ and

$$[\mathbf{J}_i] = \begin{bmatrix} \lambda_{q+i} & 1 & & \\ & \lambda_{q+i} & \ddots & 1 \\ & & \ddots & \lambda_{q+i} \end{bmatrix}_{p_i \times p_i} \tag{3.7.7}$$

so that $q + \sum_{i=1}^{p} p_i = n$, which is of the order of the matrix $[\mathbf{D}_{ss}(\lambda)]$. In the canonical form, λ_{q+i} and λ_{q+j} are not necessarily distinct when $i \neq j$. If $\sum_{i=1}^{p} p_i = 0$, then (3.7.5) degenerates to the classical non-defective form. From (3.7.5),

$$\begin{aligned}[\mathbf{J} - \lambda\mathbf{I}] &= [\mathbf{P}]^{-1}[\mathbf{B}^{-1}\mathbf{A} - \lambda\mathbf{I}][\mathbf{P}] \\ &= [\mathbf{BP}]^{-1}[\mathbf{A} - \lambda\mathbf{B}][\mathbf{P}] \\ &= [\mathbf{Q}]^{\mathrm{T}}[\mathbf{D}(\lambda)][\mathbf{P}]\end{aligned} \tag{3.7.8}$$

where $[\mathbf{Q}]^{\mathrm{T}} = [\mathbf{BP}]^{-1}$. Taking the reciprocal of the above equation, we have

$$[\mathbf{J} - \lambda\mathbf{I}]^{-1} = [\mathbf{P}]^{-1}[\mathbf{D}_{ss}(\lambda)]^{-1}[\mathbf{Q}]^{\mathrm{T}}$$

or

$$[\mathbf{D}_{ss}(\lambda)]^{-1} = [\mathbf{P}][\mathbf{J} - \lambda\mathbf{I}]^{-1}[\mathbf{Q}]^{\mathrm{T}} = [\mathbf{Z}(\lambda)] \tag{3.7.9}$$

which is the dynamic flexibility of the discrete system. Equation (3.7.9) can be expanded as

$$[\mathbf{Z}(\lambda)] = [\mathbf{P}_0][\mathbf{J}_0 - \lambda\mathbf{I}]^{-1}[\mathbf{Q}_0]^{\mathrm{T}} + \sum_{i=1}^{p} [\mathbf{P}_i][\mathbf{J}_i - \lambda\mathbf{I}]^{-1}[\mathbf{Q}_i]^{\mathrm{T}} \tag{3.7.10}$$

where, corresponding to the canonical form, Eq. (3.7.6),

$$[\mathbf{P}] = [\mathbf{P}_0, \mathbf{P}_1, \ldots, \mathbf{P}_p]$$

and

$$[\mathbf{Q}] = [\mathbf{Q}_0, \mathbf{Q}_1, \ldots, \mathbf{Q}_p] \tag{3.7.11}$$

It is evident from (3.7.8) that $[\mathbf{P}]$ and $[\mathbf{Q}]$ are the respective collections of the right and the left generalized vectors of the following eigenvalue problem:

$$[\mathbf{D}_{ss}(\lambda)]\{\mathbf{p}\} = \{\mathbf{0}\} \qquad \text{and} \qquad \{\mathbf{q}\}^{\mathrm{T}}[\mathbf{D}_{ss}(\lambda)] = \{\mathbf{0}\}^{\mathrm{T}} \tag{3.7.12}$$

Now $[\mathbf{P}_0]$ and $[\mathbf{Q}_0]$ correspond to all non-defective eigenvalues. A non-defective eigenvalue λ_r has multiplicity equal to the degeneracy of $[\mathbf{D}_{ss}(\lambda)]$. Further, $[\mathbf{P}_i]$ and $[\mathbf{Q}_i]$ are the leading eigenvectors satisfying Eq. (3.7.12). The subsequent vectors do not satisfy (3.7.12) and are called the principal vectors. The principal vectors are linearly independent of each other and of all eigenvectors with respect to $[\mathbf{B}]$ and are generated by their leading eigenvectors. The normalization conditions are

$$[\mathbf{Q}]^{\mathrm{T}}[\mathbf{B}][\mathbf{P}] = [\mathbf{I}] \tag{3.7.13}$$

and from (3.7.5)

$$[\mathbf{Q}_i]^{\mathrm{T}}[\mathbf{A}][\mathbf{P}_i] = [\mathbf{J}_i] \tag{3.7.14}$$

which is a canonical form of order p_i.

Therefore, the dynamic flexibility of the fixed interface substructure can be expressed explicitly in terms of its natural modes, as in Eq. (3.7.10). Since $[\mathbf{J}_0]$ is diagonal, so is $[\mathbf{J}_0 - \lambda\mathbf{I}]^{-1}$. However,

$$[\mathbf{J}_i - \lambda \mathbf{I}]^{-1} = \begin{bmatrix} (\lambda_{q+i} - \lambda) & -(\lambda_{q+i} - \lambda)^{-2} & \cdots & (-1)^{p_i+1}(\lambda_{q+i} - \lambda)^{-p_i} \\ & (\lambda_{q+i} - \lambda)^{-1} & \cdots & (-1)^{p_i}(\lambda_{q+i} - \lambda)^{-p_i+1} \\ & & \ddots & \vdots \\ \mathbf{0} & & & (\lambda_{q+i} - \lambda)^{-1} \end{bmatrix} \quad (3.7.15)$$

For most engineering applications, the value of p_i is about two. Thus the second term on the right-hand side of (3.7.10) involves about two vectors at a time.

In practice it is almost impossible to evaluate all the partial modes. If only a number of the lowest partial modes, in ascending order of the moduli of the eigenvalue, are available, Eq. (3.7.10) is a good approximation. The approximation is greatly improved when, as discussed later, static contributions accounting for the contribution of higher modes are employed.

A number of special non-defective cases will be discussed in this section. Consider the case when $[\mathbf{D}_{ss}(\lambda)]$ is real, that is, all constitutive relations are real and no hysteresis damping is possible. Then the roots of $\det[\mathbf{D}_{ss}(\lambda)] = 0$ must be in complex-conjugate pairs. Suppose that $\lambda_r = \sigma_r + i\omega_r$, $\{\mathbf{q}_r\} = \{\mathbf{v}_r + i\mathbf{w}_r\}$ and $\{\mathbf{p}_r\} = \{\mathbf{x}_r + i\mathbf{y}_r\}$ are the eigensolutions, then $\bar{\lambda}_r = \sigma_r - i\omega_r$, $\{\bar{\mathbf{q}}_r\} = \{\mathbf{v}_r - i\mathbf{w}_r\}$ and $\{\bar{\mathbf{p}}_r\} = \{\mathbf{x}_r - i\mathbf{y}_r\}$ are also eigensolutions, where σ_r, ω_r, $\{\mathbf{v}_r\}$, $\{\mathbf{w}_r\}$, $\{\mathbf{x}_r\}$, $\{\mathbf{y}_r\}$ are real. Grouping the complex-conjugate pairs in (3.7.10), we have, after simplification,

$$[\mathbf{Z}(\lambda)] = \sum_{\Gamma} \frac{2}{(\lambda - \sigma_r)^2 + \omega_r^2} [\mathbf{v}_r \quad \mathbf{w}_r] \begin{bmatrix} \lambda - \sigma_r & -\omega_r \\ -\omega_r & \lambda - \sigma_r \end{bmatrix} \begin{bmatrix} \mathbf{x}_r^{\mathrm{T}} \\ \mathbf{y}_r^{\mathrm{T}} \end{bmatrix} \quad (3.7.16)$$

If we specialize further, saying that $[\mathbf{D}_{ss}(\lambda)]$ is symmetrical, then $\{\mathbf{p}_r\} = \{\mathbf{q}_r\}$, and

$$[\mathbf{Z}(\lambda)] = \sum_{\Gamma} \frac{2}{(\lambda - \sigma_r)^2 + \omega_r^2} [\mathbf{x}_r \quad \mathbf{y}_r] \begin{bmatrix} \lambda - \sigma_r & -\omega_r \\ -\omega_r & \lambda - \sigma_r \end{bmatrix} \begin{bmatrix} \mathbf{x}_r^{\mathrm{T}} \\ \mathbf{y}_r^{\mathrm{T}} \end{bmatrix} \quad (3.7.17)$$

For undamped gyroscopic systems, $\sigma_r = 0$, $\{\mathbf{w}_r\} = -\{\mathbf{y}_r\}$ and $\{\mathbf{v}_r\} = -\{\mathbf{x}_r\}$, and thus

$$[\mathbf{Z}(\lambda)] = \sum_{\Gamma} \frac{2}{\lambda^2 + \omega_r^2} [\mathbf{x}_r \quad \mathbf{y}_r] \begin{bmatrix} \lambda & -\omega_r \\ -\omega_r & \lambda \end{bmatrix} \begin{bmatrix} \mathbf{x}_r^{\mathrm{T}} \\ \mathbf{y}_r^{\mathrm{T}} \end{bmatrix} \quad (3.7.18)$$

Finally, if the system is conservative and non-gyroscopic, then the classical modal analysis requires the normalization condition (3.7.13) to be modified to

$$\{\boldsymbol{\phi}_r\}^{\mathrm{T}}[\mathbf{M}(\omega_r^2)]\{\boldsymbol{\phi}_r\} = \mathbf{1} \quad (3.7.19)$$

where

$$[\mathbf{M}(\omega_r^2)] = -\frac{\mathrm{d}}{\mathrm{d}\omega^2}[\mathbf{D}(\omega_r^2)] \quad (3.7.20)$$

Therefore $\{\mathbf{p}_r\} = \{\mathbf{q}_r\} = \{\boldsymbol{\phi}_r\}$ and $\lambda = i\omega$. After simplification, we have

$$[\mathbf{Z}(i\omega)] = \sum_{\Gamma} \frac{\{\boldsymbol{\phi}_r\}\{\boldsymbol{\phi}_r\}^{\mathrm{T}}}{\omega_r^2 - \omega^2} \quad (3.7.21)$$

If the absolute values of λ_r do not increase rapidly, the convergence of (3.7.10) with respect to the modes is slow. The improvement of the convergence can be achieved in the following manner. The discussion is given for non-defective matrices and than for defective matrices.

Expand the non-defective flexibility lambda matrix in Taylor's series,

$$[\mathbf{Z}(\lambda)] = [\mathbf{Z}(0)] + \lambda[\mathbf{Z}'(0)] + \frac{\lambda^2}{2}[\mathbf{Z}''(0)] + [\mathbf{R}(\lambda)] \tag{3.7.22}$$

where $[\mathbf{R}(\lambda)]$ is the residual flexibility. However from (3.7.10),

$$[\mathbf{Z}(\lambda)] = \sum_{\Gamma}\left[\frac{1}{\lambda_r}\{\mathbf{p}_r\}\{\mathbf{q}_r\}^{\mathrm{T}} + \frac{\lambda}{\lambda_r^2}\{\mathbf{p}_r\}\{\mathbf{q}_r\}^{\mathrm{T}} + \frac{\lambda}{\lambda_r^3}\{\mathbf{p}_r\}\{\mathbf{q}_r\}^{\mathrm{T}} + \frac{\lambda^3\{\mathbf{p}_r\}\{\mathbf{q}_r\}^{\mathrm{T}}}{\lambda_r^3(\lambda-\lambda_r)}\right] \tag{3.7.23}$$

where the following identity has been employed:

$$\frac{1}{\lambda-\lambda_r} = \frac{1}{\lambda_r} + \frac{\lambda}{\lambda_r^2} + \frac{\lambda}{\lambda_r^3} + \frac{\lambda^3}{\lambda_r^3(\lambda-\lambda_r)} \tag{3.7.24}$$

Comparing Eqs (3.7.22) and (3.7.23), we have

$$[\mathbf{Z}(\lambda)] = [\mathbf{Z}(0)] + \lambda[\mathbf{Z}'(0)] + \frac{\lambda^2}{2}[\mathbf{Z}''(0)] + \lambda^3\sum_{\Gamma}\frac{\lambda^3\{\mathbf{p}_r\}\{\mathbf{q}_r\}^{\mathrm{T}}}{\lambda_r^3(\lambda-\lambda_r)} \tag{3.7.25}$$

Equation (3.2.25) converges at a rate similar to $1/\lambda_r^3(\lambda-\lambda_r)$ and this is much faster than Eq. (3.7.10) which converges like $1/(\lambda-\lambda_r)$ only. Faster convergence can be obtained by expanding Eq. (3.7.24) to more terms.

The matrices $[\mathbf{Z}'(0)]$ and $[\mathbf{Z}''(0)]$ are obtained by differentiating the following identity with respect to λ:

$$[\mathbf{D}(\lambda)][\mathbf{Z}(\lambda)] = [\mathbf{I}] \tag{3.7.26}$$

giving

$$[\mathbf{D}'(\lambda)][\mathbf{Z}(\lambda)] + [\mathbf{D}(\lambda)][\mathbf{Z}'(\lambda)] = [\mathbf{0}] \tag{3.7.27}$$

or

$$[\mathbf{Z}'(\lambda)] = -[\mathbf{Z}(\lambda)][\mathbf{D}'(\lambda)][\mathbf{Z}(\lambda)] \tag{3.7.28}$$

Differentiate (3.7.27) once more,

$$[\mathbf{D}''(\lambda)][\mathbf{Z}(\lambda)] + 2[\mathbf{D}'(\lambda)][\mathbf{Z}'(\lambda)] + [\mathbf{D}(\lambda)][\mathbf{Z}''(\lambda)] = [\mathbf{0}]$$

and therefore,

$$[\mathbf{Z}''(\lambda)] = -[\mathbf{Z}(\lambda)]([\mathbf{D}''(\lambda)][\mathbf{Z}(\lambda)] + 2[\mathbf{D}'(\lambda)][\mathbf{Z}'(\lambda)]) \tag{3.7.29}$$

On evaluating at $\lambda = 0$,

$$[\mathbf{Z}'(0)] = -[\mathbf{Z}(0)][\mathbf{D}'(0)][\mathbf{Z}(0)] \tag{3.7.30}$$

and

$$[\mathbf{Z}''(0)] = -[\mathbf{Z}(0)]([\mathbf{D}''(0)][\mathbf{Z}(0)] + 2[\mathbf{D}'(0)][\mathbf{Z}'(0)]) \tag{3.7.31}$$

If analytic differentiation of $[\mathbf{D}(\lambda)]$ is difficult, Romberg's method is recommended. No restriction on $[\mathbf{D}(\lambda)]$ is imposed so far. If we specialize $[\mathbf{D}(\lambda)]$ in a quadratic form, then (3.7.25) degenerates to the results of Leung [41], and Palazzolo, Wang and Pilkey [42].

Following a similar procedure, we can prove that for defective matrices,

$$[\mathbf{Z}(\lambda)] = [\mathbf{Z}(0)] + \lambda[\mathbf{Z}'(0)] + \frac{\lambda^2}{2}[\mathbf{Z}''(0)] + \lambda^3\sum_{r=1}^{q}\frac{\{\mathbf{p}_r\}\{\mathbf{q}_r\}^{\mathrm{T}}}{\lambda_r^3(\lambda-\lambda_r)}$$
$$+ \lambda^3\sum_{r=1}^{q}\frac{[\mathbf{P}_i][\mathbf{J}_i - \lambda\mathbf{I}]^{-1}[\mathbf{Q}_i]^{\mathrm{T}}}{\lambda_i^3} \tag{3.7.32}$$

An alternative to finite-element modelling is continuum modelling, where the dynamic stiffness matrix is derived directly from the governing differential equations [31]. However, as the number of finite elements increases, the continuous system gives increasing accuracy. If the generalized coordinates in a finite-element model are separated into nodal coordinates $\{\mathbf{u}_n\}$ and distributed (non-nodal) coordinates $\{\mathbf{u}_d\}$, after eliminating $\{\mathbf{u}_d\}$, the resulting dynamic stiffness matrix approaches that of the continuum model. When the finite-element matrix is partitioned into $\{\mathbf{u}_n\}$ and $\{\mathbf{u}_d\}$, then by definition,

$$\begin{bmatrix} \mathbf{D}_{nn} & \mathbf{D}_{nd} \\ \mathbf{D}_{dn} & \mathbf{D}_{dd} \end{bmatrix} \begin{bmatrix} \mathbf{Z}_{nn} & \mathbf{Z}_{nd} \\ \mathbf{Z}_{dn} & \mathbf{Z}_{dd} \end{bmatrix} = [\mathbf{I}] \tag{3.7.33}$$

Solving Eq. (3.7.33) in terms of the flexibilities, we have

$$\begin{bmatrix} \mathbf{Z}_{nn} & \mathbf{Z}_{nd} \\ \mathbf{Z}_{dn} & \mathbf{Z}_{dd} \end{bmatrix} = \begin{bmatrix} \mathbf{D}_n^{-1} & -\mathbf{D}_{nn}^{-1}\mathbf{D}_{nd}\mathbf{D}_d^{-1} \\ -\mathbf{D}_{dd}^{-1}\mathbf{D}_{dn}\mathbf{D}_n^{-1} & \mathbf{D}_d^{-1} \end{bmatrix} \tag{3.7.34}$$

where $[\mathbf{D}_n] = [\mathbf{D}_{nn} - \mathbf{D}_{nd}\mathbf{D}_{dd}^{-1}\mathbf{D}_{dn}]$ and $[\mathbf{D}_d] = [\mathbf{D}_{dd} - \mathbf{D}_{dn}\mathbf{D}_{nn}^{-1}\mathbf{D}_{nd}]$. However, from Eq. (3.7.9),

$$\begin{bmatrix} \mathbf{Z}_{nn} & \mathbf{Z}_{nd} \\ \mathbf{Z}_{dn} & \mathbf{Z}_{dd} \end{bmatrix} = \begin{bmatrix} \mathbf{P}_n \\ \mathbf{P}_d \end{bmatrix} [\mathbf{J} - \lambda \mathbf{I}]^{-1} \begin{bmatrix} \mathbf{Q}_n \\ \mathbf{Q}_d \end{bmatrix}^{\mathrm{T}} \tag{3.7.35}$$

Comparing (3.7.34) and (3.7.35), we have

$$[\mathbf{D}_n] = [\mathbf{Z}_{nn}][\mathbf{P}_n][\mathbf{J} - \lambda \mathbf{I}]^{-1}[\mathbf{Q}_n]^{\mathrm{T}} \tag{3.7.36}$$

Therefore, the distributed coordinates $\{\mathbf{u}_d\}$ play no part explicitly in evaluating the condensed dynamic flexibility associated with the master nodes, so long as the generalized vectors in terms of the master nodes are available. Dropping the subscripts n, we have the same Eqs (3.7.9) and (3.7.10) for both finite-element and continuum models.

Again, it is impossible to include all partial modes in a continuum model. If only a small number of the lowest partial modes, in ascending order of the moduli of the eigenvalues, are available, Eq. (3.7.36) is a good approximation. The approximation can be improved by the method of the previous section.

When all substructure matrices $[\mathbf{D}^*(\lambda)]$ are assembled according to the equilibrium and compatibility conditions, an eigenvalue problem results for the frequency parameter λ and modes $\{\psi\}$ and $\{\boldsymbol{\phi}\}$ of the system,

$$[\mathbf{D}(\lambda)]\{\boldsymbol{\phi}\} = \{\mathbf{0}\} \qquad \text{and} \qquad \{\boldsymbol{\psi}\}^{\mathrm{T}}[\mathbf{D}(\lambda)] = \{\mathbf{0}\}^{\mathrm{T}} \tag{3.7.37}$$

where $[\mathbf{D}(\lambda)]$ denotes the system matrix. If the system is conservative, solutions of Eqs (3.7.3) by the Sturm method and inverse iteration are well known. If the system is non-defective, inverse iteration with generalized Rayleigh quotient [43] is effective.

When the system is defective at an eigenvalue due to non-conservativeness, it is always possible in an engineering application to introduce a non-conservative parameter μ in the dynamic stiffness matrix $[\mathbf{D}[\lambda, \mu)]$ so that $[\mathbf{D}(\lambda, 0)]$ is conservative. Now if we take the conservative solution as an approximation, the eigenvalue λ corresponding to μ can easily be constructed by increasing μ. When $[\mathbf{D}(\lambda, \mu)]$ is defective, it is required that

$$\Delta = \det[\mathbf{D}(\lambda, \mu)] = 0 \qquad \text{and} \qquad \frac{\partial \Delta}{\partial \lambda} = 0 \tag{3.7.38}$$

from which λ, μ can be found. If an initial approximation λ_0, μ_0 is given, the Newtonian algorithm gives the improved solution

$$\lambda = \lambda_0 + d\lambda, \qquad \mu = \mu_0 + d\mu \tag{3.7.39}$$

where $d\lambda$ and $d\mu$ are solved from

$$\begin{bmatrix} \partial\Delta_0/\partial\lambda & \partial\Delta_0/\partial\mu \\ \partial^2\Delta_0/\partial\lambda^2 & \partial^2\Delta_0/\partial\mu^2 \end{bmatrix} \begin{Bmatrix} d\lambda \\ d\mu \end{Bmatrix} = - \begin{Bmatrix} \Delta_0 \\ \partial\Delta_0/\partial\lambda \end{Bmatrix} \tag{3.7.40}$$

in which subscripts denote evaluation at λ_0.

The derivatives of the determinant can be evaluated in the following manner. Decompose the matrix **[D]** into a lower triangular matrix **[L]** with a unit main diagonal and an upper triangular matrix **[U]**

$$[\mathbf{D}] = [\mathbf{L}][\mathbf{U}] \tag{3.7.41}$$

then

$$\Delta = \prod_i u_{ii} \tag{3.7.42}$$

where u_{ii} are the elements of **[U]** along the main diagonal. Differentiate (3.7.41) with respect to λ,

$$\frac{\partial[\mathbf{D}]}{\partial\lambda} = [\mathbf{D}'] = [\mathbf{L}][\mathbf{U}'] + [\mathbf{L}'][\mathbf{U}] \tag{3.7.43}$$

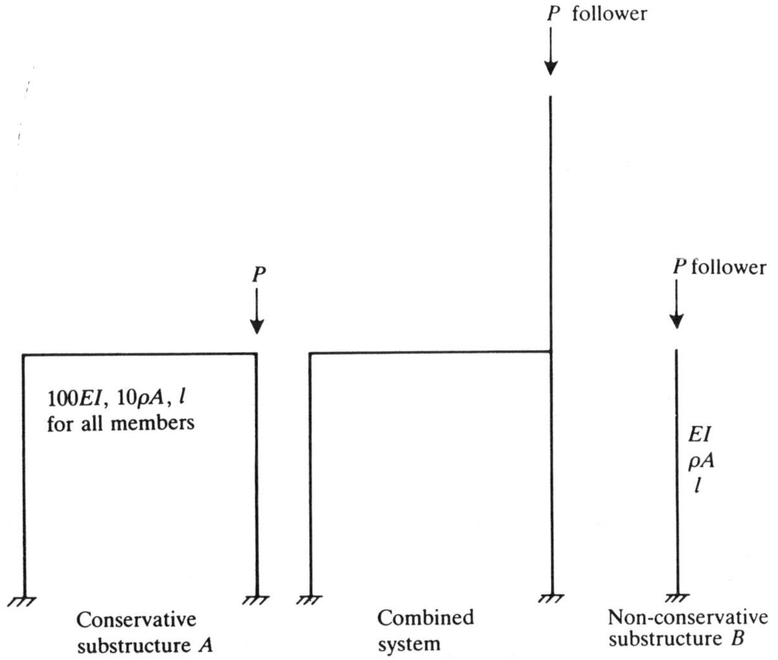

Fig. 3.7.1. A system consisting of conservative and non-conservative substructures

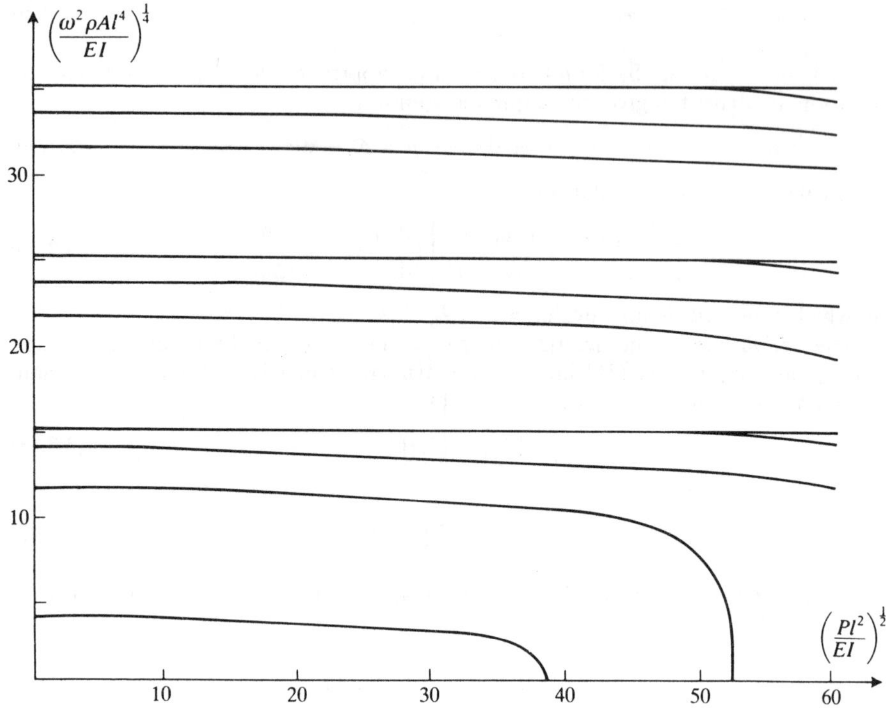

Fig. 3.7.2. Frequency diagram for the conservative substructure

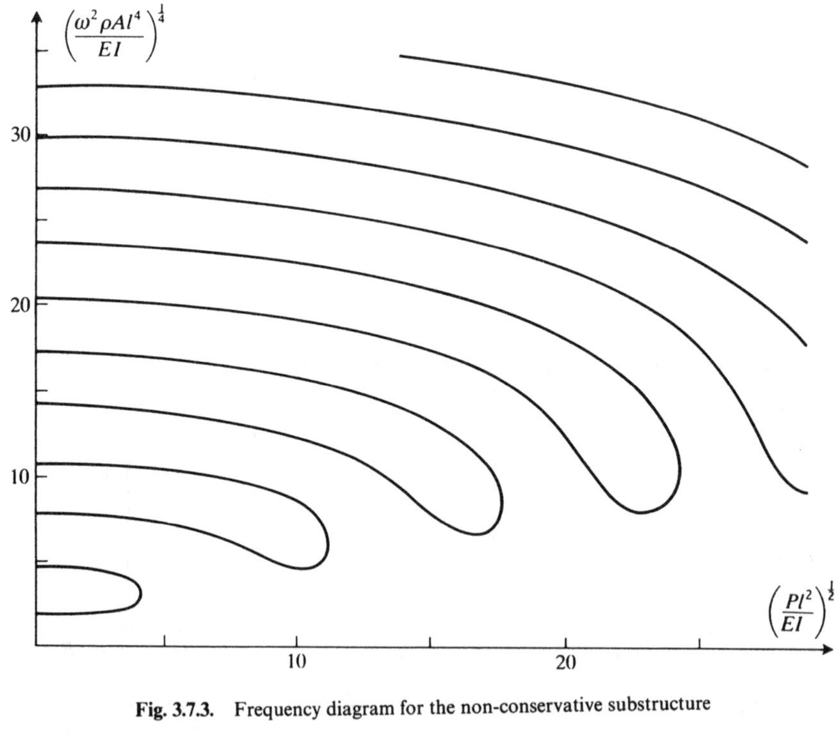

Fig. 3.7.3. Frequency diagram for the non-conservative substructure

where the elements u'_{ij} and l'_{ij} of [U′] and [L′] respectively are given by

$$\left.\begin{aligned} l'_{ij} &= \left(D'_{ij} - \sum_{k=1}^{j} l_{ik}u'_{kj} - \sum_{k=1}^{j-1} l'_{ik}u_{kj}\right)\bigg/ u_{jj}, \qquad j < i \\ u'_{ij} &= D'_{ij} - \sum_{k=1}^{i-1} l_{ik}u'_{kj} - \sum_{k=1}^{j-1} l'_{ik}u_{kj}, \qquad j \geq i \end{aligned}\right\} \tag{3.7.44}$$

and where u_{ij} and l_{ij} are elements of [U] and [L] respectively. Differentiating (3.7.42) with respect to λ, we have

$$\partial\Delta/\partial\lambda = \Delta = \sum_i u'_{ii} \prod_{j\neq i} u_{jj} \tag{3.7.45}$$

Similarly, $\partial^2\Delta/\partial\lambda^2$, $\partial^2\Delta/\partial\lambda\partial\mu$ can be obtained.

Consider a system consisting of two substructures A and B as shown in Fig. 3.7.1. Substructure A is a simple frame where the axial force P is conservative. The fixed interface natural frequencies can easily be found and are plotted in Fig. 3.7.2 against P. Substructure B is a Beck column subject to follower force P and its fixed interface natural frequencies are plotted in Fig. 3.7.3. The natural frequencies of the combined system are plotted in Fig. 3.7.4 against P. It is found that coalescence of non-adjacent modes is possible.

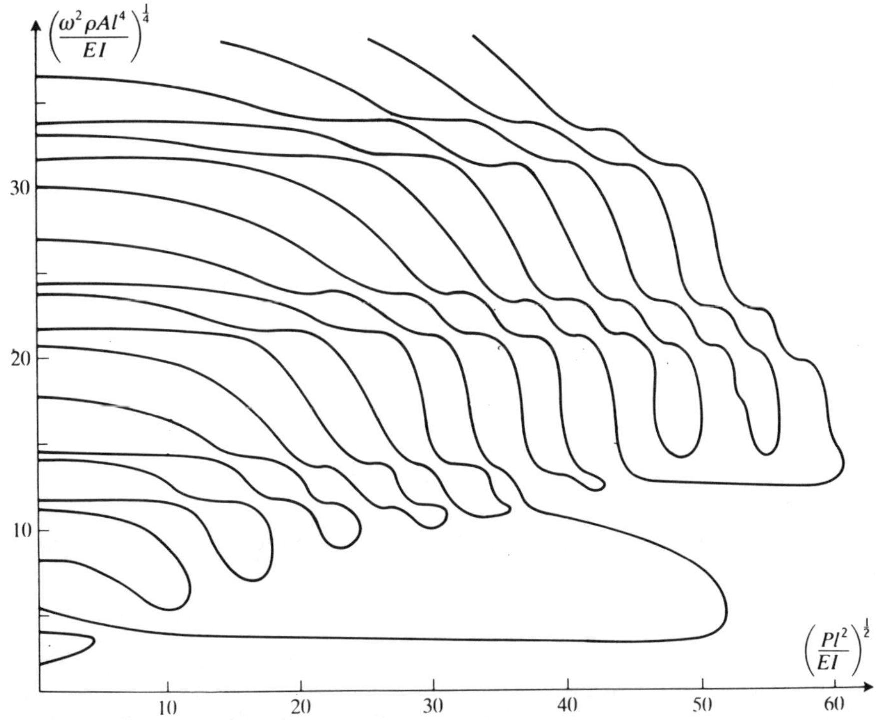

Fig. 3.7.4. Frequency diagram for the combined structure

3.8. Substructure Response

Let the undamped governing equations of a substructure be

$$[\mathbf{K}]\{\mathbf{q}\} + [\mathbf{M}]\{\ddot{\mathbf{q}}\} = \{\mathbf{Q}\} \tag{3.8.1}$$

where [**K**] and [**M**] are the finite element stiffness and mass matrices respectively, and {**q**} and {**Q**} are the nodal response and excitation vectors respectively. A dot denotes a derivative with respect to time t. In deriving Eq. (3.8.1), the distributed displacement $\{\mathbf{u}(\mathbf{x},t)\}$ is related to the nodal displacement {**q**} by the shape function matrix $[\mathbf{N}(\mathbf{x})]$

$$\{\mathbf{u}(\mathbf{x},t)\} = [\mathbf{N}(\mathbf{x})]\{\mathbf{q}(t)\} \tag{3.8.2}$$

Equation (3.8.2) is a good approximation if a sufficiently fine finite element mesh is employed. When the nodal displacement $\{\mathbf{q}(t)\}$ is partitioned according to slaves and masters, subscripts s and m respectively, Eq. (3.8.1) becomes

$$\begin{bmatrix} \mathbf{K}_{ss} & \mathbf{K}_{sm} \\ \mathbf{K}_{ms} & \mathbf{K}_{mm} \end{bmatrix} \begin{Bmatrix} \mathbf{q}_s \\ \mathbf{q}_m \end{Bmatrix} + \begin{bmatrix} \mathbf{M}_{ss} & \mathbf{M}_{sm} \\ \mathbf{M}_{ms} & \mathbf{M}_{mm} \end{bmatrix} \begin{Bmatrix} \ddot{\mathbf{q}}_s \\ \ddot{\mathbf{q}}_m \end{Bmatrix} = \begin{Bmatrix} \mathbf{0} \\ \mathbf{Q}_m(t) \end{Bmatrix} \tag{3.8.3}$$

where $\{\mathbf{Q}_s(t)\} = \{\mathbf{0}\}$ is assumed. The associated initial conditions are

$$\{\dot{\mathbf{q}}_m(0)\} = \{\dot{\mathbf{q}}_m^0\} \qquad \text{and} \qquad \{\mathbf{q}_m(0)\} = \{\mathbf{q}_m^0\} \tag{3.8.4}$$

and similar initial conditions for the slaves.

Further reduction of Eqs (3.8.3) subject to conditions (3.8.4) seems to be difficult. We can thus first proceed to steady state harmonic analysis and subsequently return to the case of general $\{\mathbf{Q}_m(t)\}$.

Let

$$\{\mathbf{Q}_m(t)\} = \{\overline{\mathbf{Q}}_m(t)\}e^{i\omega t} \qquad \text{and} \qquad \{\mathbf{q}(t)\} = \{\bar{\mathbf{q}}\}e^{i\omega t} \tag{3.8.5}$$

where ω is the excitation frequency and $i = \sqrt{-1}$. Substituting Eqs (3.8.5) into Eq. (3.8.3), we have

$$[\mathbf{D}(\omega)]\{\bar{\mathbf{q}}\} = \begin{bmatrix} \mathbf{D}_{ss}(\omega) & \mathbf{D}_{sm}(\omega) \\ \mathbf{D}_{ms}(\omega) & \mathbf{D}_{mm}(\omega) \end{bmatrix} \begin{Bmatrix} \bar{\mathbf{q}}_s \\ \bar{\mathbf{q}}_m \end{Bmatrix} = \begin{Bmatrix} \mathbf{0} \\ \overline{\mathbf{Q}}_m \end{Bmatrix} \tag{3.8.6}$$

where $[\mathbf{D}_{sm}(\omega)] = [\mathbf{K}_{sm} - \omega^2\mathbf{M}_{sm}]$, etc. From the first equation,

$$\{\bar{\mathbf{q}}_s\} = -[\mathbf{D}_{ss}(\omega)]^{-1}[\mathbf{D}_{sm}(\omega)]\{\bar{\mathbf{q}}_m\} \qquad \text{and} \qquad \{\bar{\mathbf{q}}\} = [\mathbf{T}(\omega)]\{\bar{\mathbf{q}}_m\} \tag{3.8.7}$$

where

$$[\mathbf{T}(\omega)] = \begin{bmatrix} -\mathbf{D}_{ss}^{-1}(\omega)\mathbf{D}_{sm}(\omega) \\ \mathbf{I} \end{bmatrix} \tag{3.8.8}$$

is the transformation matrix for harmonic condensation. The second of equations (3.8.6) gives

$$[\mathbf{D}_m(\omega)]\{\bar{\mathbf{q}}_m\} = [\mathbf{T}(\omega)]^{\mathrm{T}}[\mathbf{D}(\omega)][\mathbf{T}(\omega)]\{\bar{\mathbf{q}}_m\} = [\mathbf{D}_{mm} - \mathbf{D}_{ms}\mathbf{D}_{ss}^{-1}\mathbf{D}_{sm}]\{\bar{\mathbf{q}}_m\} \tag{3.8.9}$$

where the singly subscripted matrix $[\mathbf{D}_m(\omega)]$ is the condensed dynamic stiffness matrix evaluated at frequency ω. The condensed dynamic stiffness matrix can also be expressed in the following well-behaved form to reduce the numerical problem which occurs as ω tends to zero

$$[\mathbf{D}_m(\omega)] = [\mathbf{K}_m] - \omega^2[\mathbf{M}_m] - \omega^2[\mathbf{G}][\mathbf{\Lambda}][\mathbf{G}]^{\mathrm{T}} \tag{3.8.9a}$$

Here $[\mathbf{K}_m]$ and $[\mathbf{M}_m]$ are the matrices corresponding to static shape functions and $[\mathbf{G}]$ and $[\mathbf{\Lambda}]$ are as found in Sect. 3.2.

Comparing Eqs (3.8.3) and (3.8.6), we have

$$\left[\mathbf{D}\left(-\mathrm{i}\frac{\mathrm{d}}{\mathrm{d}t}\right)\right]\{\mathbf{q}\} = \{\mathbf{Q}\} \tag{3.8.10}$$

i.e. if the frequency in harmonic analysis is replaced by $-\mathrm{id}/\mathrm{d}t$, the equations for the general excitation are recovered. Corresponding to the condensed system, Eq. (3.8.9),

$$\left[\mathbf{D}_m\left(-\mathrm{i}\frac{\mathrm{d}}{\mathrm{d}t}\right)\right]\{\mathbf{q}_m(t)\} = \{\mathbf{Q}_m(t)\} \tag{3.8.11}$$

The condensed differential Eq. (3.8.11) subject to initial conditions (3.8.4) will be solved by means of the dynamic flexibility, $[\mathbf{Z}_m(\omega)]$,

$$[\mathbf{D}_m(\omega)][\mathbf{Z}_m(\omega)] = [\mathbf{I}] \tag{3.8.12}$$

by expressing $[\mathbf{Z}_m(\omega)]$ in spectral form (in terms of fixed interface modes).

In summary, for solving the distributed coordinate response $\{\mathbf{u}(\mathbf{x},t)\} = \{\bar{\mathbf{u}}(\mathbf{x},\omega)\}\mathrm{e}^{\mathrm{i}\omega t}$, two basic transformations are involved. One is the finite element interpolation,

$$\{\bar{\mathbf{u}}(\mathbf{x},\omega)\} = [\mathbf{N}(\mathbf{x})]\{\bar{\mathbf{q}}\} \tag{3.8.13}$$

and the other is harmonic condensation,

$$\{\bar{\mathbf{q}}\} = [\mathbf{T}(\omega)]\{\bar{\mathbf{q}}_m\} \tag{3.8.14}$$

so that

$$\{\bar{\mathbf{u}}(\mathbf{x},\omega)\} = [\mathbf{N}(\mathbf{x})][\mathbf{T}(\omega)]\{\bar{\mathbf{q}}_m\} \tag{3.8.15}$$

After forming the dynamic stiffness and flexibility, we replace ω by $-\mathrm{id}/\mathrm{d}t$ and obtain the condensed equations of motion in terms of masters alone.

Since Eq. (3.8.3) cannot be reduced in the time domain, we reduce Eq. (3.8.9) in the frequency domain, assemble overall substructures according to the conditions of equilibrium and compatibility, and finally obtain the governing equations in the time domain by replacing ω with $-\mathrm{id}/\mathrm{d}t$. Assembly by using the standard finite element procedure in the frequency domain results in the non-linear eigenproblem

$$[\mathbf{D}^m(\omega)]\{\bar{\mathbf{q}}^m\} = \{\mathbf{0}\} \tag{3.8.16}$$

for free vibration. The superscript m denotes global quantities with respect to the collective masters over all substructures. The following coordinate transforming from substructure $\{\bar{\mathbf{q}}_m\}$ to system $\{\bar{\mathbf{q}}^m\}$ is implied:

$$\{\bar{\mathbf{q}}_m\} = [\mathbf{C}_e]\{\bar{\mathbf{q}}^m\} \tag{3.8.17}$$

where $[\mathbf{C}_e]$ is a coordinate transformation matrix for substructure e. There are effective algorithms [17, 18] to solve Eq. (3.8.16) for non-trivial solutions of $\{\bar{\mathbf{q}}^m\}$ so that

$$[\mathbf{D}^m(\omega_j)]\{\phi_j^m\} = \{\mathbf{0}\} \tag{3.8.18}$$

where ω_j and $\{\phi_j^m\}$ are the jth mode. If condensation is not applied,

$$[\mathbf{D}(\omega_j)]\{\phi_j\} = [\mathbf{K} - \omega_j^2\mathbf{M}]\{\phi_j\} = \{\mathbf{0}\} \tag{3.8.19}$$

where $\{\phi_j\} = \{\phi_j^s\phi_j^m\}$, and **K** and **M** are the conventional stiffness and mass matrices. The orthonormality condition is

$$\{\phi_j\}^{\mathrm{T}}[\mathbf{M}]\{\phi_k\} = \delta_{jk} \tag{3.8.20}$$

where δ_{jk} is the Kronecker delta. Applying the harmonic condensation, Eq. (3.8.14), we have the orthonormality condition for the condensed modes,

$$\{\phi_j^m\}^{\mathrm{T}}[\mathbf{T}(\omega_j)]^{\mathrm{T}}[\mathbf{M}][\mathbf{T}(\omega_j)]\{\phi_k^m\} = \delta_{jk} \tag{3.8.21}$$

or

$$\{\phi_j^m\}^{\mathrm{T}}[\mathbf{M}^m(\omega_j, \omega_k)]\{\phi_k^m\} = \delta_{jk} \tag{3.8.22}$$

where the mixed mass matrix is defined as

$$[\mathbf{M}^m(\omega_j, \omega_k)] = [\mathbf{T}(\omega_j)]^{\mathrm{T}}[\mathbf{M}][\mathbf{T}(\omega_k)] \tag{3.8.23}$$

A proof is given below that

$$[\mathbf{M}^m(\omega_j, \omega_k)] = \frac{1}{\omega_k^2 - \omega_j^2}[\mathbf{D}^m(\omega_j) - \mathbf{D}^m(\omega_k)] \tag{3.8.24}$$

with which $[\mathbf{M}^m(\omega_j, \omega_k)]$ can be evaluated without using $[\mathbf{T}(\omega)]$ and $[\mathbf{M}]$.

Suppose there are two systems of harmonic excitation, $\{\mathbf{Q}_j\}e^{i\omega_j t}$ and $\{\mathbf{Q}_k\}e^{i\omega_k t}$ applied at the masters, and that the corresponding harmonic responses are $\{\mathbf{q}_j\}e^{i\omega_j t}$ and $\{\mathbf{q}_k\}e^{i\omega_k t}$ respectively. The superscript m and the overbar are omitted. The distributed coordinates $\{\mathbf{u}(\mathbf{x}, \omega_j)\}$ and $\{\mathbf{u}(\mathbf{x}, \omega_k)\}$ are given, respectively, by

$$\{\mathbf{u}_j\} = \{\mathbf{u}(\mathbf{x}, \omega_j)\} = [\mathbf{N}(\mathbf{x})][\mathbf{T}(\omega_j)]\{\mathbf{q}_j\}$$

and

$$\{\mathbf{u}_k\} = \{\mathbf{u}(\mathbf{x}, \omega_k)\} = [\mathbf{N}(\mathbf{x})][\mathbf{T}(\omega_k)]\{\mathbf{q}_k\} \tag{3.8.25}$$

The dynamic stiffness relations are

$$[\mathbf{D}_i]\{\mathbf{q}_i\} = \{\mathbf{Q}_i\} \qquad \text{and} \qquad [\mathbf{D}_k]\{\mathbf{q}_k\} = \{\mathbf{Q}_k\} \tag{3.8.26}$$

where $[\mathbf{D}_j] = [\mathbf{D}(\omega_j)]$, etc. The reciprocal theoem states that the work done by the first set of forces on the second set of displacements is equal to the work done by the second set of forces on the first set of displacements. Therefore,

$$\{\mathbf{q}_i\}^{\mathrm{T}}\{\mathbf{Q}_k\} + \omega_k^2\int\{\mathbf{u}_j\}^{\mathrm{T}}[\mathbf{m}]\{\mathbf{u}_k\}\,\mathrm{d}v = \{\mathbf{q}_k\}^{\mathrm{T}}\{\mathbf{Q}_j\} + \omega_j^2\int\{\mathbf{u}_k\}^{\mathrm{T}}[\mathbf{m}]\{\mathbf{u}_j\}\,\mathrm{d}v \tag{3.8.27}$$

where $[\mathbf{m}]$ is the inertia per unit volume and $\mathrm{d}v$ is a volume element of the system. Substituting Eqs (3.8.25) and (3.8.26) into (3.8.27), we have

$$[\mathbf{M}^m(\omega_j, \omega_k)] = [\mathbf{T}(\omega_j)]^{\mathrm{T}}[\mathbf{M}][\mathbf{T}(\omega_j)] = \frac{1}{\omega_k^2 - \omega_j^2}[\mathbf{D}_j - \mathbf{D}_k] \tag{3.8.28}$$

where the following definition of **M** is recognized:

$$[\mathbf{M}] = \int[\mathbf{N}(\mathbf{x})]^{\mathrm{T}}[\mathbf{m}][\mathbf{N}(\mathbf{x})]\,\mathrm{d}v \tag{3.8.29}$$

An obvious by-product of Eq. (3.8.28) is that the condensed mass matrix is given, when the superscript m denoting masters is re-established, by

$$[\mathbf{M}^m(\omega)] = [\mathbf{M}^m(\omega,\omega)] = [\mathbf{T}(\omega)]^{\mathrm{T}}[\mathbf{M}^m(\omega,\omega)][\mathbf{T}(\omega)] = -\frac{\partial}{\partial\omega^2}[\mathbf{D}^m(\omega)] \quad (3.8.30)$$

Since a substructure dynamic stiffness contains a greater number of modes than the number of masters retained, expansion of vectors using more modes than the order is difficult. This difficulty is alleviated by considering distributed coordinates. Let the kth distributed mode be, according to Eq. (3.8.14),

$$\{\psi_k\} = [\mathbf{N}(\mathbf{x})][\mathbf{T}(\omega_k)]\{\phi_k^m\} \quad (3.8.31)$$

To expand a vector $\{\omega(\mathbf{x})\}$ in terms of natural modes, let

$$\{\mathbf{v}(\mathbf{x})\} = \sum_k \alpha_k\{\psi_k(\mathbf{x})\} \quad (3.8.32)$$

where α_k are constants to be determined. According to Eq. (3.8.14),

$$\{\mathbf{v}(\mathbf{x})\} = [\mathbf{N}(\mathbf{x})][\mathbf{T}(0)]\{\mathbf{V}\} \quad (3.8.33)$$

where $\{\mathbf{V}\}$ is a given vector containing values of $\{\mathbf{v}(\mathbf{x})\}$ evaluated at the master nodes. Substituting Eqs (3.8.31) and (3.8.33) into (3.8.32), we have

$$[\mathbf{N}(\mathbf{x})][\mathbf{T}(0)]\{\mathbf{v}\} = \sum_k \alpha_k[\mathbf{N}(\mathbf{x})][\mathbf{T}(\omega_k)]\{\phi_k^m\} \quad (3.8.34)$$

Premultiplying by $\{\psi_j(\mathbf{x})\}^{\mathrm{T}}[\mathbf{m}]$ and integrating over the whole system yields

$$\{\phi_j^m\}^{\mathrm{T}}[\mathbf{T}(\omega_j)]^{\mathrm{T}}\int[\mathbf{N}(\mathbf{x})][\mathbf{m}][\mathbf{N}(\mathbf{x})]\,\mathrm{d}v[\mathbf{T}(0)]\{\mathbf{v}\}$$

$$= \sum_k \alpha_k\{\phi_j^m\}^{\mathrm{T}}[\mathbf{T}(\omega_j)]^{\mathrm{T}}\int[\mathbf{N}(\mathbf{x})][\mathbf{m}][\mathbf{N}(\mathbf{x})]\,\mathrm{d}v[\mathbf{T}(\omega_k)]\{\phi_j^m\}$$

or

$$\{\phi_j^m\}^{\mathrm{T}}[\mathbf{M}^m(\omega_j,0)]\{\mathbf{V}\} = \sum_k \alpha_k\{\phi_j^m\}^{\mathrm{T}}[\mathbf{M}^m(\omega_j,\omega_k)]\{\phi_j^m\}$$

However, from the orthonormality condition on the natural modes,

$$\alpha_j = \{\phi_j^m\}^{\mathrm{T}}[\mathbf{M}^m(\omega_j,0)]\{\mathbf{V}\} \quad (3.8.35)$$

Therefore, to an arbitrary vector $\{\mathbf{V}\}$ or $\{\mathbf{v}(\mathbf{x})\}$ there corresponds a unique series expansion in terms of the natural modes, as given by Eqs (3.8.32) and (3.8.35). These formulae are useful when dealing with initial conditions. In equation (3.8.35), $[\mathbf{M}(\omega,0)] = [\mathbf{K}-\mathbf{D}(\omega)]/\omega^2$, where $[\mathbf{K}] = [\mathbf{D}(0)]$ is the static stiffness.

The dynamic flexibility defined in Eq. (3.8.12) can be expressed in spectral form using global coordinates, as

$$[\mathbf{Z}^m(\omega)] = \sum_k \frac{\{\phi_k^m\}\{\phi_k^m\}^{\mathrm{T}}}{\omega_k^2-\omega^2} \quad (3.8.36)$$

In the frequency domain

$$[\mathbf{D}^m(\omega)]\{\bar{\mathbf{q}}^m\} = \{\bar{\mathbf{Q}}^m\} \qquad \text{or} \qquad \{\bar{\mathbf{q}}^m\} = [\mathbf{Z}^m(\omega)]\{\bar{\mathbf{Q}}^m\} \quad (3.8.37)$$

For time domain analysis we replace ω by $-\mathrm{i}\,\mathrm{d}/\mathrm{d}t$

$$\{\mathbf{q}^m(t)\} = [\mathbf{Z}^m(\omega)]\{\mathbf{Q}^m(t)\} = \sum_k \{\phi_k^m\}\{\phi_k^m\}^{\mathrm{T}} \frac{1}{\omega_k^2 - \omega^2}\{\mathbf{Q}^m(t)\} = \sum_k p_k(t)\{\phi_k^m\} \tag{3.8.38}$$

where

$$p_k(t) = \frac{P_k(t)}{\omega_k^2 + (\mathrm{d}/\mathrm{d}t)^2} \quad \text{and} \quad P_k(t) = \{\phi_k^m\}^{\mathrm{T}}\{\mathbf{Q}^m(t)\} \tag{3.8.39}$$

Therefore,

$$\ddot{p}_k + \omega_k^2 p_k = P_k(t) \tag{3.8.40}$$

the solution of which is given by the well known Duhamel integral,

$$p_k(t) = \int_0^t P_k(\tau)h_k(t-\tau)\,\mathrm{d}\tau + g_k(t)p_k(0) + h_k(t)\dot{p}_k(0) \tag{3.8.41}$$

where

$$h_k(t) = \omega_k^{-1}\sin\omega_k t \tag{3.8.42}$$

$$g_k(t) = \cos\omega_k t \tag{3.8.43}$$

The initial conditions for $p_k(t)$ are obtained by putting $t = 0$ in Eq. (3.8.38)

$$\{\mathbf{q}^m(0)\} = \sum_k p_k(0)\{\phi_k^m\} \tag{3.8.44}$$

According to the expansion theorem of Eqs (3.8.32) and (3.8.35), we obtain

$$p_k(0) = \{\phi_k^m\}^{\mathrm{T}}[\mathbf{M}^m(\omega_k, 0)]\{\mathbf{q}^m(0)\} \tag{3.8.45}$$

Similarly,

$$\dot{p}_k(0) = \{\phi_k^m\}^{\mathrm{T}}[\mathbf{M}^m(\omega_k, 0)]\{\dot{\mathbf{q}}^m(0)\} \tag{3.8.46}$$

Therefore, the response at the masters can be calculated as follows:

1. Evaluate the modal force $\{\phi_k^m\}\{\mathbf{Q}^m(t)\}$ from Eq. (3.8.40).
2. Find the initial conditions for the modal coordinates $p_k(t)$ from Eqs (3.8.45) and (3.8.46).
3. Integrate the Duhamel integral in Eq. (3.8.41).
4. Obtain the solution from Eq. (3.8.38).

The modal participation factors measure the relative influences of the external forces on each of the natural modes. Concentrated forces are studied first and distributed loads are considered afterwards.

Suppose the external concentrated force applies at the masters only. Modal expansion gives

$$\{\mathbf{Q}^m(t)\} = \sum_k R_k(t)\{\phi_k^m\} \tag{3.8.47}$$

where $R_k(t)$ is to be determined. Premultiplying both sides by $\{\phi_k^m\}[\mathbf{M}^m(\omega_j, \omega_k)]$ and applying the orthonormality condition, we obtain

$$R_k(t) = \{\phi_k^m\}^{\mathrm{T}}[\mathbf{M}^m(\omega_k)]\{\mathbf{Q}^m(t)\} = \{\Gamma_k\}^{\mathrm{T}}\{\mathbf{Q}^m(t)\} \tag{3.8.48}$$

where $\{\Gamma_k\}$ is the kth modal participation vector.

The problem is more complicated when forces are applied to the slaves as well.

This can be treated as a special case of a distributed load. The distributed load is considered in the frequency domain first and converted to the time domain by replacing ω by $-\mathrm{i}\,\mathrm{d}/\mathrm{d}t$ afterwards. Let the distributed harmonic force be approximated in a finite element sense, so the Eq. (3.8.14) is applicable:

$$\{\bar{\mathbf{f}}(\mathbf{x})\}\mathrm{e}^{\mathrm{i}\omega t} = [\mathbf{N}(\mathbf{x})][\mathbf{T}(\omega)]\{\bar{\mathbf{F}}^m\}\mathrm{e}^{\mathrm{i}\omega t} \tag{3.8.49}$$

Expand $\{\bar{\mathbf{f}}\}$ in modal form,

$$\{\bar{\mathbf{f}}(\mathbf{x})\} = \sum_k S_k\{\psi_k(\mathbf{x})\} = \sum_k S_k[\mathbf{N}(\mathbf{x})][\mathbf{T}(\omega_k)]\{\phi_k^m\} \tag{3.8.50}$$

where S_k is to be determined. Premultiply by $\{\psi_j\}[\mathbf{m}]$ and integrate over the whole volume, using the orthonormality condition, to obtain

$$S_k = \{\phi_k^m\}^{\mathrm{T}}[\mathbf{M}^m(\omega_k, \omega)]\{\bar{\mathbf{F}}^m\} = \{\gamma_k(\omega)\}^{\mathrm{T}}\{\bar{\mathbf{F}}^m\} \tag{3.8.51}$$

where $\{\gamma_k(\omega)\}$ is the kth distributed modal participation vector for harmonic excitation. If we replace ω by $-\mathrm{i}\,\mathrm{d}/\mathrm{d}t$, then

$$S_k = \left\{\gamma_k\left(-\mathrm{i}\frac{\mathrm{d}}{\mathrm{d}t}\right)\right\}^{\mathrm{T}}\{\mathbf{F}^m\} \tag{3.8.52}$$

where

$$\left\{\gamma_k\left(-\mathrm{i}\frac{\mathrm{d}}{\mathrm{d}t}\right)\right\}^{\mathrm{T}} = \{\phi_k^m\}^{\mathrm{T}}\left[\mathbf{M}^m\left(\omega_k, -\mathrm{i}\frac{\mathrm{d}}{\mathrm{d}t}\right)\right] \tag{3.8.53}$$

is the kth distributed modal participation vector for general excitation. Therefore, when external forces are applied at nodes other than the masters, the computation of the response is difficult. It is advisable to include all positions at which external forces are applied as masters.

3.8.1. Modal Damping

If the system is lightly damped, modal damping is a valid and convenient assumption. Suppose that the modal damping factor of the kth mode is ζ_k, then Eq. (3.8.40) is modified to

$$\ddot{p}_k + 2\zeta_k\omega_k\dot{p}_k + \omega_k^2 p_k = P_k(t) \tag{3.8.54}$$

which has the same form of solution as Eq. (3.8.41) except that

$$h_k(t) = \lambda_k^{-1}\mathrm{e}^{-\zeta_k\omega_k t}\sin\lambda_k t \qquad \text{and} \qquad g_k(t) = \mathrm{e}^{-\zeta_k\omega_k t}\cos\lambda_k t + \zeta_k\omega_k h_k(t) \tag{3.8.55}$$

where $\lambda_k^2 = (1 - \zeta_k^2)\omega_k^2$.

Since the natural modes for heavily damped systems are complex, the above-mentioned theory has to be modified substantially. This is beyond the scope of the present study.

3.9. Periodic Structures

A periodic structure consists of identical substructures coupled together in a regular manner. Depending on the arrangements, the periodicity may be linear, as in long bridges or multistorey buildings, or circular, as in domes or axisymmetrical shells, or may even be extended to two and three dimensions, as in the case of framed roofs and lattices. The dynamic analysis of such structures is greatly simplified by utilizing the periodicity property.

For two adjacent substructures coupled through only one coordinate, perhaps by suitable transformation, the method of difference calculus has been used extensively. Wah and Calcote [44] summarized this area of work in structural mechanics up to 1970. Ellington and McCallion [45] studied the free vibrations of grillages for various boundary conditions by lumped parameter models. Sundararajan and Reddy investigated the convergence of the finite strip method [46].

For linear periodic systems with multiple coupling, two categories of analysis methods have been used. One is the Holzer transfer matrix method in which displacements and forces are used as coupling coordinates; the other is the stiffness method in which displacements are employed as unknowns. Both functions lead to a set of matrix difference equations [47]. Lin and McDaniel [48] investigated the matrix difference equation associated with a periodic beam on many elastic supports and pointed out various numerical difficulties encountered in using a conventional transfer matrix approach, involving the multiplication of a chain of transfer matrices. The methods of complementary approach [48], Z-transform [49] and Leverrier algorithm [50] have been proposed to circumvent these numerical difficulties. By using displacements only as unknowns, the symmetry of substructure mass and stiffness matrices is preserved [47, 51]. The main difficulty of a stiffness approach is that a characteristic problem in the complex domain has to be solved for every harmonic frequency. However, damping effects may be considered without altering the formulation, as complex arithmetic is already employed.

The computation of the harmonic response by both approaches may be independent of the number of substructures, as the periodic structure is characterized by the propagating and non-propagating free wave motion, without any need for knowledge of the natural modes [51]. However, if all the natural modes are required, any computational algorithms claimed to be independent of the number of substructures are misleading, as the total number of natural modes equals the number of degrees of freedom per substructure times the number of substructures. While the propagation and harmonic response problems have been discussed extensively, it is interesting to note that only a few algorithms are available for multiply coupled periodic structures in natural vibration analysis. This is due to the fact that two levels of eigenvalue problem have to be handled. One is the complex characteristic problem for the matrix difference equations even for undamped vibration, and the other is for the natural vibrations. Engels and Meirovitch [50] have suggested a transfer matrix method with the Leverrier algorithm to circumvent the former eigenvalue problem. They have pointed out that the resulting frequencies cannot be determined completely, i.e. some may be missed, particularly for multiple natural frequencies. Denke, Eide and Pickard [47] have suggested a trial-and-error method, in which the resonance peak of the harmonic response is computed. Unfortunately, no guarantee is made of the completeness of the natural frequencies thus obtained from a highly complicated non-symmetrical matrix determinantal equation.

A method is now presented for studying the natural vibration of a periodic structure with multiple coupling. Non-symmetrical matrices and a complex characteristic problem are avoided. Symmetrical system mass and stiffness matrices of orders not greater than the number of coupling coordinates between two adjacent substructures are obtained explicitly. Real arithmetic only is required and the resulting symmetric eigenvalue problem of small order may be solved by many standard algorithms. Internal degrees of freedom and various boundary conditions can be handled, and the Sturm sequence property of the resulting matrices ensures the completeness of natural frequencies. Although undamped free vibration is considered, damping effects and forced vibration problems can be studied by modal analysis as the natural modes are orthonormalized. The method is restricted to substructures having a planar symmetry.

3.9.1. Theory

When an undamped substructure is driven by boundary forces $\mathbf{Q}_b \sin \omega t$ producing boundary displacements $\mathbf{q}_b \sin \omega t$ and internal displacements $\mathbf{q}_a$, the equations of motion obtained from a finite element analysis are, in matrix form,

$$\begin{pmatrix} \mathbf{D}_{bb} & \mathbf{D}_{ba} \\ \mathbf{D}_{ab} & \mathbf{D}_{aa} \end{pmatrix} \begin{Bmatrix} \mathbf{q}_b \\ \mathbf{q}_a \end{Bmatrix} = \begin{pmatrix} \mathbf{K}_{bb} & \mathbf{K}_{ba} \\ \mathbf{K}_{ab} & \mathbf{K}_{aa} \end{pmatrix} \begin{Bmatrix} \mathbf{q}_b \\ \mathbf{q}_a \end{Bmatrix} - \omega^2 \begin{pmatrix} \mathbf{M}_{bb} & \mathbf{M}_{ba} \\ \mathbf{M}_{ab} & \mathbf{M}_{aa} \end{pmatrix} \begin{Bmatrix} \mathbf{q}_b \\ \mathbf{q}_a \end{Bmatrix} = \begin{Bmatrix} \mathbf{Q}_b \\ \mathbf{0} \end{Bmatrix} \qquad (3.9.1)$$

where $\mathbf{D}_{bb}$, $\mathbf{K}_{bb}$, $\mathbf{M}_{bb}$, etc. are the dynamic stiffness, stiffness and consistent mass matrices respectively. Upon eliminating $\mathbf{q}_a$ by the Gaussian method [9], we obtain

$$\mathbf{q}_a = -\mathbf{D}_{aa}^{-1}\mathbf{D}_{ab}\mathbf{q}_b, \qquad \mathbf{Q}_a = \mathbf{D}\mathbf{q}_b \qquad (3.9.2)$$

where

$$\mathbf{D} = \mathbf{D}_{bb} - \mathbf{D}_{ba}\mathbf{D}_{aa}^{-1}\mathbf{D}_{ab} \qquad (3.9.3)$$

and the condensed mass and stiffness matrices may be obtained by differentiation [31]:

$$\mathbf{M} = -\frac{\partial \mathbf{D}}{\partial \omega^2} = \mathbf{M}_{bb} - \mathbf{D}_{ba}\mathbf{D}_{aa}^{-1}\mathbf{M}_{ab} - \mathbf{M}_{ba}\mathbf{D}_{aa}^{-1}\mathbf{D}_{ab} + \mathbf{D}_{ba}\mathbf{D}_{aa}^{-1}\mathbf{M}_{aa}\mathbf{D}_{aa}^{-1}\mathbf{D}_{ab} \qquad (3.9.4)$$

$$\mathbf{K} = \mathbf{D} + \omega^2\mathbf{M} \qquad (3.9.5)$$

If the substructure has a symmetry plane P, then some of the coordinates, $\mathbf{s}$, of $\mathbf{q}_b$ are symmetrical with respect to P and others $\mathbf{r}$, are antisymmetrical (see Fig. 3.9.1). Consider the substructure of Figure 3.9.2a where the generalized displacements are $\mathbf{s}_1, \mathbf{r}_1, \mathbf{s}_2$ and $\mathbf{r}_2$.

Here, the subscripts 1 and 2 denote the left- and right-hand sides of P respectively. Figure 3.9.2b is the mirror image of Fig. 3.9.2a, with the mirror in parallel with P. If the dynamic stiffness equations for (a) are

$$\begin{bmatrix} \mathbf{D}_{11} & \mathbf{D}_{12} & \mathbf{D}_{13} & \mathbf{D}_{14} \\ & \mathbf{D}_{22} & \mathbf{D}_{23} & \mathbf{D}_{24} \\ & & \mathbf{D}_{33} & \mathbf{D}_{34} \\ \text{sym.} & & & \mathbf{D}_{44} \end{bmatrix} \begin{Bmatrix} \mathbf{s}_1 \\ \mathbf{s}_2 \\ \mathbf{r}_1 \\ \mathbf{r}_2 \end{Bmatrix} = \begin{Bmatrix} \mathbf{S}_1 \\ \mathbf{S}_2 \\ \mathbf{R}_1 \\ \mathbf{R}_2 \end{Bmatrix} \qquad (3.9.6)$$

then the dynamic stiffness equations for (b) will be

$$\begin{bmatrix} \mathbf{D}_{11} & \mathbf{D}_{12} & \mathbf{D}_{13} & \mathbf{D}_{14} \\ & \mathbf{D}_{22} & \mathbf{D}_{23} & \mathbf{D}_{24} \\ & & \mathbf{D}_{33} & \mathbf{D}_{34} \\ \text{sym.} & & & \mathbf{D}_{44} \end{bmatrix} \begin{Bmatrix} \mathbf{s}_1 \\ \mathbf{s}_2 \\ -\mathbf{r}_1 \\ -\mathbf{r}_2 \end{Bmatrix} = \begin{Bmatrix} \mathbf{S}_1 \\ \mathbf{S}_2 \\ -\mathbf{R}_1 \\ -\mathbf{R}_2 \end{Bmatrix} \qquad (3.9.7)$$

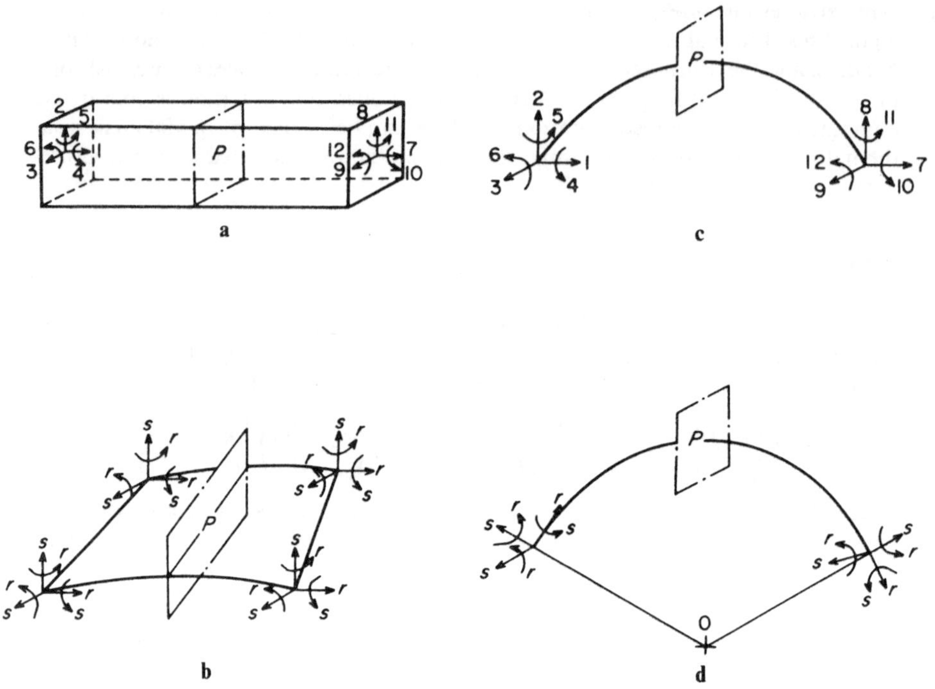

Fig. 3.9.1. The symmetrical and antisymmetrical coordinates: **a** beam type substructure; **b** shell type substructure; **c** curved substructure; **d** cyclic symmetrical substructure

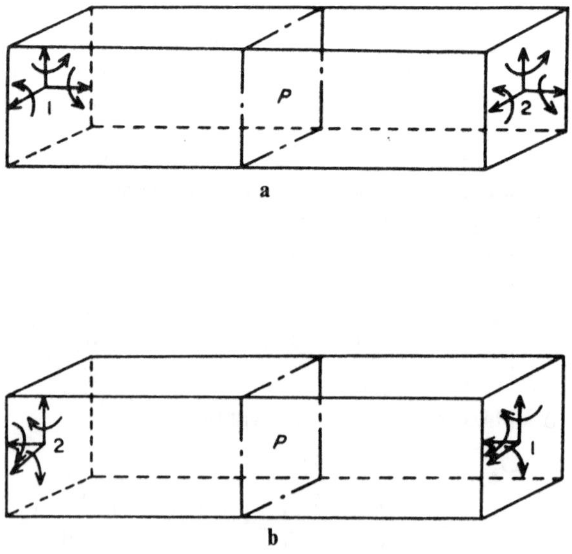

Fig. 3.9.2. **a** Substructure; **b** its mirror image

Equation (3.9.7) can be rearranged as

$$\begin{bmatrix} \mathbf{D}_{11} & \mathbf{D}_{12}^{\mathrm{T}} & -\mathbf{D}_{24} & -\mathbf{D}_{23} \\ & \mathbf{D}_{22} & -\mathbf{D}_{14} & -\mathbf{D}_{13} \\ & & \mathbf{D}_{44} & \mathbf{D}_{34}^{\mathrm{T}} \\ \text{sym.} & & & \mathbf{D}_{33} \end{bmatrix} \begin{Bmatrix} \mathbf{s}_1 \\ \mathbf{s}_2 \\ \mathbf{r}_1 \\ \mathbf{r}_2 \end{Bmatrix} = \begin{Bmatrix} \mathbf{S}_1 \\ \mathbf{S}_2 \\ \mathbf{R}_1 \\ \mathbf{R}_2 \end{Bmatrix} \tag{3.9.8}$$

Since substructures (a) and (b) are identical, then, according to reference [47],

$$\begin{aligned} &\mathbf{D}_{12} = \mathbf{D}_{12}^{\mathrm{T}}, \qquad \mathbf{D}_{34} = \mathbf{D}_{34}^{\mathrm{T}}, \qquad \mathbf{D}_{12} = \mathbf{D}_{22} \\ &\mathbf{D}_{33} = \mathbf{D}_{44}, \qquad \mathbf{D}_{13} = -\mathbf{D}_{24}, \qquad \mathbf{D}_{23} = -\mathbf{D}_{14} \end{aligned} \tag{3.9.9}$$

With these relations, n such identical substructures can be assembled by a formal finite element process to give the following difference equations for the kth interface station

$$\begin{pmatrix} \mathbf{D}_{12} & \mathbf{D}_{23} \\ -\mathbf{D}_{23}^{\mathrm{T}} & \mathbf{D}_{34} \end{pmatrix} \begin{Bmatrix} \mathbf{s}_{k-1} \\ \mathbf{r}_{k-1} \end{Bmatrix} + 2 \begin{pmatrix} \mathbf{D}_{11} & \mathbf{0} \\ \mathbf{0} & \mathbf{D}_{23} \end{pmatrix} \begin{Bmatrix} \mathbf{s}_k \\ \mathbf{r}_k \end{Bmatrix} + \begin{pmatrix} \mathbf{D}_{12} & -\mathbf{D}_{23} \\ \mathbf{D}_{23}^{\mathrm{T}} & \mathbf{D}_{34} \end{pmatrix} \begin{Bmatrix} \mathbf{s}_{k+1} \\ \mathbf{r}_{k+1} \end{Bmatrix}$$

$$= \begin{Bmatrix} \mathbf{S}_k \\ \mathbf{R}_k \end{Bmatrix} \qquad k = 1, 2, \ldots, n-1 \tag{3.9.10}$$

$$\begin{pmatrix} \mathbf{D}_{11} & \mathbf{D}_{13} \\ \mathbf{D}_{13}^{\mathrm{T}} & \mathbf{D}_{33} \end{pmatrix} \begin{Bmatrix} \mathbf{s}_0 \\ \mathbf{r}_0 \end{Bmatrix} + \begin{pmatrix} \mathbf{D}_{12} & -\mathbf{D}_{23} \\ \mathbf{D}_{23}^{\mathrm{T}} & \mathbf{D}_{34} \end{pmatrix} \begin{Bmatrix} \mathbf{s}_1 \\ \mathbf{r}_1 \end{Bmatrix} = \begin{Bmatrix} \mathbf{S}_0 \\ \mathbf{R}_0 \end{Bmatrix} \tag{3.9.11}$$

$$\begin{pmatrix} \mathbf{D}_{12} & \mathbf{D}_{23} \\ -\mathbf{D}_{23}^{\mathrm{T}} & \mathbf{D}_{34} \end{pmatrix} \begin{Bmatrix} \mathbf{s}_{n-1} \\ \mathbf{r}_{n-1} \end{Bmatrix} + \begin{pmatrix} \mathbf{D}_{11} & -\mathbf{D}_{13} \\ -\mathbf{D}_{13}^{\mathrm{T}} & \mathbf{D}_{33} \end{pmatrix} \begin{Bmatrix} \mathbf{s}_n \\ \mathbf{r}_n \end{Bmatrix} = \begin{Bmatrix} \mathbf{S}_n \\ \mathbf{R}_n \end{Bmatrix} \tag{3.9.12}$$

These difference equations are solved for some periodic boundary conditions in the following section and for more general boundary conditions at stations $k = 0$ and $k = n$ later. In the following analysis, it is understood that there is a mass matrix and a stiffness matrix associated with every **D** considered, according to Eqs (3.9.4) and (3.9.5).

Equations (3.9.10)–(3.9.12) may be combined into the familiar form

$$\bar{\mathbf{D}}\mathbf{q} = \mathbf{Q} \tag{3.9.13}$$

where

$$\mathbf{q} = [\mathbf{s}_0^{\mathrm{T}}\mathbf{r}_0^{\mathrm{T}}\mathbf{s}_1^{\mathrm{T}}\mathbf{r}_1^{\mathrm{T}}\ldots\mathbf{s}_n^{\mathrm{T}}\mathbf{r}_n^{\mathrm{T}}]^{\mathrm{T}}, \qquad \mathbf{Q} = [\mathbf{S}_0^{\mathrm{T}}\mathbf{R}_0^{\mathrm{T}}\mathbf{S}_1^{\mathrm{T}}\mathbf{R}_1^{\mathrm{T}}\ldots\mathbf{S}_n^{\mathrm{T}}\mathbf{R}_n^{\mathrm{T}}]^{\mathrm{T}} \qquad (3.9.14, 15)$$

3.9.2. Periodic Boundary Conditions

Equations (3.9.10) are to be solved for a system having an infinite number of identical substructures arranged in a uniform manner such that the planes of symmetry of the substructures are parallel to each other. The system is studied by isolating n consecutive substructures.

Let

$$\mathbf{s}_k = \sum_j \mathbf{e}_j \sin jk\alpha, \qquad \mathbf{r}_k = \sum_j \mathbf{g}_j \cos jk\alpha \tag{3.9.16}$$

where $\alpha = \pi/n$, $\mathbf{e}_i$ and $\mathbf{g}_i$ are vectors to be determined, and the summation is understood to range from 0 to n. The summation operator $\sum^*$ is defined by

$$\sum_k {}^*a_k = \tfrac{1}{2}(a_0 + a_n) + \sum_{k=1}^{n=1} a_k \tag{3.9.17}$$

Then it can be proved that

$$\begin{aligned}\sum_k {}^*a_k \sin ik\alpha \sin jk\alpha &= 0 \quad \text{when } i \neq j,\\ &= n/2 \quad \text{when } i = j\end{aligned} \tag{3.9.18}$$

$$\left.\begin{aligned}\sum_k {}^*a_k \cos ik\alpha \cos jk\alpha &= 0 \quad \text{when } i \neq j,\\ &= n/2 \quad \text{when } i = j \neq 0 \text{ or } n\\ &= n \quad \text{when } i = j = 0 \text{ or } n\end{aligned}\right\} \tag{3.9.19}$$

With the aid of expressions (3.9.18) and (3.9.19), Eqs (3.9.16) transform equation (3.9.13) to

$$\mathbf{Dx} = \mathbf{X} \tag{3.9.20}$$

where $\mathbf{D} = \operatorname{diag}[\mathbf{D}_j]$, $\mathbf{x} = \operatorname{col}\{\mathbf{x}_j\}$, $\mathbf{X} = \operatorname{col}\{\mathbf{X}_j\}$, $j = 0, 1, \ldots, n$,

$$\mathbf{D}_j = \begin{pmatrix} \mathbf{D}_{11} + \mathbf{D}_{12}\cos j\alpha & \mathbf{D}_{23}\sin j\alpha \\ \mathbf{D}_{23}^{\mathrm{T}}\sin j\alpha & \mathbf{D}_{33} + \mathbf{D}_{34}\cos j\alpha \end{pmatrix} \mathbf{x}_j = \begin{Bmatrix} \mathbf{e}_j \\ \mathbf{g}_j \end{Bmatrix}, \qquad \mathbf{X}_j = \begin{Bmatrix} \mathbf{E}_j \\ \mathbf{G}_j \end{Bmatrix}$$

$$\mathbf{E}_j = \frac{1}{n}\sum_k {}^*\mathbf{S}_k \sin jk\alpha, \qquad \mathbf{G}_j = \frac{1}{n}\sum_k {}^*\mathbf{R}_k \cos jk\alpha,$$

$$\mathbf{G}_0 = \frac{2}{n}\sum_k {}^*\mathbf{R}_k, \qquad \mathbf{G}_n = \frac{2}{n}\sum_k {}^*(-1)^k\mathbf{R}_k$$

Since $\mathbf{D}$ has been decomposed into block diagonal form, Eq. (3.9.20) may be written in the uncoupled form

$$\mathbf{D}_j\mathbf{x}_j = \mathbf{X}_j, \qquad j = 0, 1, \ldots, n \tag{3.9.21}$$

These uncoupled equations may be solved directly for displacements in a harmonic analysis, or static analysis when $\omega = 0$. If free vibration is considered, i.e. when $\mathbf{X} = \mathbf{0}$, many efficient algorithms are available. Here it is suggested that the eigenvalue problem

$$\mathbf{D}_j(\omega)\mathbf{x}_j = \mathbf{0}, \qquad j = 0, 1, \ldots, n \tag{3.9.22}$$

be solved in the following stages: locate an approximate natural frequency ω_0 by an infallible Sturm sequence method [32]; obtain the corresponding mode shape by inverse iteration; improve the natural frequency by Rayleigh's quotient [33]. If an eigenvector $\mathbf{x}_j$ of Eq. (3.9.22) is obtained, then the corresponding eigenvector of Eq. (3.9.20) is

$$\mathbf{x} = [\mathbf{0}^{\mathrm{T}}\mathbf{0}^{\mathrm{T}}\ldots\mathbf{0}^{\mathrm{T}}\mathbf{x}_j^{\mathrm{T}}\mathbf{0}^{\mathrm{T}}\ldots\mathbf{0}^{\mathrm{T}}]^{\mathrm{T}} \tag{3.9.23}$$

and, for the same mode,

$$\mathbf{s}_k = \mathbf{e}_j \sin jk\alpha, \qquad \mathbf{r}_k = \mathbf{g}_j \sin jk\alpha \tag{3.9.24}$$

If the generalized mass of $\mathbf{x}$ is normalized to unity, it can be shown that, under the transformation (3.9.16), the generalized mass of $\mathbf{g}$ in equation (3.9.13) is n.

To the author's knowledge, Eqs (3.9.21) and (3.9.22) are new. If it is not required to eliminate the internal degrees of freedom, $\mathbf{D}_j = \mathbf{K}_j - \omega^2\mathbf{M}_j$, where $\mathbf{K}_j$ and $\mathbf{M}_j$ are constant symmetrical matrices, and Eqs (3.9.22) may be solved by many standard algorithms. When complete solutions for the natural modes of the original system are required, n such eigenvalue problems have to be solved. However, since the substructure matrices are obtained by approximating the continuous system by one with a finite number of degrees of freedom, such as in the finite element method, very high frequency modes are meaningless. If a fixed number of modes are required, the present algorithm is independent of the number of substructures.

Equation (3.9.13) may also be uncoupled in a similar manner into a number of smaller systems for another class of periodic boundary conditions where $\mathbf{s}_0 = \mathbf{s}_n$ and $\mathbf{r}_0 = \mathbf{r}_n$.

Example 3.9.1

As the number of elements does not affect the computational effort greatly and the roundoff errors are not significant, an immediate application of the theory is to study the convergence of some finite elements with respect to the number of elements.

Consider a rod element whose stiffness matrix is

$$\mathbf{K} = \frac{EA}{l}\begin{pmatrix} 1 & -1 \\ -1 & 1 \end{pmatrix}$$

and whose mass matrices are

$$\mathbf{M}_c = \frac{EA}{\omega^2 l}\lambda\varepsilon^2\begin{pmatrix} 2 & 1 \\ 1 & 2 \end{pmatrix}, \qquad \mathbf{M}_l = \frac{EA}{\omega^2 l}\lambda\varepsilon^2\begin{pmatrix} 3 & 0 \\ 0 & 3 \end{pmatrix}, \qquad \mathbf{M}_a = \frac{1}{2}(\mathbf{M}_c + \mathbf{M}_l)$$

for consistent mass, lumped mass and averaged mass models respectively, where $\lambda\varepsilon^2 = pl^2\omega^2/6E$, $\varepsilon = j\alpha = j\pi/n$, and ρ, A, l, ω and E have their usual meanings. If the rod considered is simply supported at the ends and is represented by n elements, then corresponding to Eq. (3.9.20) the only non-vanishing matrices (scalars, in this case) are $D_{33} = EA(1 - 2\lambda\varepsilon^2)/l$ and $D_{34} = -EA(1 + \lambda\varepsilon^2)/l$ for the consistent mass model. From the determinant $D_j = 0$ for the natural frequencies, λ is determined from

$$1 - 2\lambda\varepsilon^2 - (1 + \lambda\varepsilon^2)\cos\varepsilon = 0$$

Similarly, frequency equations for lumped mass and averaged mass models are, respectively,

$$1 - 3\lambda\varepsilon^2\cos\varepsilon = 0, \qquad 1 - \frac{5\lambda\varepsilon^2}{2} - (1 + \tfrac{1}{2}\lambda\varepsilon^2)\cos\varepsilon = 0$$

The lowest solutions of these equations are, respectively,

$$\lambda_c = \left(1 + \frac{\varepsilon^2}{12} + \frac{\varepsilon^4}{360} - \frac{17\varepsilon^6}{30\,240} + \cdots\right)\Big/6$$

$$\lambda_l = \left(1 - \frac{\varepsilon^2}{12} + \frac{\varepsilon^4}{360} - \frac{\varepsilon^6}{20\,160} + \cdots\right)\Big/6$$

$$\lambda_a = \left(1 - \frac{\varepsilon^4}{240} - \frac{\varepsilon^6}{6048} + \cdots\right)\Big/6$$

Compared with the exact value $\lambda = \frac{1}{6}$, one has the following estimates for the relative errors of the frequency parameters,

Table 3.9.1. Percentage errors in λ of rod models

Mode j	Consistent mass			Lumped mass			Averaged mass		
	(1)	(2)	(3)	(1)	(2)	(3)	(1)	(2)	(3)
1	0.09168	0.9139	0.09142	−0.09134	−0.09139	−0.09135	−0.00005	−0.00005	−0.00005
2	0.36605	0.36554	0.36607	−0.36499	−0.36554	−0.36501	−0.00081	−0.00080	−0.00080
3	0.82509	0.82247	0.82512	−0.81978	−0.82247	−0.81977	−0.00411	−0.00406	−0.00407
4	1.4706	1.4622	1.4704	−1.4536	−1.4622	−1.4536	−0.01290	−0.01283	−0.01292
5	2.3049	2.2846	2.3044	−2.2638	−2.2846	−2.2639	−0.03164	−0.03132	−0.03166
6	3.3313	3.2899	3.3299	−3.2469	−3.2899	−3.2469	−0.06592	−0.06494	−0.06596
10	9.4269	9.1385	9.3984	−8.8109	−9.1385	−8.8110	−0.52102	−0.50108	−0.52288
15	21.585	20.562	21.408	−18.943	−20.562	−18.945	−2.7316	−2.5367	−2.7851
20	37.956	36.554	37.154	−31.608	−36.554	−31.628	−8.8110	−8.0172	−9.4127

(1) Using 30 finite elements.
(2) Using the first terms of Eqs (3.9.9)–(3.9.11).
(3) Using up to ε^6 terms of Eqs (3.9.9)–(3.9.11).

$$\delta_c = \frac{\varepsilon^2}{12} + \frac{\varepsilon^4}{360} - \frac{17\varepsilon^6}{30\,240} + \cdots \tag{3.9.25}$$

$$\delta_l = -\frac{\varepsilon^2}{12} + \frac{\varepsilon^4}{360} - \frac{\varepsilon^6}{20\,160} + \cdots \tag{3.9.26}$$

$$\delta_a = -\frac{\varepsilon^4}{240} - \frac{\varepsilon^6}{6048} + \cdots \tag{3.9.27}$$

These formulae are confirmed by a full finite element analysis for a uniform rod with $l = \rho = A = E = 1$ and $n = 30$, and the results are compared in Table 3.9.1. The computation was done on an ICL 1904s computer using 22-digit arithmetic. The first terms of expressions (3.9.25) and (3.9.26) are in agreement with those given in Refs. [52, 53] by an exact dynamic stiffness method, and differ in sign from those obtained in Ref. [54] by the equivalent energy method. The higher order terms of the error estimates have not appeared elsewhere.

Przemieniecki's model [55], where

$$\mathbf{K} = \frac{EA}{l}\begin{pmatrix} 1 & -1 \\ -1 & 1 \end{pmatrix} - \frac{\omega^4\rho^2Al^3}{360E}\begin{pmatrix} 8 & 7 \\ 7 & 8 \end{pmatrix}, \qquad \mathbf{M} = \frac{\rho Al}{6}\begin{pmatrix} 2 & 1 \\ 1 & 2 \end{pmatrix} - \frac{\omega^4\rho^2Al^3}{180E}\begin{pmatrix} 8 & 7 \\ 7 & 8 \end{pmatrix}$$

can be studied similarly. The lowest frequency parameter λ and the relative error are determined to be

$$\lambda_p = \left(1 + \frac{\varepsilon^4}{120} - \frac{17\varepsilon^6}{5800} + \cdots\right)\Big/6, \qquad \delta_p = \frac{\varepsilon^4}{120} - \frac{7\varepsilon^6}{5800} + \cdots$$

Therefore, this complicated model, involving frequency-dependent matrices, is not necessarily more accurate than the much simpler averaged mass model.

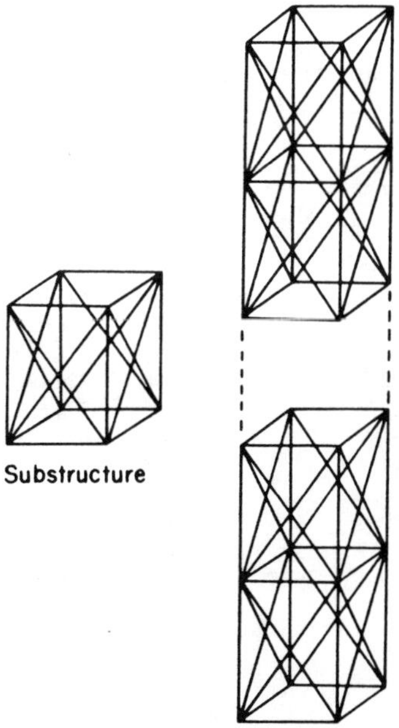

Fig. 3.9.3. A simply supported system with ten identical substructures

Example 3.9.2

Consider, as a second example, the frame system shown in Fig. 3.9.3 consisting of ten identical substructures. Each beam element is a steel rod with diameter $0.02\,\mathrm{m}^2$. The lengths of the beam elements with axes in the X, Y and Z directions are 2.0 m, 1.8 m and 2.5 m respectively. The inclining members have compatible lengths so that all the members are initially stress free at rest. The unconstrained system has 264 d.o.f. when the boundary stations are simply supported. All the 264 modes were computed with matrices of order 24 and the results are listed in Table 3.9.2. The first 36 modes were calculated by a full finite element analysis to give the comparison of the natural frequencies shown in Table 3.9.3. Most of the results are very encouraging. Severe

Table 3.9.2. The 264 natural frequencies ($\mathrm{rad\,s^{-1}}$) of Example 3.9.2

0	0	0	0	37.52	42.35	137.1	141.2	140.0	152.7	174.1	182.6
264.0	265.7	288.9	299.9	300.0	300.3	302.4	302.9	304.7	306.2	307.1	308.1
314.4	314.9	316.0	318.3	320.5	322.2	326.7	327.6	331.5	332.8	333.5	336.8
338.6	339.2	339.3	339.7	343.0	344.1	346.0	347.0	348.4	348.9	350.1	352.8
354.5	356.4	359.1	359.8	360.0	360.3	363.3	363.3	363.8	369.3	396.5	369.7
377.5	384.7	391.1	393.3	396.2	398.9	400.2	401.3	401.8	404.1	406.9	408.1
410.9	413.6	415.0	416.1	416.4	422.4	423.2	432.3	432.4	436.2	438.7	438.9
466.5	448.1	454.1	459.7	460.2	461.7	463.2	464.4	465.6	466.3	470.1	471.3
481.4	484.0	485.7	494.7	530.2	539.7	540.6	540.7	542.6	543.3	547.9	553.9
557.4	560.0	560.4	561.0	563.1	564.1	569.8	570.8	571.0	574.1	574.4	575.1
581.3	583.4	586.5	586.5	591.9	593.9	600.6	601.0	605.9	606.2	607.1	614.4
620.9	622.4	623.2	633.7	649.8	656.9	662.3	675.7	678.9	679.5	687.0	704.6
724.8	750.0	768.9	783.7	792.2	807.2	808.6	812.8	816.8	831.5	832.2	852.5
892.6	929.7	933.9	950.7	976.9	996.7	1061	1082	1123	1145	1187	1199
1225	1237	1261	1307	1327	1331	1334	1359	1362	1427	1491	1513
1592	1645	1694	1737	1786	1789	1855	1978	2014	2179	2184	2185
2339	2382	2402	2422	2450	2465	2470	2526	2546	2553	2558	2570
2573	2576	2577	2603	2604	2606	2612	2617	2618	2625	2669	2685
2691	2692	2693	2697	2704	2719	2761	2796	2868	2896	2898	2959
2969	2998	3040	3057	3058	3062	3062	3063	3064	3083	3108	3118
3119	3126	3132	3140	3149	3152	3155	3163	3181	3186	3191	3204
3229	3252	3256	3268	3273	3320	3371	3404	3415	3464	3660	3731

Table 3.9.3. Comparison of the first 36 modes

Mode	(1)	(2)	Mode	(1)	(2)	Mode	(1)	(2)
1	0	–[a]	13	264.0418	264.0418	25	314.4099	314.4099
2	0	–	14	265.7299	265.7299	26	314.8707	315.0227
3	0	–	15	288.9278	288.9278	27	316.0244	316.0250
4	0	–	16	299.9190	299.9199	28	318.2979	318.3000
5	37.5177	37.5177	17	300.0254	300.0255	29	320.4968	320.4976
6	42.3533	42.3533	18	300.2778	300.2779	30	322.2253	323.1162
7	137.1328	127.9503[b]	19	302.4169	302.4180	31	326.6653	327.2928
8	138.6693	–	20	302.9405	302.9483	32	327.5650	–
9	139.9730	139.9730	21	304.6617	304.6617	33	331.5418	331.6011
10	152.7073	146.9168[b]	22	306.2060	306.2060	34	332.8089	332.8587
11	174.1168	169.0613[b]	23	307.6627	307.6757	35	333.4717	333.5282
12	182.6064	182.6064	24	308.1406	–	36	336.7568	336.7851

[a] Rigid body modes or partial modes.
[b] Modes with large axial displacements.
(1) Natural frequencies calculated by present method.
(2) Natural frequencies calculated by finite element.

discrepancy exists for some modes. A careful comparison of the modal shapes reveals that, in order to simulate the simply supported boundary conditions in the full analysis, additional axial restrictions on the boundary have to be imposed to eliminate the rigid body modes, and as a result, some of the axial modes are distorted. Unfortunately, it has not been possible for the author to simulate the perfectly simply supported conditions for the finite element model without violating the axial and/or torsional freedoms by adding springs. The modal density, i.e. the number of modes per frequency, is plotted in Fig. 3.9.4. This is useful for statistical vibration analysis and is not sensitive to boundary conditions.

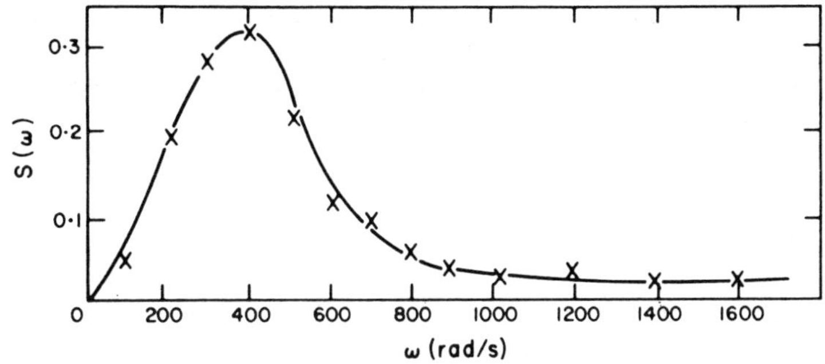

Fig. 3.9.4. Modal density

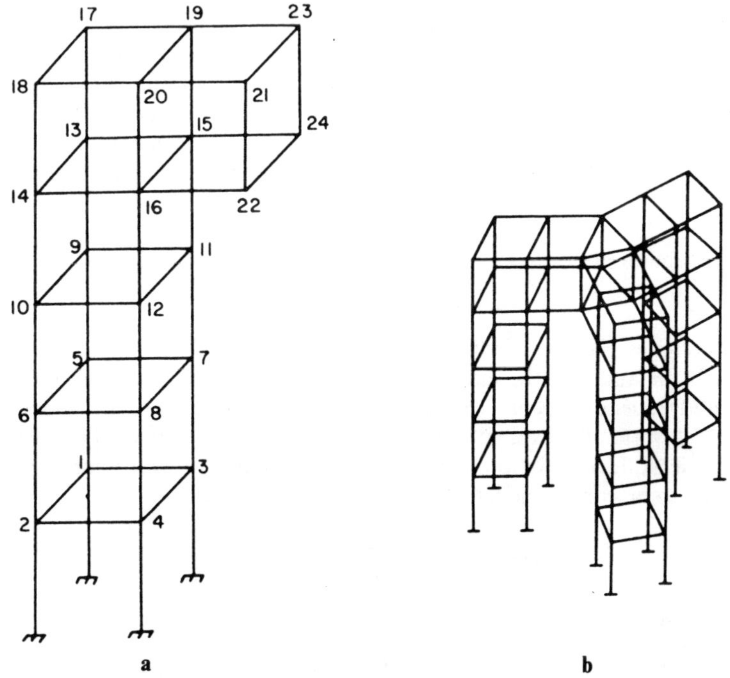

Fig. 3.9.5. Example 3.9.3: **a** substructure; **b** system with three substructures

It is noted that the results for very high modes are not necessarily realistic due to the finite element approximation.

Example 3.9.3

With some minor modifications, a set of equations similar to that of expression (3.9.20) is obtained for circular symmetrical structures. Consider a substructure as shown in Fig. 3.9.5a. The internal d.o.f. are first eliminated according to the method in Ref. [9] or [4], and the 24 generalized coordinates corresponding to nodes 21–24 are retained. If n such substructures are arranged axisymmetrically as shown in Fig. 3.9.5b for $n = 3$, then the lowest natural frequencies may be computed by the present method with matrices of order 12 independent of n. The first six natural frequencies for $n = 3(2)11$ have been calculated by a full finite element analysis and the results are identical to those listed.

3.9.3. Finite Systems

So far periodic systems with an infinite number of substructures have been considered and also systems with periodic boundary conditions. A system with periodic boundary conditions at its end stations is called a simply supported system. Although it is possible, by numerical methods such as embedding [56], to solve a system of matrix difference equations with general boundary conditions, it is not a simple task. In what follows here the system solution is to be obtained by using eigenfunctions of the corresponding simply supported system. System boundary conditions may be obtained either by relaxing or constraining the end displacements of the corresponding simply supported system. An efficient method of relaxing the boundary displacements of a structural system has been developed by the author [9]. Constraining of the boundary displacements will be considered here.

Let the normal modes of the simply supported system be ψ and λ^2, where

$$\overline{\mathbf{K}}\psi = \overline{\mathbf{M}}\psi\lambda^2 \tag{3.9.28}$$

and ψ is normalized, so that

$$\psi^{\mathrm{T}}\overline{\mathbf{K}}\psi = \lambda^2, \qquad \psi^{\mathrm{T}}\overline{\mathbf{M}}\psi = I \tag{3.9.29}$$

Here ψ includes the modal displacements $\mathbf{q}_b = [\mathbf{sr}]^{\mathrm{T}}$ (cf. Eq. (3.9.2)), at stations $k = 0, 1, \ldots, n$, and also the modal displacements $\mathbf{q}$ within all the substructures. $\overline{\mathbf{K}}$ and $\overline{\mathbf{M}}$ are frequency independent as no internal coordinate has been eliminated. It is shown below that the explicit forms of $\overline{\mathbf{K}}$ and $\overline{\mathbf{M}}$ are not required. The corresponding dynamic stiffness relation is $\overline{\mathbf{D}}\mathbf{q} = \mathbf{Q}$. If some of the coordinates $\mathbf{q}_m$ of $\mathbf{q} = [\mathbf{q}_m\mathbf{q}_s]^{\mathrm{T}}$ are to be constrained, then

$$\overline{\mathbf{D}}\mathbf{q} = \begin{pmatrix} \mathbf{D}_{mm} & \mathbf{D}_{ms} \\ \mathbf{D}_{sm} & \mathbf{D}_{ss} \end{pmatrix} \begin{Bmatrix} \mathbf{q}_m \\ \mathbf{q}_s \end{Bmatrix} = \begin{Bmatrix} \mathbf{Q}_m \\ \mathbf{0} \end{Bmatrix} \tag{3.9.30}$$

Eliminating $\mathbf{q}_s$ gives

$$\mathbf{D}^{-1}\mathbf{Q}_m = \mathbf{q}_m \tag{3.9.31}$$

where

$$\mathbf{D} = \mathbf{D}_{mm} - \mathbf{D}_{ms}\mathbf{D}_{ss}^{-1}\mathbf{D}_{sm} \tag{3.9.32}$$

is a frequency-dependent matrix. If $\mathbf{q}_m$ is constrained to zero, the resulting eigenvalue problem is

$$\mathbf{D}^{-1}\mathbf{Q}_m = \mathbf{0} \tag{3.9.33}$$

From this, the frequency determinant is

$$\det[\mathbf{D}^{-1}(\omega)] = 0 \tag{3.9.34}$$

and the corresponding solution for $\mathbf{Q}_m$ is the modal clamping force required to give $\mathbf{q}_m = 0$. Since $\mathbf{D}_{ss}$ is a large order matrix, the inversion $\mathbf{D}_{ss}^{-1}$ required for each frequency considered in solving Eq. (3.9.34) is expensive and should be avoided. Expanding $\mathbf{D}^{-1}$ by Taylor's series up to ω^4 gives

$$\mathbf{D}^{-1}(\omega) = \mathbf{K}_0^{-1} + \omega^2\mathbf{K}_0^{-1}\mathbf{M}_0\mathbf{K}_0^{-1} + \omega^4\mathbf{R}(\omega) \tag{3.9.35}$$

where $\mathbf{R}$ is a remainder, $\mathbf{K}_0 = \mathbf{D}(0)$, and $\mathbf{M}_0 = \mathbf{M}(0)$. Also, from Eqs (3.9.29), $[\bar{\mathbf{D}}]^{-1} = \psi[\lambda^2 - \omega^2\mathbf{I}]\psi^{\mathrm{T}}$, or

$$\begin{pmatrix} \mathbf{D}_{mm} & \mathbf{D}_{ms} \\ \mathbf{D}_{sm} & \mathbf{D}_{ss} \end{pmatrix}^{-1} \begin{Bmatrix} \mathbf{Q}_m \\ \mathbf{0} \end{Bmatrix} = \begin{Bmatrix} \psi_m \\ \psi_s \end{Bmatrix} [\lambda^2 - \omega^2\mathbf{I}]^{-1} \begin{Bmatrix} \psi_m \\ \psi_s \end{Bmatrix}^{\mathrm{T}} \begin{Bmatrix} \mathbf{Q}_m \\ \mathbf{0} \end{Bmatrix} = \begin{Bmatrix} \mathbf{q}_m \\ \mathbf{q}_s \end{Bmatrix}$$

Eliminating $\mathbf{q}_s$ gives

$$\mathbf{D}^{-1}\mathbf{Q}_m = \mathbf{q}_m \tag{3.9.36}$$

$$\mathbf{D}^{-1} = \psi_m[\lambda^2 - \omega^2\mathbf{I}]\psi_m^{\mathrm{T}} \tag{3.9.37}$$

However, $1/(\lambda_i^2 - \omega^2) = \lambda_i^{-2} + \omega^2\lambda_i^{-4} + \omega^4/[\lambda_i^4(\lambda_i^2 - \omega^2)]$, so that

$$\mathbf{D}^{-1} = \psi_m\lambda^{-2}\psi_m^{\mathrm{T}} + \omega^2\psi_m\lambda^{-4}\psi_m^{\mathrm{T}} + \omega^4\psi_m[\lambda_i^{-4}(\lambda_i^2 - \omega^2)]\psi_m^{\mathrm{T}} \tag{3.9.38}$$

where $\lambda^2 = \mathrm{diag}[\lambda_j^2]$. Comparing like powers of ω in Eqs (3.9.35) and (3.9.38) gives

$$\mathbf{D}^{-1} = \mathbf{K}_0^{-1} + \omega^2\mathbf{K}_0^{-1}\mathbf{M}_0\mathbf{K}_0^{-1} + \omega^4\psi_m[\lambda_i^{-4}(\lambda_i^2 - \omega^2)^{-1}]\psi_m^{\mathrm{T}} \tag{3.9.39}$$

Note that in this exact form all matrices are of reduced size and the convergence rate of the last term is very fast with respect to the number of modes. Since only the active components ψ_m at the end stations of ψ are required, internal modal displacements of the system need not be determined explicitly.

With $\mathbf{Z} = \mathbf{D}^{-1}$, the eigenvalue problem to be solved is

$$\mathbf{Z}\mathbf{Q}_m = \mathbf{0} \tag{3.9.40}$$

where $\mathbf{Q}_m$ is the clamping force eigenvector. Differentiating $\mathbf{Z}$ with respect to ω^2 gives

$$\mathbf{Y} = \frac{\partial \mathbf{Z}}{\partial \omega^2} = -\mathbf{ZMZ} \tag{3.9.41}$$

Since $\mathbf{M}$ is positive-definite, so is $\mathbf{Y}$, for non-vanishing $\mathbf{Z}$. If an approximate eigenvalue ω_0 is obtained from a natural frequency bracketing method [57], the force eigenvector may be computed by the inverse iteration process, $\mathbf{Z}(\omega_0)\mathbf{Q}_m^{(i+1)} = \rho\mathbf{Y}(\omega_0)\mathbf{Q}_m^{(i)}$, where ρ is a normalization factor such that $\mathbf{Q}_m^{\mathrm{T}}\mathbf{Y}\mathbf{Q}_m = 1$, and an improved natural frequency $\bar{\omega}$ is then given by $\bar{\omega}^2 = \omega_0^2 + \rho$.

Example 3.9.4

Consider a simply supported beam whose normal modes are

$$\varphi_i(x) = \sqrt{\frac{2}{\rho Al}}\sin\left(\frac{i\pi x}{l}\right), \qquad i = 1, 2, \ldots, \qquad \omega_i^2 = \frac{(i\pi)^4}{\lambda}, \qquad \lambda = \frac{\rho Al^4}{EI}$$

where E, I, l, ρ and A have their usual meanings. If the end rotations are chosen as active coordinates, then the normal modes are represented by

$$\psi_m = \frac{2}{\rho Al}\frac{\pi}{l}\begin{pmatrix} 1 & 2 & \ldots & i & \ldots \\ -1 & 2 & \ldots & (-1)^i i & \ldots \end{pmatrix}$$

If the beam is idealized by two finite elements, then

$$\bar{\mathbf{K}} = \frac{8EI}{l^2}\begin{bmatrix} l^2 & & & \text{sym.} \\ -3l & 24 & & \\ \frac{1}{2}l^2 & 0 & 2l^2 & \\ 0 & 3l & \frac{1}{2}l^2 & l^2 \end{bmatrix}$$

With the first and the last coordinates as active, $\bar{\mathbf{K}}$ is condensed to

$$\mathbf{K}_0 = \mathbf{K}_{mm} - \mathbf{K}_{ms}\mathbf{K}_{ss}^{-1}\mathbf{K}_{sm} = \frac{EI}{l}\begin{pmatrix} 4 & 2 \\ 2 & 4 \end{pmatrix}$$

which is identical to the stiffness matrix for a beam idealized by one finite element. Therefore, $\mathbf{K}_0$ remains unchanged regardless of the number of finite elements if the same active coordinates are used. Similarly, the consistent mass matrix is

$$\mathbf{M}_0 = \frac{\rho Al^3}{420}\begin{pmatrix} 4 & -3 \\ -3 & 4 \end{pmatrix}$$

Therefore,

$$\mathbf{Z} = \mathbf{D}^{-1} = \frac{l}{EI}\left(\frac{1}{6}\begin{pmatrix} 2 & -1 \\ -1 & 2 \end{pmatrix} + \frac{\omega^2\lambda}{15\,120}\begin{pmatrix} 32 & -31 \\ -31 & 32 \end{pmatrix}\right)$$
$$+ \omega^4\psi_m[\omega_i^{-4}(\omega_i^2 - \omega^2)^{-1}]\psi_m^{\mathrm{T}}$$

Table 3.9.4. Convergence to the natural frequency parameter; figures in parentheses are percentage errors

Mode	No. of terms	Clamped–hinged beam λ_i^2	Clamped–clamped beam λ_i^2	Example 5 ω_i
1	1	15.461 (0.28)	22.405 (0.16)	78.094 (0.54)
	2	15.419 (0.00)	22.390 (0.08)	77.685 (0.09)
	3	15.418 (0.00)	22.373 (0.00)	77.678 (0.00)
	Exact	15.418	22.373	77.678
2	2	50.316 (0.70)	61.861 (0.28)	86.270 (0.02)
	3	49.981 (0.04)	61.754 (0.12)	86.256 (0.00)
	4	49.965 (0.00)	61.685 (0.00)	86.255 (0.00)
	Exact	49.965	61.685	86.255
3	3	105.29 (1.00)	121.57 (0.54)	150.59 (0.75)
	4	104.33 (0.08)	120.93 (0.00)	149.74 (0.18)
	5	104.24 (0.00)	120.91 (0.00)	149.52 (0.03)
	Exact	104.24	120.91	149.47

The natural frequencies of a clamped–clamped beam and a clamped–hinged beam can now be calculated, respectively, by letting $\det \mathbf{Z} = 0$ and by equating the first element of $\mathbf{Z}$ to zero. The convergence for the first three modes is tabulated in Table 3.9.4, where $\lambda_i^4 = \omega_i^4 \rho A l^2/EI$ and the numbers in parentheses are the percentage errors. It is shown that the convergence is very rapid and that even higher modes may be determined accurately.

Example 3.9.5

Consider the case when the boundaries at junctions $j = 0$ and n of the frame system in Example 3.9.2 are clamped. The condensed matrices $\mathbf{K}_0$ and $\mathbf{M}_0$ of the corresponding simply supported system may be determined by the fact that given harmonic forces with amplitudes $\mathbf{R}_0$ and $\mathbf{R}_n$ and frequency ω, the corresponding boundary displacement amplitudes $\mathbf{r}_0$ and $\mathbf{r}_n$ may be computed by the method mentioned in Sect. 3.9.3, i.e. a relation of the form

$$[\mathbf{D}(\omega)]\begin{Bmatrix}\mathbf{r}_0\\ \mathbf{r}_n\end{Bmatrix} = \begin{Bmatrix}\mathbf{R}_0\\ \mathbf{R}_n\end{Bmatrix} \tag{3.9.42}$$

may be assumed. With the unit force amplitude vectors, $[1\ \ 0\ \ 0\ \ \ldots\ \ 0]^T$, $[0\ \ 1\ \ 0\ \ \ldots\ \ 0]^T$, etc., applied to the right-hand side of Eq. (3.9.42), and the force vectors collected in matrix form, Eq. (3.9.42) gives

$$\mathbf{D}(\omega)\mathbf{Z}(\omega) = \mathbf{I} \tag{3.9.43}$$

where $\mathbf{Z}(\omega)$ is the collection of the response vectors for each loading case. Now, if $\omega = 0$, $\mathbf{Z}(0) = \mathbf{K}_0^{-1}$, then in the limit

$$\lim_{\omega\to 0} \frac{1}{\omega^2}(\mathbf{Z}(\omega) - \mathbf{K}_0^{-1}) = \mathbf{K}_0^{-1}\mathbf{M}_0\mathbf{K}_0^{-1} \tag{3.9.43}$$

Therefore, $\mathbf{K}_0^{-1}$ and $\mathbf{K}_0^{-1}\mathbf{M}_0\mathbf{K}_0^{-1}$ are obtained without actually carrying out the condensation process. With $\mathbf{K}_0^{-1}$, $\mathbf{K}_0^{-1}\mathbf{M}_0\mathbf{K}_0^{-1}$ and the natural modes for the corresponding simply supported conditions obtained previously, Eq. (3.9.39) may be written out explicitly. The natural frequencies computed by the Sturm sequence method are listed in Table 3.9.5 for different modes and different number of terms, the results being compared with those from a full finite element analysis. It is noted that the convergence behaviour is similar to that for the case of a simple beam.

It is well known that the distribution of natural frequencies in the high frequency

Table 3.9.5. Natural frequencies (rad s^{-1}) of Example 3.9.5

Mode	No. of substructures					
	3	5	7	9	11	13
1	6.3036[a]	6.3035[a]	6.3018[a]	6.2995[a]	6.2987[a]	6.2964[a]
2	7.3238	7.2167	7.1348	7.0667	7.0420	6.9795
3	17.1685[a]	14.4003[a]	12.7491[a]	11.1987[a]	9.8887[a]	9.0669[a]
4	21.1520[a]	16.9637[a]	14.2772[a]	13.2058[a]	11.9378[a]	11.0671[a]
5	21.5640	19.7850[a]	16.8932[a]	14.2292[a]	13.4855[a]	12.4956[a]
6	24.1010	21.1694	18.9847[a]	16.8594[a]	14.2043[a]	13.7489[a]

[a] Repeated natural frequencies of order two.

range is not sensitive to the boundary conditions and it is expected that some of the natural frequencies of the modified system are very close to those of the original system. Therefore, a direct application of Eq. (3.9.39) may lead to numerical problems as $\omega \to \lambda_i$. However, for most structural response analyses, the first few lowest frequency modes are adequate (see, e.g. [58]).

3.9.4. Periodicity in Two Dimensions

Although only systems with one-dimensional periodicity have been considered so far, systems with higher dimensional periodicity may also be analysed. In such cases, care must be taken to rearrange the order of the generalized coordinates as the symmetry is dependent on the direction considered. The following example is designed to illustrated the process.

Example 3.9.6

Consider the natural vibration of a simply supported square plate with unit sides as shown in Fig. 3.9.6. Sixteen d.o.f. elements are employed so that the generalized displacements are w, $\mathrm{d}w/\mathrm{d}x$, $\mathrm{d}w/\mathrm{d}y$, $\mathrm{d}^2w/\mathrm{d}x\,\mathrm{d}y$ at each node.

Taking n elements on each side and rearranging the generalized coordinates according to the symmetry in one direction gives $\mathbf{x}_j = [q_1 q_2 q_3 q_4 q_5 q_6 q_7 q_8]_j^{\mathrm{T}}$ and $\mathbf{D}_j \mathbf{x}_j = \mathbf{0}$. Then, interchanging the orders in $\mathbf{x}_j$ according to the symmetry in the other direction gives $\mathbf{x}_{ij} = [q_1 q_2 q_3 q_4]_{ij}^{\mathrm{T}}$. The corresponding uncoupled dynamic stiffness equations are $\mathbf{D}_{ij}\mathbf{x}_{ij} = \mathbf{0}$, $i, j = 0, 1, \ldots, n$, where

$$\mathbf{D}_{ij} = \mathbf{K}_{ij} - \lambda \varepsilon^4 \mathbf{M}_{ij}, \qquad \varepsilon^2 = \tfrac{1}{2}(\varepsilon_i^2 + \varepsilon_j^2), \qquad \varepsilon_i = \frac{i\pi}{n}, \qquad \varepsilon_j = \frac{j\pi}{n},$$

$$\lambda = \frac{151\,200 \mu \omega^2}{176\,400 \pi^4 (i^2 + j^2)^2 D}$$

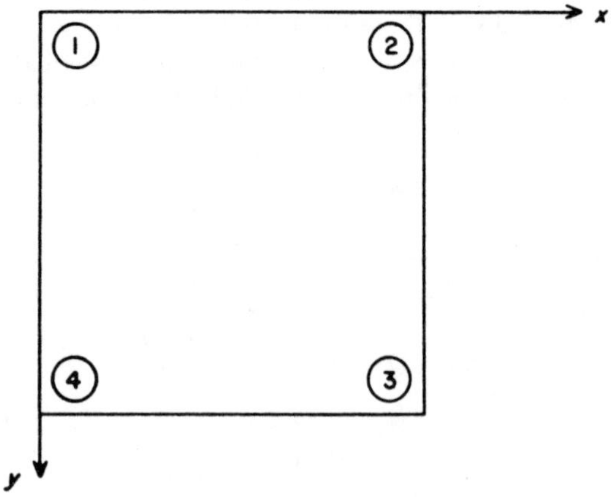

Fig. 3.9.6. A square plate element

$$\mathbf{K}_{ij} = \begin{bmatrix} 445\,823 - 219\,024cj - (219\,024 + 7775cj)ci & & & \text{sym.} \\ -79\,272sj - 34\,127sjci & 72\,575 + 21\,815cj + (3023 + 15\,983cj)ci & & \\ -(79\,271 + 24\,127cj)si & -13\,283sjsi & 72\,575 + 3023cj + (21\,815 + 15\,983cj)ci & \\ -13\,283sjsi & (1512 + 4211cj)si & 1512sj + 4211sjci & 4223 - 696cj - (695 + 995cj)ci \end{bmatrix}$$

$$\mathbf{M}_{ij} = \begin{bmatrix} 24\,336 + 8424cj + (8424 + 2196cj)ci & & & \text{sym.} \\ 2027sj + 702sjci & 623 - 478cj + (216 - 162cj)ci & & \\ 2027si + 702cjsi & 168sjsi & 623 + 216cj - (468 + 162cj)ci & \\ 168sjsi & 51si - 38cjsi & 51sj - 38sjci & 15 - 11cj - (11 - 9cj)ci \end{bmatrix}$$

$$si = \sin \varepsilon_i, \quad sj = \sin \varepsilon_j, \quad ci = \cos \varepsilon_i, \quad cj = \cos \varepsilon_j$$

Here μ is the mass density per unit area, D is the flexural rigidity and $v = 0.3$ has been used. Retaining q_1 for the lowest natural frequencies and expanding the frequency determinant up to ε^4, we have

$$\lambda = \frac{37\,812}{44\,100}, \qquad \omega_{ij}^2 = 1.000\,317\pi^4(i^2 + j^2)\frac{D}{\mu}$$

Thus, compared with the exact result, $\omega_{ij}^2 = \pi^4(i^2 + j^2)D/\mu$, the result from this element has a formulation error of about $0.000\,317 = 0.016\%$ when n tends to infinity.

For two-dimensional periodic structures with arbitrary boundary conditions, the method of relaxing the boundaries has not been very successful. This is due to the fact that the number of boundary coordinates increases rapidly with the number of substructures involved, and the solution of the resulting large order non-linear eigenvalue problem is not a simple task because of the presence of roundoff errors.

3.10. Derivatives of Substructure

Without loss of generality, we consider the one-parameter system after the finite element discretization

$$[\mathbf{D}(\lambda)]\{\mathbf{q}\} = [\mathbf{A} + \lambda\mathbf{B}]\{\mathbf{q}\} = \{\mathbf{Q}\} \tag{3.10.1}$$

where $\mathbf{A}$ and $\mathbf{B}$ are matrices independent of λ. In a vibration problem, $\mathbf{A} = \mathbf{K}$ is the stiffness matrix, $\mathbf{B} = \mathbf{M}$ is the mass matrix and $\lambda = -\omega^2$ is the frequency parameter. In a dynamic buckling problem $\mathbf{A} = \mathbf{K} - \omega^2\mathbf{M}$ is the dynamic stiffness matrix, $\mathbf{B} = \mathbf{G}$ is the geometric matrix and λ is the load factor. Note that, in all cases, $\mathbf{A}$ and $\mathbf{B}$ are independent of λ. Now, partition Eq. (3.10.1) according to the slave coordinates $\{\mathbf{q}_s\}$ and the master coordinates $\{\mathbf{q}_m\}$ and assume $\{\mathbf{Q}_s\} = \{\mathbf{0}\}$ for simplicity

$$\begin{bmatrix} \mathbf{D}_{ss} & \mathbf{D}_{sm} \\ \mathbf{D}_{ms} & \mathbf{D}_{mm} \end{bmatrix} \begin{Bmatrix} \mathbf{q}_s \\ \mathbf{q}_m \end{Bmatrix} = \left(\begin{bmatrix} \mathbf{A}_{ss} & \mathbf{A}_{sm} \\ \mathbf{A}_{ms} & \mathbf{A}_{mm} \end{bmatrix} + \lambda \begin{bmatrix} \mathbf{B}_{ss} & \mathbf{B}_{sm} \\ \mathbf{B}_{ms} & \mathbf{B}_{mm} \end{bmatrix} \right) \begin{Bmatrix} \mathbf{q}_s \\ \mathbf{q}_m \end{Bmatrix} = \begin{Bmatrix} 0 \\ \mathbf{Q}_m \end{Bmatrix} \tag{3.10.2}$$

From the first of Eq. (3.10.2)

$$\{\mathbf{q}\} = \begin{bmatrix} -\mathbf{D}_{ss}^{-1}\mathbf{D}_{sm} \\ \mathbf{I} \end{bmatrix} \{\mathbf{q}_m\} = [\mathbf{T}(\lambda)]\{\mathbf{q}_m\} \tag{3.10.3}$$

we obtain the condensed equation

$$[\mathbf{D}^*(\lambda)]\{\mathbf{q}_m\} = [\mathbf{A}^*(\lambda) + \lambda\mathbf{B}^*(\lambda)]\{\mathbf{q}_m\} = \{\mathbf{Q}_m\} \tag{3.10.4}$$

where

$$\left.\begin{aligned} [\mathbf{A}^*(\lambda)] &= [\mathbf{T}(\lambda)]^{\mathrm{T}}[\mathbf{A}][\mathbf{T}(\lambda)] \\ [\mathbf{B}^*(\lambda)] &= [\mathbf{T}(\lambda)]^{\mathrm{T}}[\mathbf{B}][\mathbf{T}(\lambda)] \\ [\mathbf{D}^*(\lambda)] &= [\mathbf{T}(\lambda)]^{\mathrm{T}}[\mathbf{D}][\mathbf{T}(\lambda)] \end{aligned}\right\} \tag{3.10.5}$$

We require to prove that

$$\frac{\partial}{\partial\lambda}[\mathbf{D}^*(\lambda)] = [\mathbf{B}^*(\lambda)] \tag{3.10.6}$$

Obviously,

$$[\mathbf{T}'(\lambda)] = \frac{\partial}{\partial\lambda}[\mathbf{T}(\lambda)] = \frac{\partial}{\partial\lambda}\begin{bmatrix} -\mathbf{D}_{ss}^{-1}\mathbf{D}_{sm} \\ \mathbf{I} \end{bmatrix} = -\begin{bmatrix} -\mathbf{D}_{ss}^{-1}\mathbf{B}_{ss}\mathbf{D}_{ss}^{-1}\mathbf{D}_{sm} + \mathbf{D}_{ss}^{-1}\mathbf{B}_{sm} \\ \mathbf{0} \end{bmatrix} \tag{3.10.7}$$

Then, from Eq. (3.10.5)

$$\begin{aligned} \frac{\partial}{\partial\lambda}[\mathbf{D}^*(\lambda)] &= [\mathbf{T}'(\lambda)]^{\mathrm{T}}[\mathbf{D}(\lambda)][\mathbf{T}(\lambda)] \\ &\quad + [\mathbf{T}(\lambda)]^{\mathrm{T}}[\mathbf{D}'(\lambda)][\mathbf{T}(\lambda)] \\ &\quad + [\mathbf{T}(\lambda)]^{\mathrm{T}}[\mathbf{D}(\lambda)][\mathbf{T}'(\lambda)] \end{aligned} \tag{3.10.8}$$

However,

$$[\mathbf{D}(\lambda)][\mathbf{T}(\lambda)] = \begin{bmatrix} \mathbf{0} \\ \mathbf{D}^* \end{bmatrix} \tag{3.10.9}$$

and from Eqs (3.10.7) and (3.10.9)

$$[\mathbf{T}'(\lambda)]^{\mathrm{T}}[\mathbf{D}(\lambda)][\mathbf{T}(\lambda)] = [\mathbf{0}] \tag{3.10.10}$$

Therefore, Eq. (3.10.8) becomes

$$\begin{aligned} \frac{\partial}{\partial\lambda}[\mathbf{D}^*(\lambda)] &= [\mathbf{T}(\lambda)]^{\mathrm{T}}[\mathbf{D}'(\lambda)][\mathbf{T}(\lambda)] \\ &= [\mathbf{T}(\lambda)]^{\mathrm{T}}[\mathbf{B}][\mathbf{T}(\lambda)] = [\mathbf{B}^*(\lambda)] \end{aligned}$$

as required. Similarly, if Eq. (3.10.1) is written as

$$[\mathbf{D}(\boldsymbol{\lambda})]\{\mathbf{q}\} = \left[\sum_i \lambda_i \mathbf{D}_i\right]\{\mathbf{q}\} = \{\mathbf{Q}\} \tag{3.10.11}$$

for a number of parameters λ_i, after condensation,

$$[\mathbf{D}^*(\lambda)]\{\mathbf{q}_m\} = \left[\sum_i \lambda_i \mathbf{D}_i^*(\lambda)\right]\{\mathbf{q}_m\} = \{\mathbf{Q}_m\} \tag{3.10.12}$$

we can prove that

$$[\mathbf{D}_i^*(\lambda)] = \frac{\partial}{\partial \lambda_i}[\mathbf{D}^*(\lambda)] \tag{3.10.13}$$

Although we have assumed in Eq. (3.10.1) that both matrices [**A**] and [**B**] are independent of λ, this assumption is never used in the derivation of Eqs (3.10.6) and (3.10.13). Therefore, the extension of the formulae to multilevel substructuring is straightforward.

It is noted that no assumption about the symmetry of the matrices [**D**], [**A**] and [**B**] in Eq. (3.10.1) was made. For a flutter problem involving follower force, matrix [**B**] is symmetric but [**A**] is not. The symmetry is preserved after condensation, i.e. $[\mathbf{B}^*(\lambda)]$ is still symmetrical.

References

1. RED Bishop, DC Johnson 1960. The mechanics of vibration. Cambridge University Press
2. WT Thomson 1972. Theory of vibration with applications. Prentice-Hall, Englewood Cliffs, NJ
3. GB Warburtion 1976. The dynamical behaviour of structure, 2nd edn. Pergamon, Oxford
4. YT Leung 1978. An accurate method of dynamic condensation in structural vibration analysis. Int J Num Meth Engng 12, 1705
5. R Guyan 1975. Reduction of stiffness and mass matrices. AIAA J 3, 380
6. BM Irons 1965. Structural eigenvalue problem: elimination of unwanted variables. AIAA J 3, 961–962
7. T Bamford et al 1971. Dynamic analysis of large structural system. In: Proc ASME synthesis of vibrating systems, Washington, DC
8. RD Henshell, JH Ong 1975. Automatic masters for eigenvalue economization. Earthquake Engng Struct Dyn 3, 375–383
9. AYT Leung 1979. An accurate method of dynamic substructuring with simplified computation. Int J Num Meth Engng 14, 1241–1256
10. M Paz 1984. Dynamic condensation. Am Inst Aeronaut J 22, 724–726
11. A Simpson 1980. The Kron methodology and practical algorithm for eigenvalue, sensitivity and response analysis of large scale structural systems. Aeronaut J 84, 417–433
12. WC Hurty 1969. Dynamic analysis by dynamic partitioning, AIAA J 7, 1152–1154
13. RR Craig, MCC Bampton 1968. Coupling of structures for dynamic analysis. AIAA J 6, 1313–1319
14. DN Herting et al 1977. Development of an automated mullet-stage modal synthesis system for NASTRAN. In: 6th NASTRAN user's colloquium, NASA CP 2018
15. RL Goldman 1969. Vibration analysis by dynamic partitioning. AIAA J 7, 1152–1154
16. SN Hou 1969. Review of modal synthesis techniques and a new approach. Shock Vib Bull 40(4), 25–39
17. WA Benfield, RF Hruda 1971. Vibration analysis of structures by component mode substitution. AIAA J 9, 1255–1261
18. TL Wilson 1977. A NASTRAN DMAP alter for the coupling of modal and physical coordinates substructures. In: 6th NASTRAN user's colloquium, NASA CP 2018
19. EL Wilson, EP Bayo 1986. Use of 4 special Ritz vectors in dynamic substructure analysis. Am Soc Civ Engrs J Struct Engng 112, 1944–1954
20. X Lu, Y Chen, J Chen 1986. Dynamic substructures analysis using Lanczos vectors. In: First world congress in computational mechanics, Austin, Texas
21. GH Sotiropoulos 1984. Comment on the substructure synthesis methods. J Sound Vib 98, 150–153
22. AYT Leung 1988. A simple dynamic substructure method. Earthquake Engng Struct Dyn 16, 827–837

23. AYT Leung 1988. Damped dynamic substructures. Int J Num Meth Engng 26, 2355–2365
24. TK Hasselman 1976. Modal coupling in lightly damped structures. AIAA J 14, 1627–1628
25. TK Hasselman 1976. Damping synthesis from substructures test. AIAA J 14, 1409–1418
26. AL Hale 1984. Substructure synthesis and its iterative improvement for large non-conservative vibrator systems. AIAA J 22, 265–272
27. AL Hale, LA Bergman 1985. The dynamic synthesis of general non-conservative structures from separately identified substructure models. J Sound Vib 98, 431–446
28. AYT Leung 1990. Non-conservative dynamic substructures. Dyn Stab Syst 5, 45–47
29. RR Craig 1977. Methods of component mode synthesis. Shock Vib Dig 9, 3–10
30. L Meirovitch, AL Hale 1981. On the substructure synthesis method. AIAA J 19, 940–947
31. TH Richards, AYT Leung 1977. An accurate method in structural vibrating analysis. J Sound Vib 55, 363–376
32. WH Wittrick, FW Williams 1971. A general algorithm for computing natural frequencies of elastic structures. Q J Mech Appl Math 24, 263–284
33. G Peters, JH Wilkinson 1971. $\mathbf{Ax} = \lambda\mathbf{Bx}$ and the general eigenvalue problems. SIAM J Num Anal 7, 479
34. KJ Bathe, EL Wilson 1972. Large eigenvalue problems in dynamic analysis. ASCE Proc. Paper 9433
35. T Hopper, FW Williams 1977. Mode finding in nonlinear eigenvalue calculations. J Struct Mech 5, 255–278
36. AL Hale, L Meirovitch 1982. Procedure for improving substructures representation in dynamic synthesis. J Sound Vib 84, 269–287
37. WC Hurty 1960. Vibrations of structural systems by component mode synthesis. J Engng Mech Div ASCE 86 (EM4), 51–69
38. W Romberg 1955. Vereinfachte numerische integration. Der Kgl Norske Vid Selsk Forh 28, 30–36
39. V. Kolousek 1973 Dynamics in engineering structures. Butterworth, London
40. AS Deif 1982. Advanced matrix theory for scientists and engineers. Abacus Press, London
41. AYT Leung 1979. Accelerated convergence of dynamic flexibility in series form. Engng Struct 1, 203–206
42. AB Palazzolo, BP Wang, WD Pilkey 1982. A receptance formula for general second degrees square lambda matrices. Int J Num Meth Engng 18, 829–843
43. AYT Leung 1987. Inverse iteration for damped natural vibration. J Sound Vib 118, 193–198
44. T Wah, LR Calcote 1970. Structural analysis by finite difference calculus. Van Nostrand Reinhold, New York
45. JP Ellington, H McCallion 1959. The free vibrations of grillages. Am Soc Mech Engrs J Appl Mech 26, 603–607
46. C Sundararajan, DV Reddy 1975. Finite strip-difference calculus technique for plate vibration problems. Int J Solids Struct 11, 425–435
47. PH Denke, GR Eide, J Pickard 1975. Matrix difference equations analysis of vibrating periodic structures. Am Inst Aeronaut Astronaut J 13, 160–166
48. YK Lin, TJ McDaniel 1969. Dynamics of beam-type periodic structures. Am Soc Mech Engrs, J Engng Indust 91, 1133–1141
49. L Meirovitch, RC Engels 1977. Response of periodic structures by Z transform method. Am Inst Aeronaut Astronaut J 15, 167–174
50. RC Engels, L Meirovitch 1978. Response of periodic structures by modal analysis. J Sound Vib 56, 481–493
51. DJ Mead 1973. A general theory of harmonic wave propagation in linear periodic systems with multiple coupling. J Sound Vib 27, 235–260
52. JE Walz, RE Fulton, NJ Cyrus, RT Eppink 1970. Accuracy of finite element approximations to structural problems. NASA TN D-5728
53. AA Liepins 1978. Rod and beam finite element matrices and their accuracy. Am Inst Aeronaut Astronaut J 16, 531–534
54. RH MacNeal 1972. The NASTRAN theoretical manual. NASA SP-221 (01)
55. JS Przemieniecki 1968. Theory of matrix structural analysis. McGraw-Hill. New York
56. J Casti, R Kalaba 1973. Imbedding methods in applied mathematics. Addison-Wesley, Palo Alto, CA
57. A Simpson 1974. Scanning Kron's determinant. Q J Mech Appl Math 27, 27–43
58. YT Leung 1980. Accelerated converging methods in structural response analysis. Presented at the conference on advanced structural dynamics, University of Southampton

Chapter 4

Dynamic Stiffness

Finite elements are related to continuous elements by means of Simpson's hypothesis (Sect. 2.8). If the non-essential coordinates (slaves) are eliminated by means of dynamic substructure methods, dynamic stiffnesses result. We shall extend the formulation of Chap. 2 to include follower forces, parametrically excited axial forces, in-plane moments and response analysis. A general formulation will be given in Chap. 5 where curved members will be considered.

4.1. Follower Force

An externally applied force which changes the direction of application (but not the magnitude) is called a follower force. The dynamic stability of equilibrium of elastic systems in the presence of follower forces has been reviewed and studied [1]. Due to the difficulty in solving the resulting non-linear and non-conservative eigenvalue problem by means of the frequency determinant and the compatibility and equilibrium requirements, the method has not been applied to complex structures such as frames. In this section the method is reformulated according to the finite element concept using a frequency-dependent shape function. Thus the compatibility and equilibrium requirements at the common nodes of the constituent members are easily satisfied. The resulting non-linear non-conservative eigenvalue problem is then solved by means of a newly developed parametric inverse iteration with the intensity of the follower forces taken as an iteration parameter, the derivatives of the dynamic stiffness matrix being approximated by a Romberg algorithm [2]. The flutter frequency and load are computed simultaneously by means of the Newtonian method, if the load–frequency plots are required. A Timoshenko column is used as an example.

The lateral displacement $v(x, t)$ of a Timoshenko column subject to constant axial force P is given by

$$v(x,t) = [\mathbf{N}(x,\omega)]\{\mathbf{q}\}e^{i\omega t} \tag{4.1.1}$$

where ω is the vibration frequency, $[\mathbf{N}]$ the frequency-dependent shape function and $\{\mathbf{q}\}$ the generalized nodal displacements. Following the standard procedure to find the dynamic stiffness matrix $[\mathbf{D}]$ for a conservative member [3–5], we have

$$[\mathbf{D}(\omega, P)]\{\mathbf{q}\} = \{\mathbf{Q}\} \tag{4.1.2}$$

where $\{\mathbf{Q}\}$ is the generalized nodal force vector. The elements of $[\mathbf{D}(\omega, P)]$ can be found in Chap. 2.

Now, without loss of generality, if the axial force at node 2 ($x = l$) is a follower force, then, the nodal shear force $Q_3 e^{i\omega t}$ is modified to

$$\bar{Q}_3 e^{i\omega t} = -EI\frac{\partial^3 v}{\partial x^3}(l, t) = Q_3 e^{i\omega t} - P\frac{\partial v}{\partial x}(l, t) \tag{4.1.3}$$

Therefore, the third row of the dynamic stiffness equation has to be modified accordingly to

$$[\bar{\mathbf{D}}(\omega, P)]\{\mathbf{q}\} = \left[\mathbf{D}(\omega, P) + \begin{Bmatrix} 0 \\ 0 \\ P\partial[\mathbf{N}(x, \omega)]/\partial x \\ 0 \end{Bmatrix}\right]\{\mathbf{q}\} = \{\bar{\mathbf{Q}}\} \tag{4.1.4}$$

A similar form is obtained for a follower axial force acting at node 1.

In a follower force analysis of a frame structure vibrating at frequency ω, the dynamic stiffness matrices of the individual beam members are formed, the matrices are transformed in line with the global coordinates according to the orientations, then the system matrix is assembled according to the usual finite element procedure. Concentrated masses and stiffnesses are added to the main diagonal of the system matrix at the appropriate positions.

Consider the following non-linear non-conservative eigenvalue problem

$$\{\mathbf{f}(\lambda, \phi)\} = [\mathbf{D}(\lambda) + \mu\mathbf{A}(\lambda)]\{\boldsymbol{\phi}\} = \{\mathbf{0}\} \tag{4.1.5}$$

Here $[\mathbf{D}(\lambda)]$ and $[\mathbf{A}(\lambda)]$ are the conservative and non-conservative parts respectively, and μ is a control parameter $0 < \mu < 1$. When $\mu = 0$, the non-linear eigenvalue problem (4.1.5) is conservative; the eigenvalue λ can be solved with certainty by the Sturm sequence method (e.g. [3]), and the corresponding eigenvector $\{\boldsymbol{\phi}\}$ obtained by the inverse iteration method [6, 7]. In performing the inverse iteration, the derivative of the dynamic stiffness matrix with respect to the eigenvalue is required. Since in the Sturm sequence bisection searching for the eigenvalue, the dynamic stiffness matrix is evaluated at regular intervals, the Romberg algorithm [2] can evaluate the dynamic stiffness derivative by simple arithmetic and is highly recommended here.

When the solution of the corresponding conservative eigenvalue problem is in hand, λ_0 and $\{\boldsymbol{\phi}_0\}$ say, the solution of the non-conservative problem (4.1.5) is assumed to be

$$\lambda = \lambda_0 + d\lambda \quad \text{and} \quad \{\phi\} = \{\phi_0\} + \{d\phi\} \tag{4.1.6}$$

Substitute Eq. (4.1.6) into Eq. (4.1.5) and neglect higher order terms:

$$\begin{aligned}\{\mathbf{f}(\lambda_0 + d\lambda, \boldsymbol{\phi}_0 + d\boldsymbol{\phi})\} &= [D(\lambda_0) + D'(\lambda_0)d\lambda + \mu\mathbf{A}(\lambda_0) + \mu\mathbf{A}'(\lambda_0)d\lambda]\{\boldsymbol{\phi} + d\boldsymbol{\phi}\} \\ &= \{0\}\end{aligned}$$

or

$$[D(\lambda_0) + \mu\mathbf{A}(\lambda_0)]\{\boldsymbol{\phi}_0 + d\boldsymbol{\phi}\} = -[\mathbf{D}'(\lambda_0) + \mu\mathbf{A}'(\lambda_0)]\{\boldsymbol{\phi}_0\}d\lambda \tag{4.1.7}$$

where a prime denotes differentiation with respect to λ. Equation (4.1.7) represents an inverse iteration algorithm. That is, when an approximate eigenvector $\{\boldsymbol{\phi}_0\}$ is given on the right-hand side, an improved eigenvector $\{\boldsymbol{\phi}_0 + \mathrm{d}\boldsymbol{\phi}\}$ is obtained on the left-hand side within a scaling constant.

The eigenrow $\{\boldsymbol{\psi}\}$ is obtained similarly by letting $\{\boldsymbol{\psi}_0\} = \{\boldsymbol{\phi}_0\}$ $(=\{\boldsymbol{\phi}_0^*\})$, the complex conjugate, if conservative gyroscopic forces are present), and the inverse iteration results in

$$\{\boldsymbol{\psi}_0 + \mathrm{d}\boldsymbol{\psi}\}^{\mathrm{T}}[\mathbf{D}(\lambda_0) + \mu\mathbf{A}(\lambda_0)] = -\{\boldsymbol{\psi}_0\}^{\mathrm{T}}[\mathbf{D}'(\lambda_0) + \mu\mathbf{A}'(\lambda_0)]\,\mathrm{d}\lambda \tag{4.1.8}$$

When both $\{\boldsymbol{\phi}\}$ and $\{\boldsymbol{\psi}\}$ converge, the Newtonian algorithm for the following equation in λ, $g(\lambda) = \{\boldsymbol{\psi}\}^{\mathrm{T}}[\mathbf{D}(\lambda) + \mu\mathbf{A}(\lambda)]\{\boldsymbol{\phi}\} = 0$, suggests the generalized Rayleigh quotient

$$\lambda = \lambda_0 - \frac{g(\lambda_0)}{g'(\lambda_0)} = \lambda_0 - \frac{\{\boldsymbol{\psi}\}^{\mathrm{T}}[\mathbf{D}(\lambda_0) + \mu\mathbf{A}(\lambda_0)]\{\boldsymbol{\phi}\}}{\{\boldsymbol{\psi}\}^{\mathrm{T}}[\mathbf{D}'(\lambda_0) + \mu\mathbf{A}(\lambda_0)]\{\boldsymbol{\phi}\}} \tag{4.1.9}$$

to give the improved eigenvalue.

If the inverse iterations (4.1.7) and (4.1.8) do not converge within five cycles, the control parameter μ must be reduced, or the iteration will converge to other undesirable modes. In the case of a follower force, the parameter μ is taken to be the axial force and the eigenvalue's path can easily be traced.

When the follower force is near critical, the inverse iteration will switch between the two modes of concern. Since all quantities in Eq. (4.1.9) are real, it is impossible to predict an unstable post-flutter complex frequency. Therefore, when the follower force is near critical, the classical method to find the critical load p and frequency λ from the following two conditions is adopted,

$$\Delta(\lambda, p) = 0 \qquad \text{and} \qquad \Delta'(\lambda, p) = 0 \tag{4.1.10}$$

where Δ is the determinant of the total dynamic stiffness matrix $[\overline{\mathbf{D}}]$ and a prime denotes a derivative with respect to λ. From the initial approximations obtained using the inverse iteration λ_0, p_0, an improved critical solution, $\lambda = \lambda_0 + \mathrm{d}\lambda$, $p = p_0 + \mathrm{d}p$, is found from the Newtonian algorithm

$$\begin{bmatrix} \Delta'_0 & \dot{\Delta}_0 \\ \Delta''_0 & \dot{\Delta}'_0 \end{bmatrix} \begin{Bmatrix} \mathrm{d}\lambda \\ \mathrm{d}p \end{Bmatrix} = -\begin{Bmatrix} \Delta_0 \\ \Delta'_0 \end{Bmatrix} \tag{4.1.11}$$

where the subscript 0 denotes evaluation at λ_0, p_0, and dots denote derivatives with respect to p. If the matrices $[\overline{\mathbf{D}}'] = (\partial/\partial\lambda)[\overline{\mathbf{D}}]$, $[\overline{\mathbf{D}}''] = (\partial^2/\partial\lambda^2)[\overline{\mathbf{D}}]$, $[\dot{\overline{\mathbf{D}}}] = (\partial/\partial p)[\overline{\mathbf{D}}]$, $[\dot{\overline{\mathbf{D}}}'] = (\partial^2/\partial\lambda\partial p)[\overline{\mathbf{D}}]$ evaluated at λ_0, p_0 by the Romberg algorithm, then Δ'_0, $\dot{\Delta}_0$, $\dot{\Delta}'_0$, Δ''_0 are computed as in Sect. 3.9.

Example 4.1.1

Consider the classical problem of a follower force (see Fig. 4.1.1) as the first example. The eigenvalue λ, where $\lambda^4 = \omega^2\rho A l^4/EI$ is plotted against the intensity Pl^2/EI for the first ten modes in Fig. 4.1.2. The first six critical follower forces have been calculated and are listed in Table 4.1.1. The critical follower forces are approximately $P = (2n - 0.5)^2\pi^2 EI/l^2$ to within 5%. The influence of shear and rotatory inertia effects can then be considered by taking as the slenderness ratio squared, $r^2 = 0(0.2)1.4$ and $s^2 = (E/kG)r^2 = 3.6r^2$. The first two natural modes are plotted

Fig. 4.1.1. Follower force

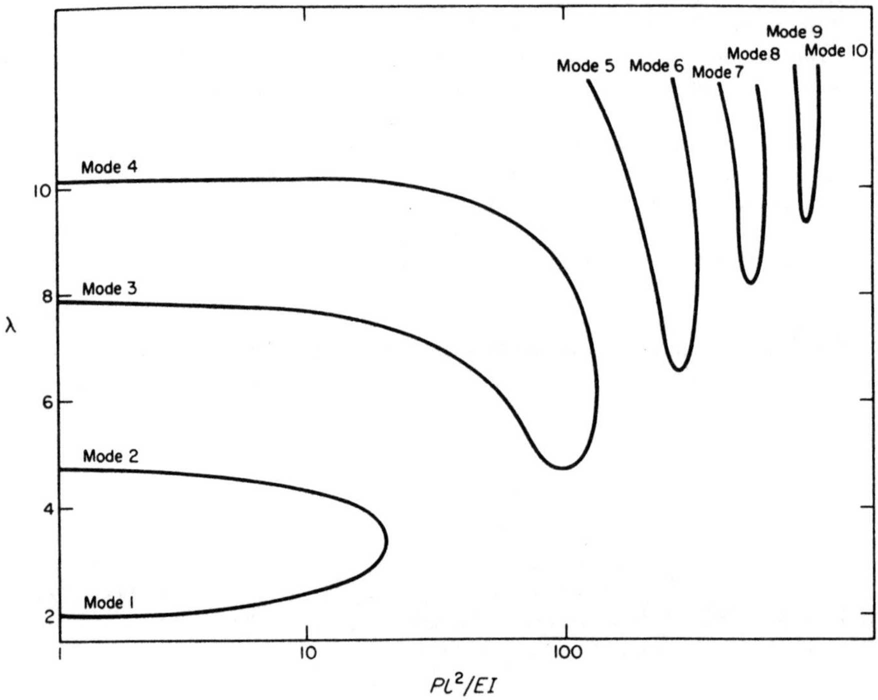

Fig. 4.1.2. Frequency parameter vs. axial load parameter

Table 4.1.1. First six critical follower forces

Flutter mode	λ	Pl^2/EI
1	3.3188	20.0509
2	6.0872	127.811
3	2.2190	317.981
4	10.056	588.715
5	11.706	939.368
6	13.223	1369.63

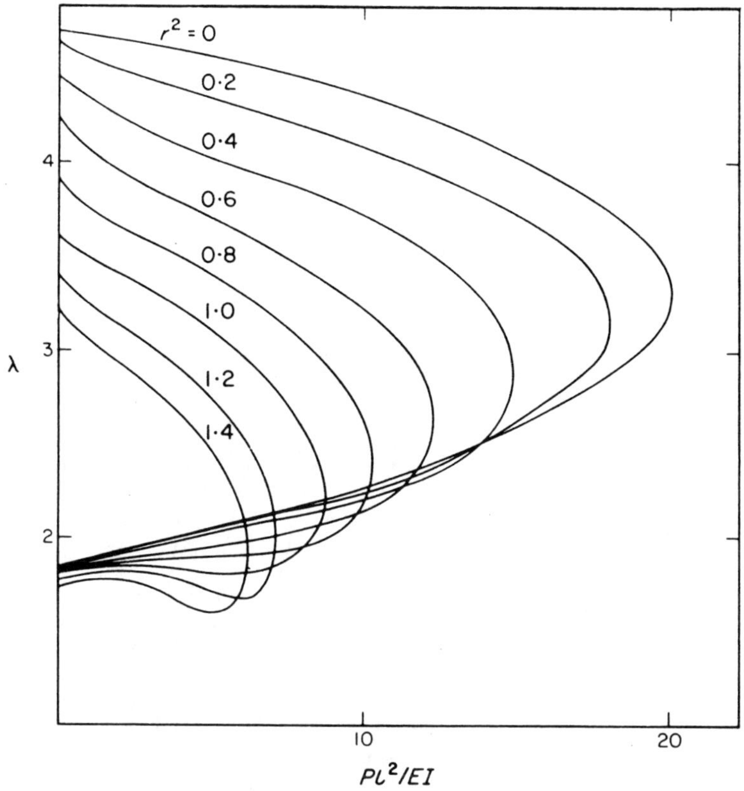

Fig. 4.1.3. Frequency parameter vs. axial load parameter with different radii of gyration

against Pl/EI in Fig. 4.1.3. It is seen that the influence of shear and rotatory inertia is substantial (about 70% reduction in Pl^2/EI) and the shapes of the parametric curves are altered.

Example 4.1.2

Next, consider the influence of a concentrated mass M as shown in Fig. 4.1.4, such that $MEI/\rho Al^4 = 0, 1$. The parametric curves are shown in Fig. 4.1.5. The reduction of critical follower forces on all modes is observed.

Example 4.1.3

The influence of the foundation stiffness K (see Fig. 4.1.6) is considered as the third example. For $KEI/l = 0, 1, 10, 100, \infty$ the parametric curves are plotted in Fig. 4.1.7. It is found that when $k = 0$ the system is unstable in the beginning and divergence instability only is possible. When $KEI/l = 100$, the parametric curves are indistinguishable from those for the rigid foundation considered in Example 1.

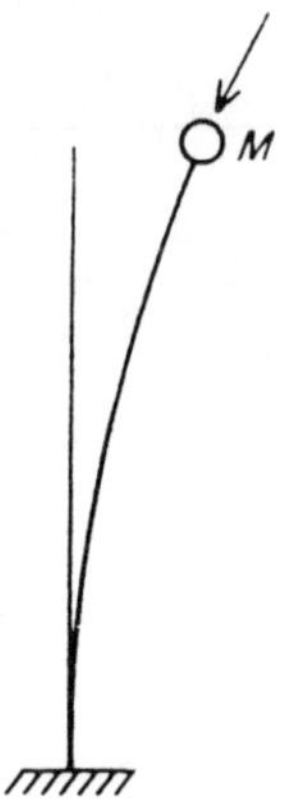

Fig. 4.1.4. A follower force system with end mass

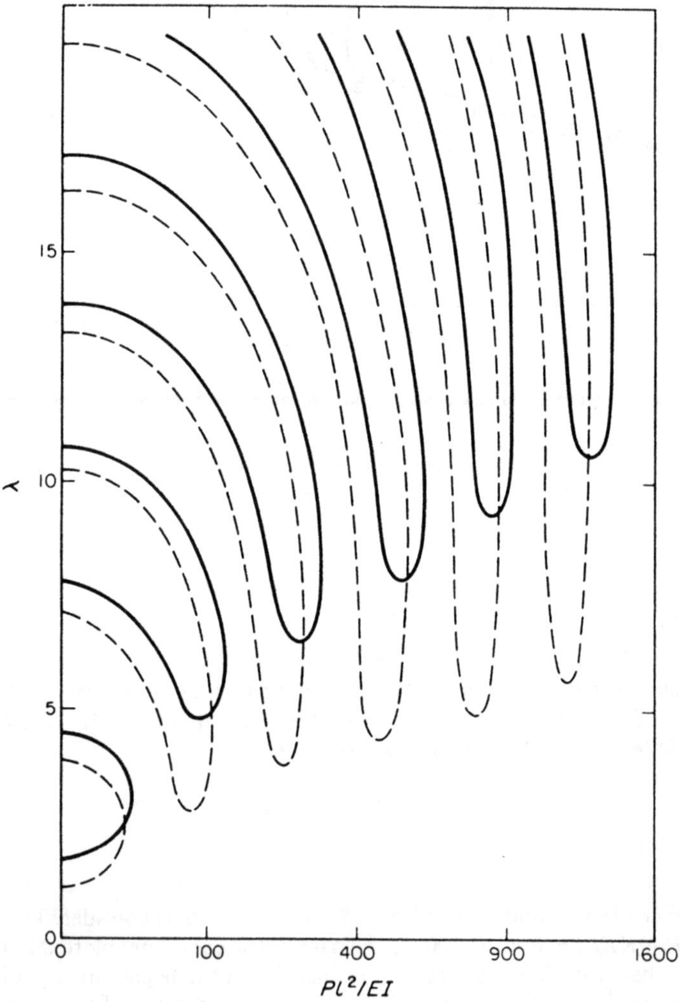

Fig. 4.1.5. Frequency diagram for Example 4.1.2: solid lines, $M = 0$; dashed lines $MEI/\rho Al^4 = 1$

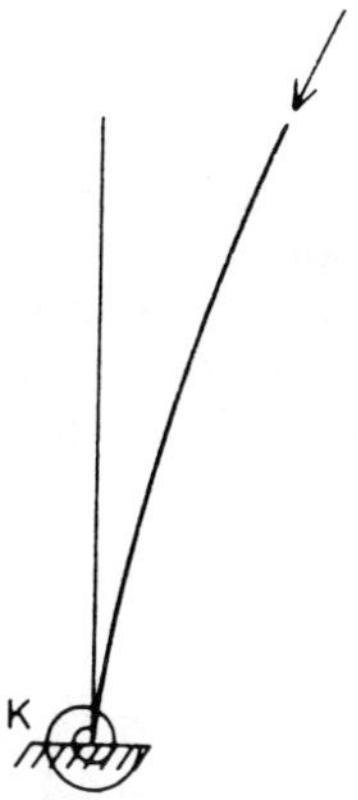

Fig. 4.1.6. Example 4.1.3

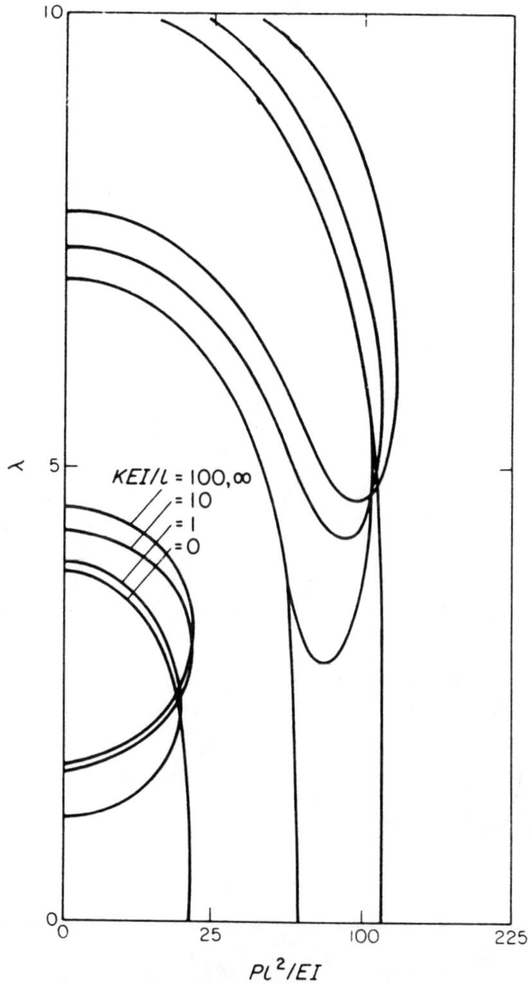

Fig. 4.1.7. Frequency diagram for Example 4.1.3

Example 4.1.4

The non-uniform step beam shown in Fig. 4.1.8 has been studied by means of two elements. The parametric curves are given in Fig. 4.1.9. It is interesting to note that modes 11 and 14 emanating from the conservative system can coalesce. There is no obvious evidence to predict which modes may coalesce under the influence of the follower force. The influence of shear and rotatory inertia is also depicted in Fig. 4.1.10. The same phenomena observed in Example 4.1.1 are noted.

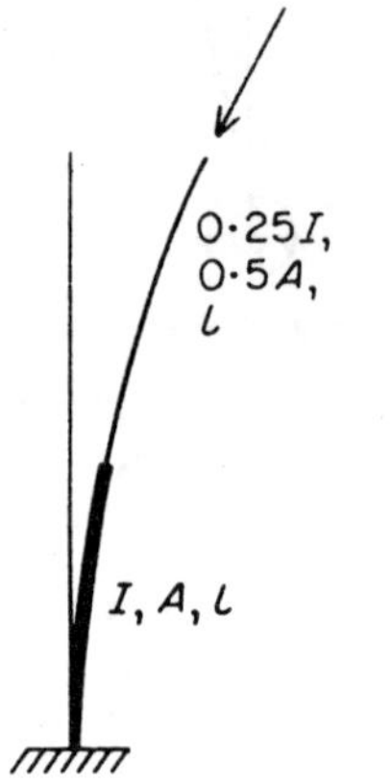

Fig. 4.1.8. Example 4.1.4

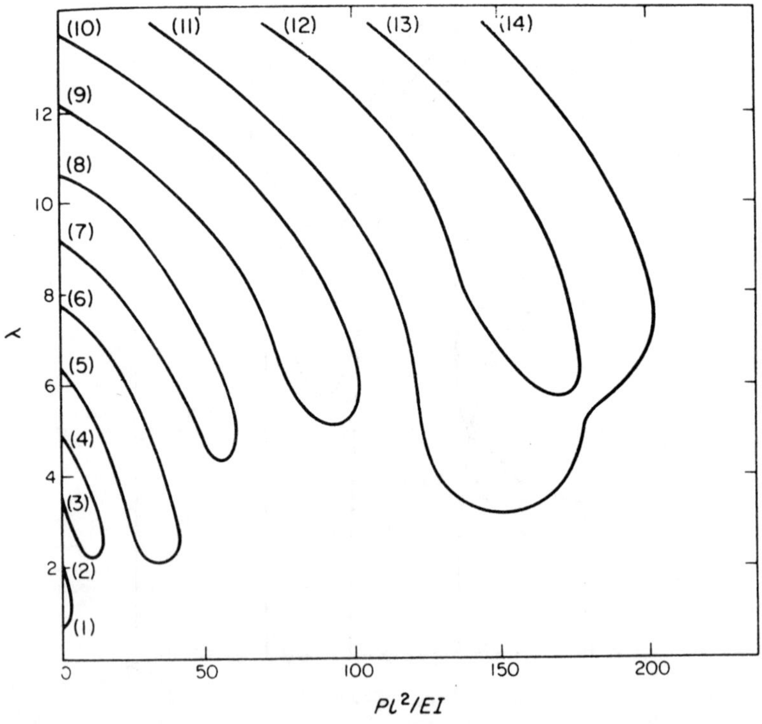

Fig. 4.1.9. Frequency diagram for Example 4.1.4. Numbers in brackets are the conservative modes

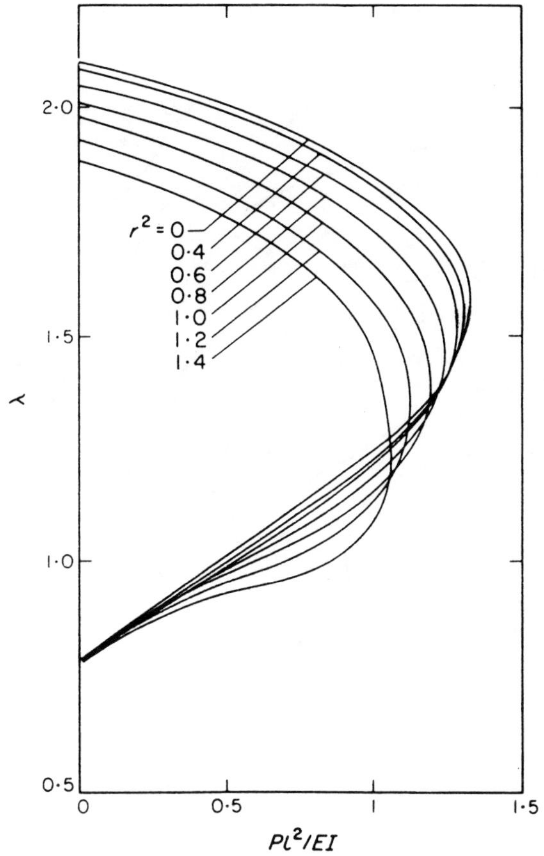

Fig. 4.1.10. Frequency diagram for Example 4.1.4 with the radii of gyration as parameter

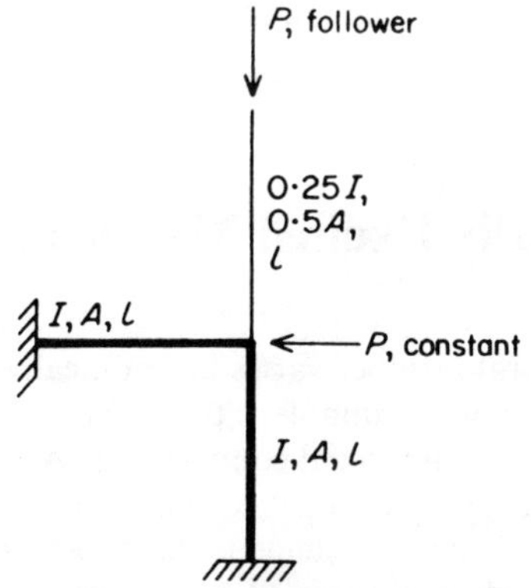

Fig. 4.1.11. Example 4.1.5

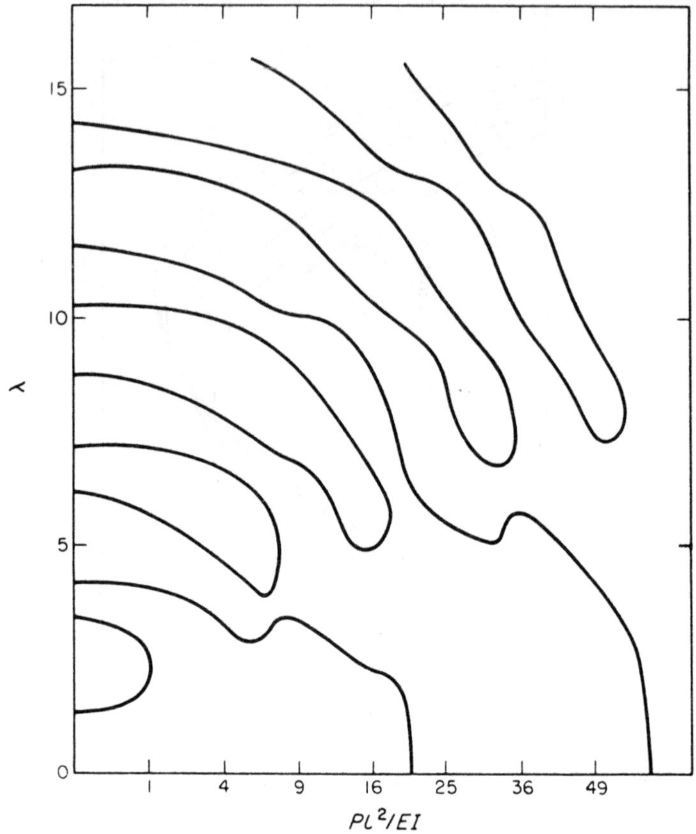

Fig. 4.1.12. Frequency diagram for Example 4.1.5

Example 4.1.5

As a final example, consider the simple frame shown in Fig. 4.1.11. Due to the artificial hinge at the common node, the lower member can experience divergence instability. The characteristic curves, shown in Fig. 4.1.12, are not as smooth as those in the previous cases due to the uneven distribution of natural frequencies.

4.2. Parametrically Excited Members

If the stiffness of a structural member varies harmonically with time, the member is parametrically excited. A typical example is the stiffness of a straight beam member affected by a harmonically applied axial compression. A structure becomes dynamically unstable if the frequency of the axial compression reaches certain values.

The dynamic stability of elastic systems is discussed extensively by Bolotin [8], where the solution methods are confined to modal analysis. While the Fourier expansion is effective for time discretization, the finite element method is also successful for spatial discretization [9, 10]. The spatial discretization is alternatively achieved by frequency-dependent shape functions, which are the exact solutions of each element under mono-frequency vibration. Since the shape functions vary according to vibrating frequency, the method is exact for mono-frequency vibration and is able to predict an infinite number of vibration modes using very few degrees of freedom. However, even if the parametric excitation is mono-frequency, the Fourier transformation results in a set of coupled ordinary differential equations in the spatial variable with various frequencies of interest. A mixed frequency formulation is introduced here to obtain the dynamic stiffness matrix, the determinant equation of which has roots corresponding to the stability boundaries. The Galerkin approximation is effectively introduced. A convergence study is carried out by taking more elements then necessary. Since the mono-frequency shape functions are very close to the exact solutions for multiple frequency vibrations, the frequency components of which are clearly distinct in the case of parametric excitation, it is found that the calculated stability boundaries are not affected by the number of elements being used in the numerical examples.

Consider a beam member isolated from the system. The governing partial differential equation for lateral vibration $v(x, \tau)$ under parametric excitation of period $T = 2\pi/\theta$ is

$$EI\frac{\partial^4 v}{\partial x^4} + (P + 2Q\cos\tau)\frac{\partial^2 v}{\partial x^2} + m\theta^2\frac{\partial^2 v}{\partial \tau^2} = 0 \qquad (4.2.1)$$

where EI is the flexural rigidity, $P + 2Q\cos\tau$ is the time-varying axial force, $\tau = \theta t$ is the non-dimensional time, and m is the mass per unit length. Since the set of trigonometric functions is complete, for steady state periodic vibration, the following Fourier expansion in terms of the unknown functions $a_k(x)$ and $b_k(x)$ is admissible:

$$v(x, \tau) = \tfrac{1}{2}b_0(x) + \sum_k [a_k(x)\sin(k\tau/2) + b_k(x)\cos(k\tau/2)] \qquad (4.2.2)$$

Here the summation over k ranges from 1 to n, the number of terms under consideration. The dynamic stability boundaries are determined from the non-trivial solutions of $a_k(x)$ and $b_k(x)$, having periods T and $2T$, respectively. Non-trivial solutions with period T are given by Eq. (4.2.2) when k is even, and those with period $2T$ when k is odd.

Substituting Eq. (4.2.2) into Eq. (4.2.1), multiplying by $\sin(j\tau/2)$ and $\cos(j\tau/2)$ respectively and integrating with respect to τ from 0 to 2π, we have

$$EI\frac{d^4\{\mathbf{a}^0\}}{dx^4} + [P\mathbf{I} + Q\mathbf{J}]\frac{d^2\{\mathbf{a}^0\}}{dx^2} - Qa_1\{\mathbf{e}_1\} - [\mathbf{M}_a^0]\{\mathbf{a}^0\} = \{0\} \qquad (4.2.3)$$

$$EI\frac{d^4\{\mathbf{b}^0\}}{dx^4} + [P\mathbf{I} + Q\mathbf{J}]\frac{d^2\{\mathbf{b}^0\}}{dx^2} - Qb_1\{\mathbf{e}_1\} - [\mathbf{M}_b^0]\{\mathbf{b}^0\} = \{0\} \qquad (4.2.4)$$

$$EI\frac{d^4\{\mathbf{a}^e\}}{dx^4} + [P\mathbf{I} + Q\mathbf{J}]\frac{d^2\{\mathbf{a}^e\}}{dx^2} - [\mathbf{M}_a^e]\{\mathbf{a}^e\} = \{0\} \qquad (4.2.5)$$

$$EI\frac{d^4\{\mathbf{b}^e\}}{dx^4} + [P\mathbf{I} + Q\mathbf{J}]\frac{d^2\{\mathbf{b}^e\}}{dx^2} - [\mathbf{M}_b^e]\{\mathbf{b}^e\} = \{0\} \qquad (4.2.6)$$

where

$$\{\mathbf{a}^0\} = [a_1, a_3, a_5, \ldots]^T, \qquad \{\mathbf{b}^0\} = [b_1, b_3, b_5, \ldots]^T$$
$$\{\mathbf{a}^e\} = [a_2, a_4, a_6, \ldots]^T, \qquad \{\mathbf{b}^e\} = [\tfrac{1}{2}b_0, b_2, b_4, b_6, \ldots]^T$$
$$[\mathbf{I}] = \mathrm{diag}\,[1, 1, \ldots] \qquad \{\mathbf{e}_1\} = [1, 0, 0, \ldots]^T$$
$$[\mathbf{M}_a^0] = m\theta^2 \,\mathrm{diag}[1^2, 2^2, 3^2, \ldots], \quad [\mathbf{M}_a^0] = m\theta^2 \,\mathrm{diag}[0^2, 1^2, 2^2, \ldots]$$

and

$$[\mathbf{J}] = \begin{bmatrix} 0 & 1 & & & \\ 1 & 0 & 1 & & \\ & 1 & 0 & 1 & \\ & & 1 & 0 & . \\ & & & . & . \end{bmatrix}$$

It is difficult to solve the set of ordinary differential equations (4.2.3)–(4.2.6). Since the primary vibration frequencies for the unknowns $a_k(x)$ and $b_k(x)$ are $k\theta/2$, which will not be coincident or close to each other for different k, it is reasonable to express the unknowns as

$$a_k(x) = [\mathbf{N}_k(R, x)]\{\mathbf{p}_k\} \quad \text{and} \quad b_k(x) = [\mathbf{N}_k(R, x)]\{\mathbf{q}_k\} \tag{4.2.7}$$

where $[\mathbf{N}_k(R, x)]$ is a four-element shape function matrix. Each of the elements is a solution for the vibration beam,

$$EI\frac{d^4\phi}{dx^4} + R\frac{d^2\phi}{dx^2} - m\left(\frac{k\theta}{2}\right)^2\phi = 0 \tag{4.2.8}$$

with the following four sets of boundary conditions, respectively:

1. $\phi(0) = 1, \quad \phi'(0) = \phi(l) = \phi'(l) = 0$
2. $\phi'(0) = 1, \quad \phi(0) = \phi(l) = \phi'(l) = 0$
3. $\phi(l) = 1, \quad \phi(0) = \phi'(0) = \phi'(l) = 0$
4. $\phi'(l) = 1, \quad \phi(0) = \phi'(0) = \phi(l) = 0$

Here R is equal to $P - Q$, $P + Q$, or P, l is the length of the beam concerned and a prime denotes a derivative with respect to x. The shape functions can be found, for example, from references [11] and [12]. The nodal vectors $\{\mathbf{p}_k\}$ and $\{\mathbf{q}_k\}$ in Eqs (4.2.7) can be considered as the generalized nodal displacements for the kth Fourier component.

Premultiplying Eqs (4.2.3)–(4.2.6) by $\{\mathbf{a}^0\}^T$, $\{\mathbf{b}^0\}^T$, $\{\mathbf{a}^e\}^T$ and $\{\mathbf{b}^e\}^T$, respectively, and integrating over x from zero to l, we have, respectively,

$$[\mathbf{D}_a^0]\{\mathbf{p}^0\} = \{\mathbf{0}\}, \quad [\mathbf{D}_a^e]\{\mathbf{p}^e\} = \{\mathbf{0}\}, \quad [\mathbf{D}_b^0]\{\mathbf{q}^0\} = \{\mathbf{0}\} \quad \text{and}$$
$$[\mathbf{D}_b^e]\{\mathbf{q}^e\} = \{\mathbf{0}\} \tag{4.2.9}$$

where

$$[\mathbf{D}_a^0] = \begin{bmatrix} \mathbf{D}(P-Q, \theta/2) & \mathbf{G}_{13} & & \\ \mathbf{G}_{31} & \mathbf{D}(P, 3\theta/2) & \mathbf{G}_{35} & \\ & \mathbf{G}_{53} & \mathbf{D}(P, 5\theta/2) & . \\ & & . & . \end{bmatrix} \tag{4.2.10}$$

$$[\mathbf{D}_b^0] = \begin{bmatrix} \mathbf{D}(P+Q,\theta/2) & \mathbf{G}_{13} & & \\ \mathbf{G}_{31} & \mathbf{D}(P,3\theta/2) & \mathbf{G}_{35} & \\ & \mathbf{G}_{53} & \mathbf{D}(P,5\theta/2) & . \\ & & . & . \end{bmatrix} \tag{4.2.11}$$

$$[\mathbf{D}_a^e] = \begin{bmatrix} \mathbf{D}(P,\theta) & \mathbf{G}_{24} & & \\ \mathbf{G}_{42} & \mathbf{D}(P,2\theta) & \mathbf{G}_{46} & \\ & \mathbf{G}_{64} & \mathbf{D}(P,3\theta) & . \\ & & . & . \end{bmatrix} \tag{4.2.12}$$

$$[\mathbf{D}_b^e] = \begin{bmatrix} \mathbf{D}(P,0) & \mathbf{G}_{02} & & \\ \mathbf{G}_{20} & \mathbf{D}(P,\theta) & \mathbf{G}_{24} & \\ & \mathbf{G}_{42} & \mathbf{D}(P,2\theta) & . \\ & & . & . \end{bmatrix} \tag{4.2.13}$$

The mixed frequency matrix $[\mathbf{G}_{ij}]$ is defined as

$$[\mathbf{G}_{ij}] = [\mathbf{G}(\theta_i, \theta_j)] = Q \int_0^1 [\mathbf{N}_i(R_i, x)]^{\mathrm{T}} [\mathbf{N}_j''(R_j, x)] \, \mathrm{d}x \tag{4.2.14}$$

where $[\mathbf{N}_i(R, x)]$ is given by Eqs (4.2.7) and primes denote derivatives with respect to x. The value of the equivalent axial force R is equal to $P - Q$, $P + Q$, or P, depending on the shape functions being used. It should be noted that $[\mathbf{G}_{13}]$, etc., in Eqs (4.2.10) and (4.2.11) are not the same, because in Eq. (4.2.10), $[\mathbf{N}_1] = [\mathbf{N}_1(P - Q, x)]$ and in Eq. (4.2.11), $[\mathbf{N}_1] = [\mathbf{N}_1(P + Q, x)]$.

The matrices $[\mathbf{D}]$ and $[\mathbf{G}_{ij}]$ in Eqs (4.2.10)–(4.2.13) are assembled element by element according to the finite element procedure. The dynamic stability boundaries determined by the conditions

$$\begin{aligned} \det[\overline{\mathbf{D}}_a^0(P, Q, \theta)] &= 0, \qquad \det[\overline{\mathbf{D}}_b^0(P, Q, \theta)] = 0 \\ \det[\overline{\mathbf{D}}_a^e(P, Q, \theta)] &= 0, \qquad \det[\overline{\mathbf{D}}_b^e(P, Q, \theta)] = 0 \end{aligned} \tag{4.2.15}$$

where $[\overline{\mathbf{D}}_a^0]$, etc., are assembled global matrices corresponding to those given in Eqs (4.2.10)–(4.2.13).

Before solving Eqs (4.2.15), a brief account of the finite element formulated with frequency-independent shape functions is advantageous. Results, similar to Eqs (4.2.10)–(4.2.13), obtained by conventional methods [8–10], are respectively,

$$[\overline{\mathbf{D}}_a^0] = \begin{bmatrix} \mathbf{K} - (P-Q)\mathbf{G} - \theta^2\mathbf{M}/4 & -Q\mathbf{G} & & \\ -Q\mathbf{G} & \mathbf{K} - P\mathbf{G} - 9\theta^2\mathbf{M}/4 & -Q\mathbf{G} & \\ & -Q\mathbf{G} & \mathbf{K} - P\mathbf{G} - 25\theta^2\mathbf{M}/4 & . \\ & & . & . \end{bmatrix}$$

$$[\overline{\mathbf{D}}_b^0] = \begin{bmatrix} \mathbf{K} - (P+Q)\mathbf{G} - \theta^2\mathbf{M}/4 & -Q\mathbf{G} & & \\ -Q\mathbf{G} & \mathbf{K} - P\mathbf{G} - 9\theta^2\mathbf{M}/4 & -Q\mathbf{G} & \\ & -Q\mathbf{G} & \mathbf{K} - P\mathbf{G} - 25\theta^2\mathbf{M}/4 & . \\ & & . & . \end{bmatrix}$$

$$[\overline{\mathbf{D}}_a^e] = \begin{bmatrix} \mathbf{K} - P\mathbf{G} - \theta^2\mathbf{M} & -Q\mathbf{G} & & \\ -Q\mathbf{G} & \mathbf{K} - P\mathbf{G} - 4\theta^2\mathbf{M} & -Q\mathbf{G} & \\ & -Q\mathbf{G} & \mathbf{K} - P\mathbf{G} - 9\theta^2\mathbf{M} & . \\ & & . & . \end{bmatrix}$$

and

$$[\overline{\mathbf{D}}_b^e] = \begin{bmatrix} \mathbf{K} - P\mathbf{G} & -Q\mathbf{G} & & & \\ -Q\mathbf{G} & \mathbf{K} - P\mathbf{G} - \theta^2\mathbf{M} & -Q\mathbf{G} & & \\ & -Q\mathbf{G} & \mathbf{K} - P\mathbf{G} - 4\theta^2\mathbf{M} & -Q\mathbf{G} & \\ & & -Q\mathbf{G} & \mathbf{K} - P\mathbf{G} - 9\theta^2\mathbf{M} \end{bmatrix}$$

Here **K**, **G**, **M** are the usual stiffness, geometric and mass matrices, respectively. The relation between P, Q and θ describing the dynamic stability boundaries is typified by

$$\det[\overline{\mathbf{D}}(P, Q, \theta)] = 0 \tag{4.2.16}$$

where $[\overline{\mathbf{D}}]$ denotes $[\overline{\mathbf{D}}_a^0]$, $[\overline{\mathbf{D}}_a^e]$, $[\overline{\mathbf{D}}_b^0]$ or $[\overline{\mathbf{D}}_b^e]$. For given values of P and Q, Eq. (4.2.16) is reduced to a linear symmetrical algebraic eigenvalue problem with $\lambda = \theta^2$

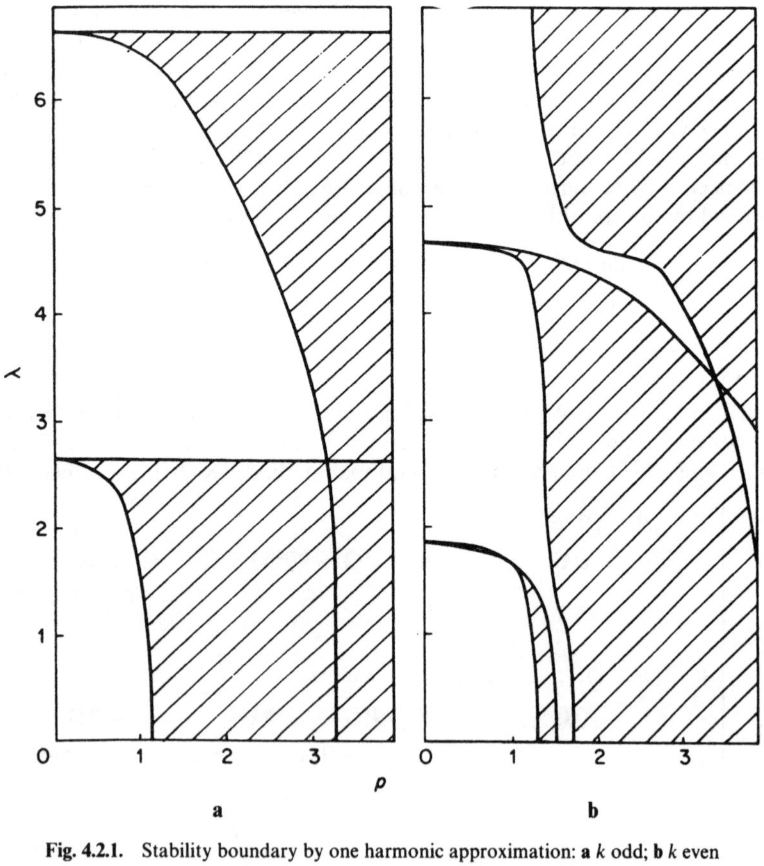

Fig. 4.2.1. Stability boundary by one harmonic approximation: **a** k odd; **b** k even

as an eigenvalue. Since $[\overline{\mathbf{D}}]$ can be rewritten as $[\overline{\mathbf{D}}(\lambda)] = [\overline{\mathbf{K}}] - \lambda[\overline{\mathbf{M}}]$, Sylvester's inertia theorem [12, 13] states that for positive-definite $[\overline{\mathbf{K}}]$ and $[\overline{\mathbf{M}}]$, the number of eigenvalues λ which are smaller than a specific value λ^* is equal to the number of negative elements on the main diagonal of $[\overline{\mathbf{D}}(\lambda^*)]$, which has been triangulated by Gaussian elimination without interchanges. If $[\overline{\mathbf{K}}]$ is not positive-definite, then the number of negative elements on the main diagonal of the triangulated $[\overline{\mathbf{D}}(\lambda^*)]$ must be modified by the number of negative elements on the main diagonal of the matrix $[\overline{\mathbf{K}}]$ triangulated by the Gaussian method without interchanges. Therefore, by counting the numbers, one is able to locate the eigenvalues with confidence. This counting technique has been generalized to frequency-dependent matrices and is known as the Wittrick–Williams algorithm. By varying P and Q, one can compute θ with confidence.

Consider a cantilever excited parametrically by a harmonic axial force $P(1 + 2\cos\theta t)$. Using just one element for the analysis, one obtains four matrices of order $2n$, corresponding to Eqs (4.2.10)–(4.2.13), where n is the number of harmonic terms considered. The roots of the matrix determinants for θ as P varies give the dynamic stability boundaries. Three harmonic terms are considered and the results for Eqs (4.2.10), (4.2.11), (4.2.12) and (4.2.13) are superimposed respectively in Figs 4.2.1–4.2.6. Figure 4.2.1a, b shows the approximations obtained by taking just one

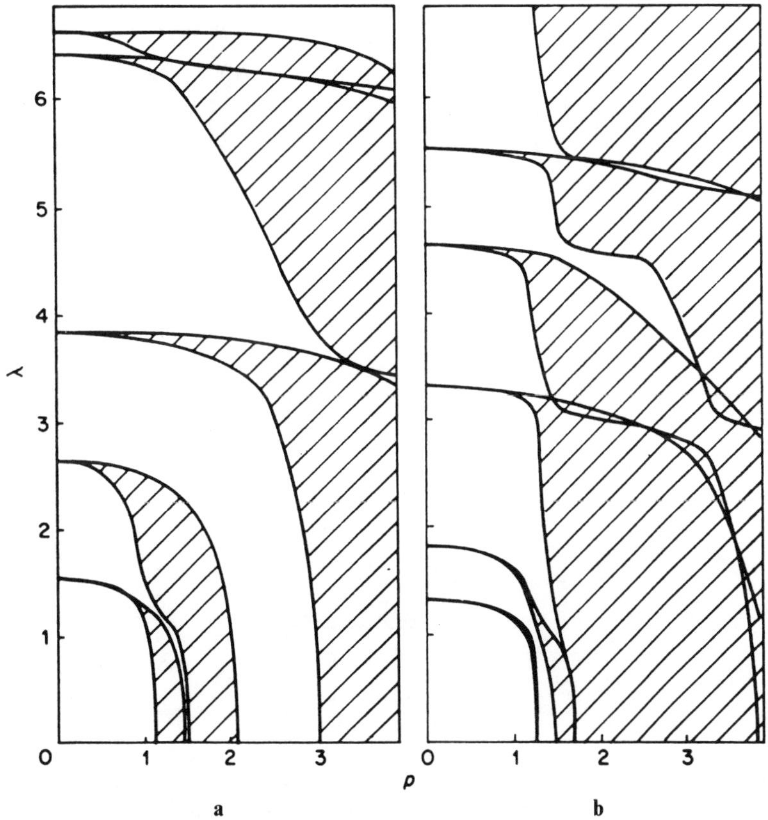

Fig. 4.2.2. As Fig. 4.2.1 but using two harmonic terms

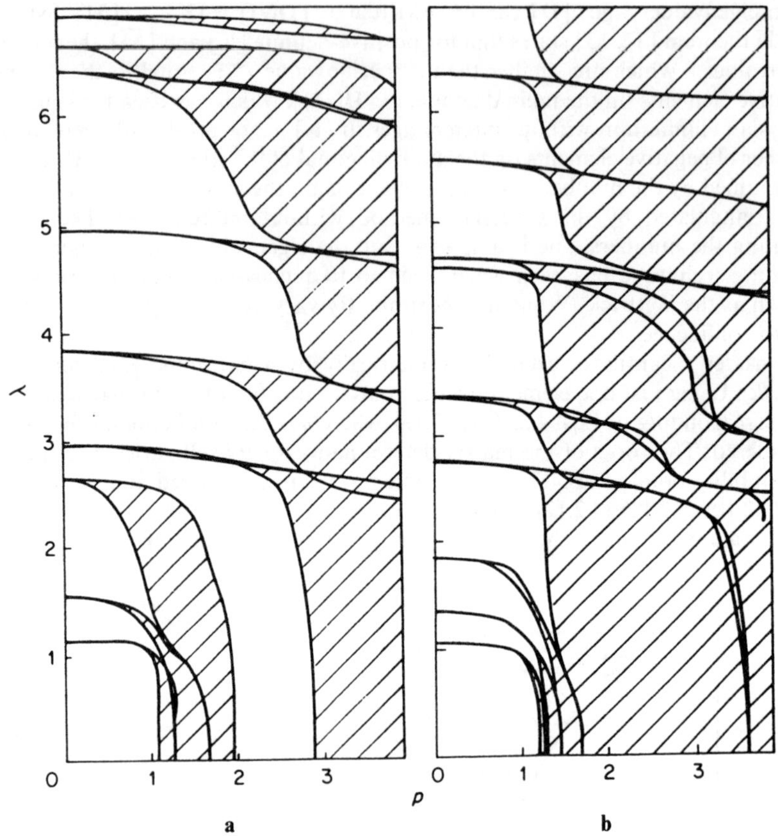

Fig. 4.2.3. As Fig. 4.2.1 but using three harmonic terms

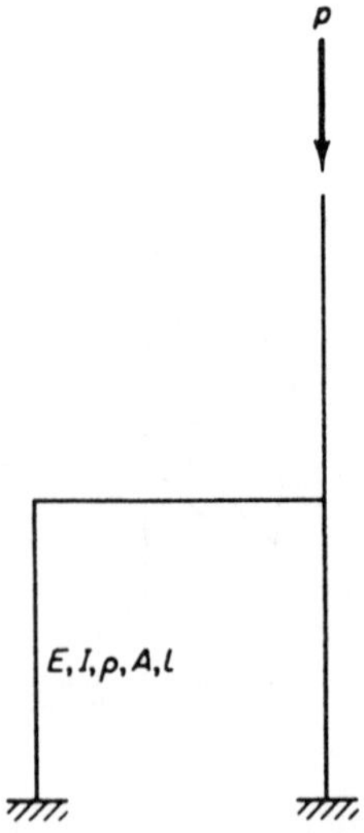

Fig. 4.2.4. An example frame

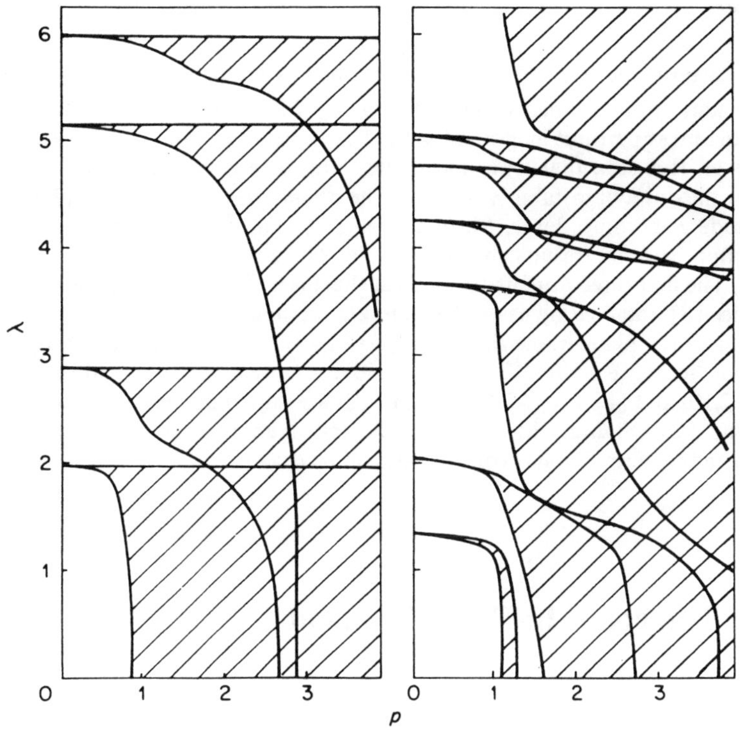

a b

Fig. 4.2.5. Stability boundaries of frame by one harmonic term: **a** k odd; **b** k even

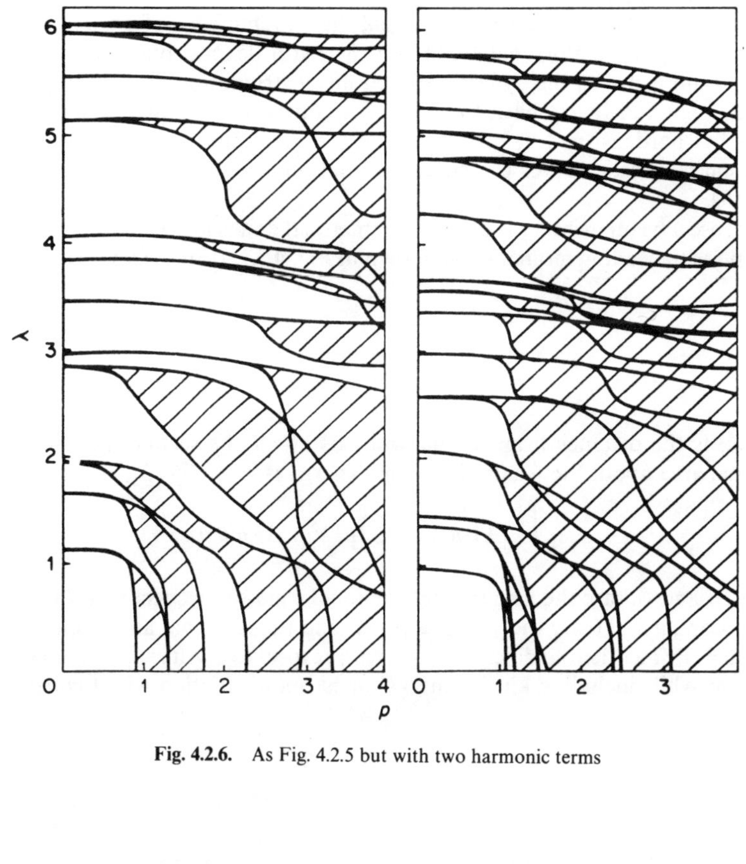

a b

Fig. 4.2.6. As Fig. 4.2.5 but with two harmonic terms

harmonic term and Fig. 4.2.2a, b those obtained by taking two harmonic terms. The coordinates are non-dimensionalized to $p = \sqrt{Pl^2 EI}$ and $\lambda = \sqrt[4]{\theta^2 \rho A l^4/EI}$, where ρA is the mass per unit length, EI is the flexural rigidity, and l is the length of the cantilever. It is seen that fewer harmonic terms give reasonably good approximations at the onset of instability, and that the boundaries vary substantially when the amplitude P becomes large. More elements are taken subsequently; however, no distinction from the results shown is noted. The shaded area denotes an unstable region. The first three principal regions at the onset of instability agree with those of reference [8] obtained by using the Galerkin method.

Consider the plane frame shown in Fig. 4.2.4 as a second example. Five d.o.f. per harmonic term are enough. Two harmonic terms are considered and the results for Eqs (4.2.10) and (4.2.11), (4.2.12) and (4.2.13) are superimposed respectively in Figs 4.2.5 and 4.2.6. Figure 4.2.5a, b shows the approximations obtained by taking just one harmonic term. Similar conclusions can be drawn.

4.3. Effects of In-Plane Moment

The effects of constant moments leading to lateral buckling are studied here by means of the dynamic stiffness method. The study is restricted to beam members whose shear centres coincide with their centroids.

A collection of papers on static torsional–flexural buckling of a beam member can be found in [14]. Unfortunately, in contrast with flexural buckling, the effects of constant moments on the dynamic behaviour of structures are rarely reported. For a single member, it has been found that a constant moment softens the flexural modes but hardens the torsional modes at the same time. Therefore, the convexity of the characteristic curves is lost, an unusual phenomenon for a conservative system [15].

Suppose a beam member is doubly symmetrical and is loaded initially by an equal and opposite bending moment L as shown in Fig. 4.3.1; then the initial stress is given by

$$\sigma_0 = \frac{L\eta}{I} \tag{4.3.1}$$

where η is the distance from the neutral axis and I the second moment of cross-sectional area about the y-axis. The strain energy is given by [3],

$$\frac{1}{2}\int_0^1 [EI(u'')^2 + GJ(\theta')^2 - 2Lu'\theta']\,\mathrm{d}z \tag{4.3.2}$$

where EI and GJ are the flexural and torsional rigidities respectively, l is the length of the beam, u and θ are the x displacement and the torsional angle, and warping is neglected. Primes denote differentiation with respect to z. After applying Hamilton's principle, while including kinetic energy for harmonic motion with frequency ω, we have

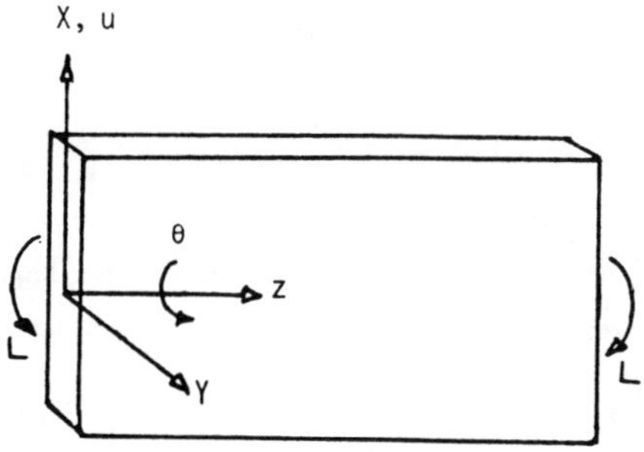

Fig. 4.3.1. A doubly symmetrical cross-section beam

$$EIu^{\text{iv}} + L\theta'' = m\omega^2 u \tag{4.3.3}$$

$$Lu'' - GJ\theta'' = mr^2\omega^2\theta \tag{4.3.4}$$

together with the natural boundary conditions

$$Q\delta u + R\delta u' + S\delta\theta = 0 \tag{4.3.5}$$

where

$$Q(z) = EIu''' + L\theta', \quad R(z) = -EIu'', \quad S(z) = Lu' - GJ\theta' \tag{4.3.6}$$

m is the translational inertia per unit length and mr^2 the torsional inertia per unit length. If L is a follower at one end, $z = l$ say, then

$$\bar{Q}(l) = EIu''', \quad \bar{R}(l) = -EIu'' \quad \text{and} \quad \bar{S}(l) = -GJ\theta' \tag{4.3.7}$$

The generalized displacement vector for the member is

$$\{\mathbf{q}\}^{\mathrm{T}} = [u(0), u'(0), \theta(0), u(l), u'(l), \theta(l)] \tag{4.3.8}$$

and the corresponding generalized force vector is

$$\{\mathbf{Q}\}^{\mathrm{T}} = [Q(0), R(0), S(0), -Q(l), -R(l), -S(l)] \tag{4.3.9}$$

Let $\lambda = m\omega^2 l^4/EI$, $\mu = L(l^2/EI)r$, $\phi = GJ(l^2/EI)r^2$ and $v = r\theta$, then Eqs (4.3.3) and (4.3.4) can be written as

$$\begin{bmatrix} \partial^4 & \mu\partial^2 \\ \mu\partial^2 & -\phi\partial^2 \end{bmatrix} \begin{Bmatrix} u \\ v \end{Bmatrix} = \lambda \begin{Bmatrix} u \\ v \end{Bmatrix} \tag{4.3.10}$$

where $\partial \equiv \mathrm{d}/\mathrm{d}\zeta$ and $\zeta = z/l$. To solve Eq. (4.3.10), one assumes $\{u, v\} = \mathrm{e}^{\alpha\zeta}\{\bar{u}, \bar{v}\}$, where α is a characteristic root, thus

$$\begin{bmatrix} \alpha^4 - \lambda & \mu\alpha^2 \\ \mu\alpha^2 & -\phi\alpha^2 - \lambda \end{bmatrix} \begin{Bmatrix} \bar{u} \\ \bar{v} \end{Bmatrix} = \{\mathbf{0}\} \tag{4.3.11}$$

For non-trivial solutions, the determinant of Eq. (4.3.11) must be zero, thus

$$\phi x^3 + (\lambda + \mu^2)x^2 - \lambda\phi x - \lambda^2 = 0 \tag{4.3.12}$$

where $x = \alpha^2$. Letting $\xi = (\lambda + \mu)/3\phi$ and $y = x + \xi$, Eq. (4.3.12) is reduced to the standard form,

$$y^3 - 3fy' + 2g = 0 \tag{4.3.13}$$

where $3f = 3\xi^2 + \lambda$ and $2g = 2\xi^3 + \lambda\xi - \lambda^2/\phi$. The discriminant $\Delta = g^2 - f^3$ of Eq. (4.3.13) determines the nature of the roots. Now,

$$\frac{27\Delta}{\lambda^2} = -\lambda\left(1 - \frac{\lambda}{\phi^2}\right)^2 - \frac{\mu^2}{\phi^2}\left[5\lambda + \frac{3\lambda^2 + 3\lambda\mu^2 + \mu^4}{\phi^2}\right] < 0 \tag{4.3.14}$$

since $\lambda = m\omega^2 l^4/EI > 0$. Therefore, all the three roots of Eq. (4.3.13), and hence Eq. (4.3.12), are real. The three roots of Eq. (4.3.12) are given by

$$x_i = -2\sqrt{q}\cos\left(\delta + \frac{2\pi i}{3}\right) - \xi, \qquad i = 0, 1, 2 \tag{4.3.15}$$

Since x_i are real, the characteristic roots of Eq. (4.3.11) must be either purely imaginary or real depending on whether x_i is negative or positive, and the characteristic solutions must be of the form $\cos \alpha_i\zeta$, $\sin \alpha_i\zeta$ and $\cosh \alpha_i\zeta$, $\sinh \alpha_i\zeta$, respectively. From Eq. (4.3.11),

$$B_i = \frac{v_i}{u_i} = \frac{\mu x_i}{\phi x_i + \lambda} = \frac{\lambda - x_i^2}{\mu x_i}, \qquad i = 0, 1, 2 \tag{4.3.16}$$

where the last two terms must be equal to each other providing a numerical check when implementing. Therefore, the solutions of Eq. (4.3.10) are

$$u(\zeta) = A_1 c(\alpha_0\zeta) + A_2 s(\alpha_0\zeta) + A_3 c(\alpha_1\zeta) + A_4 s(\alpha_1\zeta) + A_5 c(\alpha_2\zeta) + A_6 s(\alpha_2\zeta) \tag{4.3.17}$$

$$v(\zeta) = A_1 B_0 c(\alpha_0\zeta) + A_2 B_0 s(\alpha_0\zeta) + A_3 B_1 c(\alpha_1\zeta) + A_4 B_1 s(\alpha_1\zeta) + A_5 B_2 c(\alpha_2\zeta) + A_6 B_2 s(\alpha_2\zeta) \tag{4.3.18}$$

where $\alpha_i = \sqrt{|x_i|}$,

$$c(\alpha_i\zeta) = \cosh \alpha_i\zeta, \qquad s(\alpha_i\zeta) = \sinh \alpha_i\zeta \quad \text{if } x_i > 0$$

$$c(\alpha_i\zeta) = \cos \alpha_i\zeta, \qquad s(\alpha_i\zeta) = \sin \alpha_i\zeta \quad \text{if } x_i < 0$$

and A_k, $k = 1, 2, \ldots, 6$ are the six integration constants to be determined in terms of the six boundary conditions at the two ends of the beam. The derivatives of Eqs (4.3.17) and (4.3.18) are often required. It is convenient to denote

$$\frac{d}{d\zeta}c(\alpha_i\zeta) = j\alpha_i s(\alpha_i\zeta) \quad \text{and} \quad \frac{d}{d\zeta}s(\alpha_i\zeta) = \alpha_i c(\alpha_i\zeta) \tag{4.3.19}$$

where $j = 1$ if $x_i > 0$, and $j = -1$ if $x_i < 0$. By evaluating $u(z)$, $u'(z)$, and $\theta(z)$ at $z = 0$ and l we have

$$\begin{Bmatrix} u(0) \\ u'(0) \\ \theta(0) \\ u(l) \\ u'(l) \\ \theta(l) \end{Bmatrix} = \begin{bmatrix} 1 & 0 & 1 & 0 & 1 & 0 \\ 0 & \alpha_0/l & 0 & \alpha_1/l & 0 & \alpha_2/l \\ B_0/r & 0 & B_1/r & 0 & B_2/r & 0 \\ c_0 & s_0 & c_1 & s_1 & c_2 & s_2 \\ j\alpha_0 s_0/l & \alpha_0 c_0/l & j\alpha_1 s_1/l & \alpha_1 c_1/l & j\alpha_2 s_2/l & \alpha_2 c_2/l \\ B_0 c_0/r & B_0 s_0/r & B_1 c_1/r & B_1 s_1/r & B_2 c_2/r & B_2 s_2/r \end{bmatrix} \begin{Bmatrix} A_1 \\ A_2 \\ A_3 \\ A_4 \\ A_5 \\ A_6 \end{Bmatrix} \tag{4.3.20}$$

or, from Eq. (4.3.5),

$$\{\mathbf{q}\} = [\mathbf{C}]\{\mathbf{A}\} \tag{4.3.21}$$

where $[\mathbf{C}]$ and $\{\mathbf{A}\}$ are the corresponding matrix and vector in Eq. (4.3.20).

The relation between the nodal forces and the coefficients $\{\mathbf{A}\}$ can be obtained by substituting Eqs (4.3.17) and (4.3.18) into (4.3.6) and (4.3.9),

$$\{\mathbf{Q}\} = \frac{EI}{l^3}[\mathbf{Y}_0, \mathbf{Y}_1, \mathbf{Y}_2]\{\mathbf{A}\} = [\mathbf{Y}]\{\mathbf{A}\} \tag{4.3.22}$$

where

$$[\mathbf{Y}_i] = \begin{bmatrix} 0 & \alpha_i(j\alpha_i^2 + \mu B_i) \\ -j\alpha_i^2 l & 0 \\ 0 & \alpha_i r(\mu - \phi B_i) \\ -\alpha_i s_i(\alpha_i^2 + j\mu B_i) & -\alpha_i c_i(j\alpha_i^2 + \mu B_i) \\ j\alpha_i^2 c_i l & j\alpha_i^2 s_i l \\ -j\alpha_i s_i r(\mu - \phi B_i) & -\alpha_i c_i r(\mu - \phi B_i) \end{bmatrix}$$

and $c_i = c(\alpha_i)$, $s_i = s(\alpha_i)$, $i = 0, 1, 2$. Therefore, from Eqs (4.3.21) and (4.3.22), one has the dynamic stiffness relation,

$$\{\mathbf{Q}\} = [\mathbf{Y}][\mathbf{C}]^{-1}\{\mathbf{q}\} = [\mathbf{D}]\{\mathbf{q}\} \tag{4.3.23}$$

We can prove by a lengthy algebraic process that the dynamic stiffness $[\mathbf{D}]$ is symmetrical when Eqs (4.3.12) and (4.3.16) are observed. Therefore, it is computationally more efficient to write

$$[\mathbf{D}] = [\mathbf{C}^{-T}\mathbf{Y}^{T}] \tag{4.3.24}$$

where $[\mathbf{C}]$ and $[\mathbf{Y}]$ are given by Eqs (4.3.21) and (4.3.22).

If L is a follower moment at $z = l$, then $Q(l)$, $R(l)$ and $S(l)$ in Eq. (4.3.9) will be replaced by $\bar{Q}(l)$, $\bar{R}(l)$ and $\bar{S}(l)$ of Eq. (4.3.7). The last three rows of $[\mathbf{Y}_i]$ in Eq. (4.3.22) must be modified accordingly.

The influence of the initial stress σ_0 on the natural frequency ω_0 is usually shown on a characteristic diagram using σ and ω as ordinate and abscissa respectively. For a conservative system, the characteristic curves meet the ordinate and abscissa at right angles. It was proved by Huseyin [15] that the characteristic curve varies from zero slope at $\omega = 0$ to unit slope at $\sigma = 0$ monotonically and therefore is always convex for a conservative system. However, the monotonicity no longer applies if the initial stress is produced by an initial moment which softens the flexural modes but hardens the torsional modes simultaneously.

Let us consider a beam member which is fixed at $z = l$ and is flexurally restrained but torsionally free at $z = 0$ (Fig. 4.3.2). The initial constant moment is produced by an offset at the supports. Therefore, $u(0) = u'(0) = u(l) = u'(l) = \theta(l) = 0$, and there is only one d.o.f. $\theta(0) \neq 0$. The characteristic diagrams for $r/l = 0.02$ and 0.04 are shown in Fig. 4.3.3a, b. The ordinate is $\mu^{1/2}/\phi^{1/4} = (Ll/\sqrt{EI\,GJ})^{1/2}$ and the abscissa

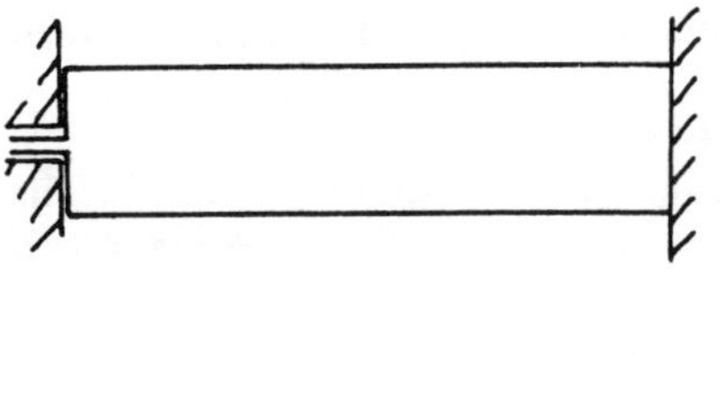

Fig. 4.3.2. A beam free to rotate at one end only

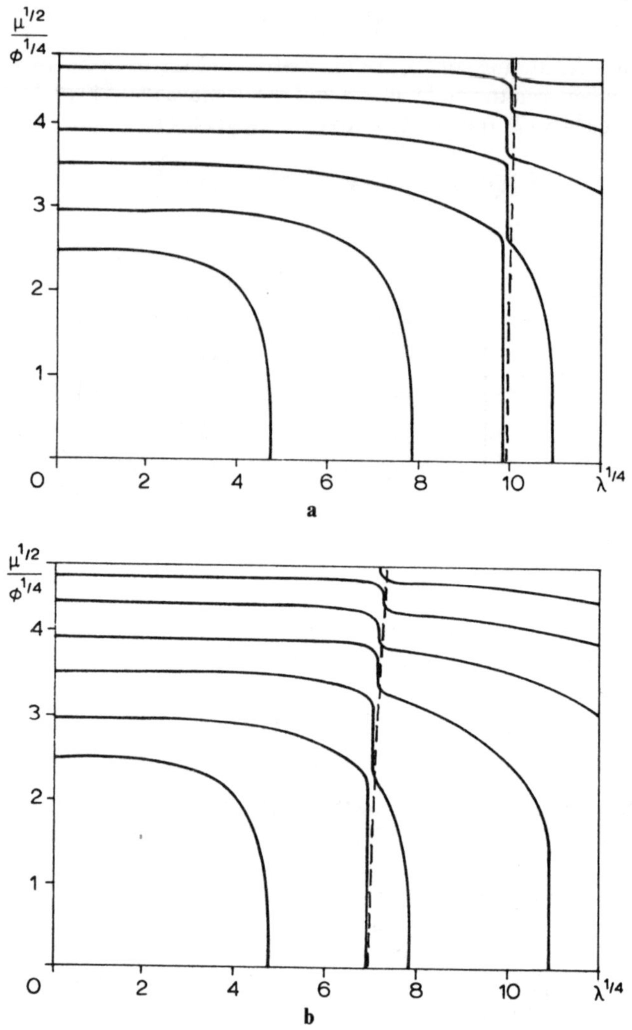

Fig. 4.3.3. Characteristic curves for a restrained beam: **a** $r/l = 0.02$; **b** $r/l = 0.04$

is $\lambda^{1/4}$ so that the buckling modes and unstressed natural modes are distributed almost uniformly along the coordinate axes. The first three unstressed flexural modes are approximately $\lambda^{1/4} = 4.73$, 4.78 and 11.00. The first unstressed torsional modes for $r/l = 0.02$ and 0.04 are $\lambda^{1/4} = 9.8699$ and 6.9791 respectively, for $E = 2.6G$, and $J = 4I$. The first buckling mode [16] is given by $L = 2\pi\sqrt{EIGJ}/l$, or $\mu^{1/2}/\phi^{1/4} = (\pi/2)^2 = 2.5066$. All these points are shown in Fig. 4.3.3. If the flexural deformation is neglected, then the dotted curves for torsional response result. It is clear that the initial moment softens the flexural modes but hardens the torsional modes. The monotonicity of slope variation in the characteristic curves is lost due to the interaction of the flexural and torsional modes.

Consider a cantilever subject to a follower moment L as another example. The unstressed flexural modes are approximately given by $\lambda^{1/4} = 1.87$, $(n + \frac{1}{2})\pi$, $n > 1$;

and the unstressed torsional modes are given by $\lambda^{1/4} = 1.1137\sqrt{\pi(2n-1)/2}\,\sqrt{l/r}$, $n = 1, 2, \ldots$. The characteristic curves are plotted in Fig. 4.3.4a–j for $r/l = 0.02$ to 0.20, where the unstressed flexural modes and torsional modes are labelled f and t respectively. The first conservative lateral buckling mode when $\lambda = 0$ is given by [16] $\mu/\phi^{1/2} = Ll/\sqrt{EIGJ} = \pi/2$, or $\mu^{1/2}/\phi^{1/4} = 1.2533$ which is lower than the first lateral flutter mode at [1] $\mu/\phi^{1/2} = 1.43\pi$ or $\mu^{1/2}/\phi^{1/4} = 2.1195$. It is seen that flutter

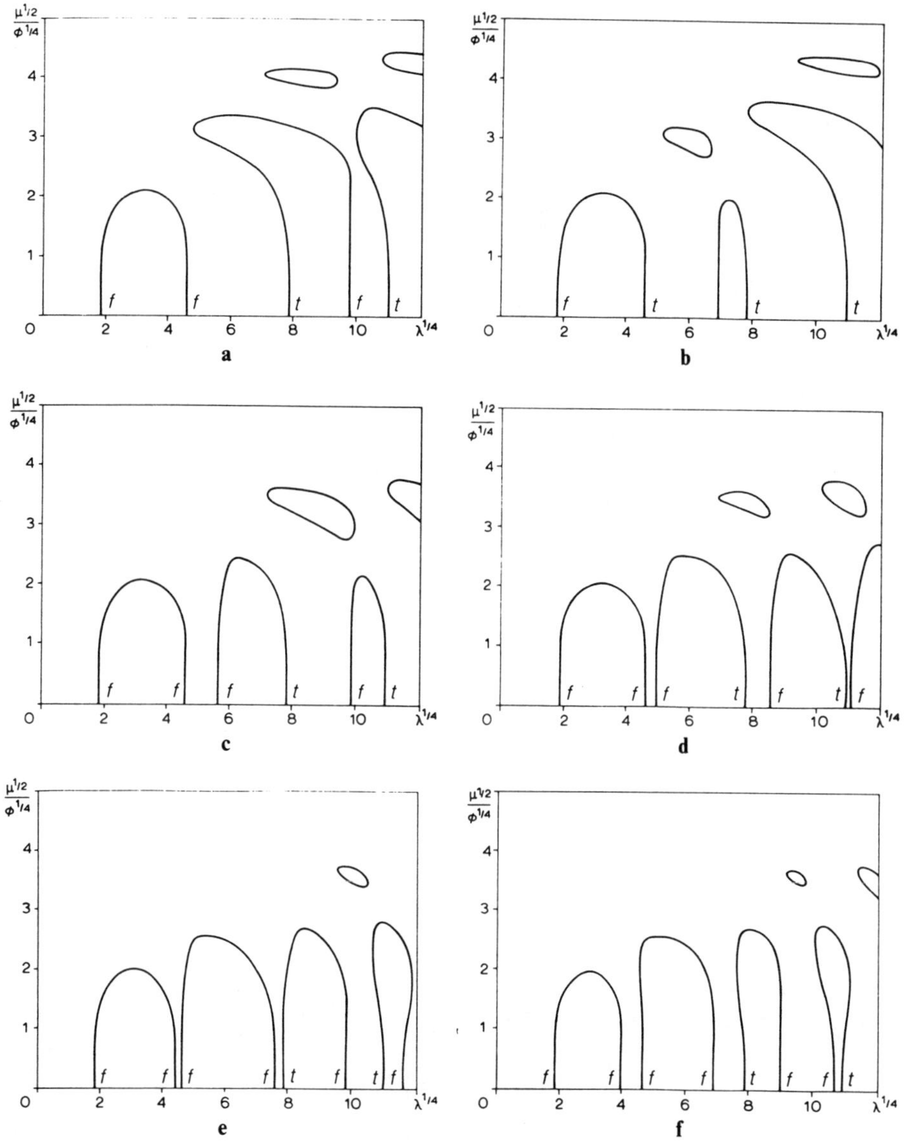

Fig. 4.3.4. Characteristic curves for a cantilever beam: **a** $r/l = 0.02$; **b** $r/l = 0.04$; **c** $r/l = 0.06$; **d** $r/l = 0.08$; **e** $r/l = 0.10$; **f** $r/l = 0.12$; **g** $r/l = 0.14$; **h** $r/l = 0.16$; **i** $r/l = 0.18$; **j** $r/l = 0.20$; **k** $r/l = 0.028$; **l** $r/l = 0.030$; **m** $r/l = 0.032$; **n** $r/l = 0.034$ *(continued on page 156)*

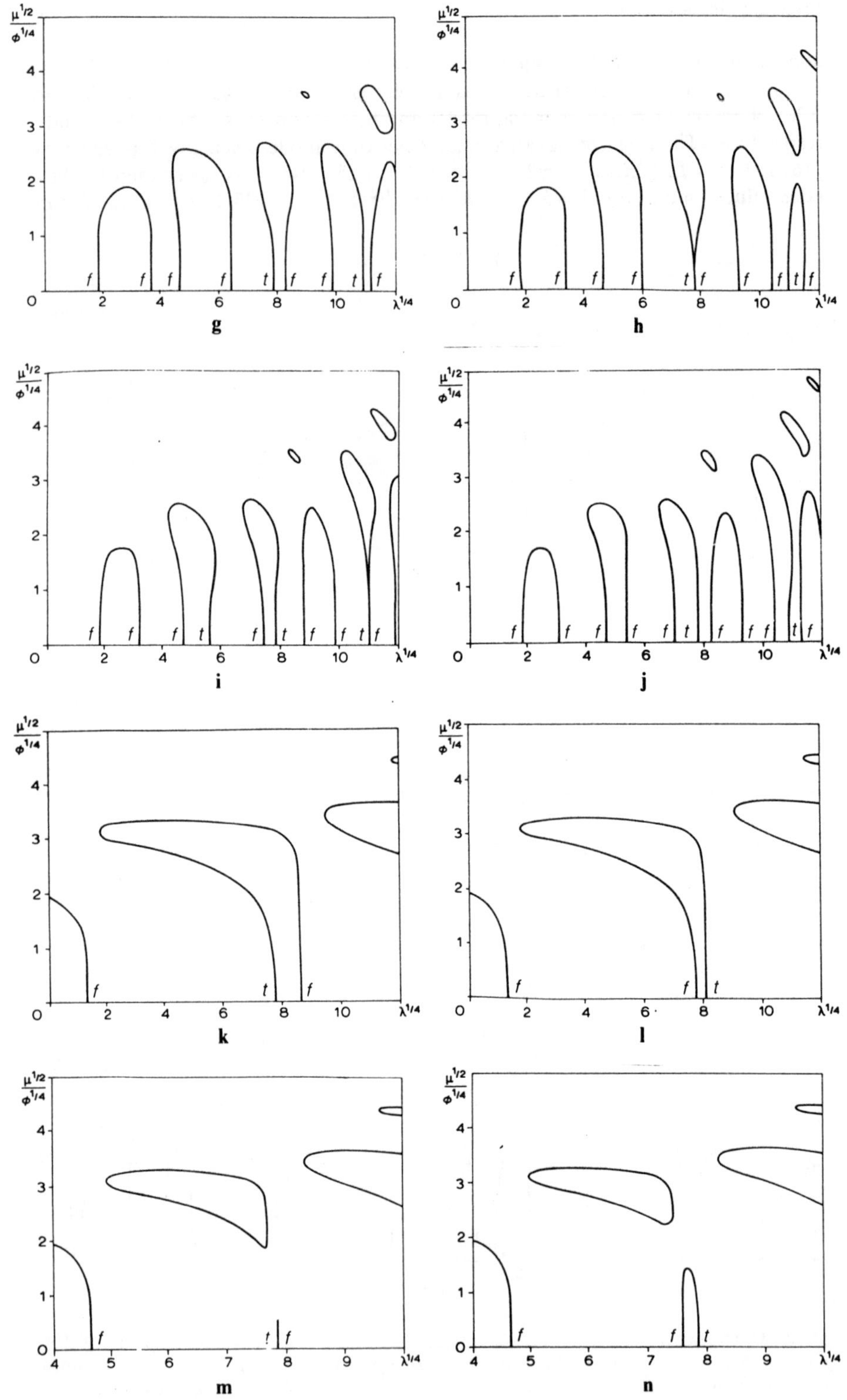

Fig. 4.3.4 *(continued)*

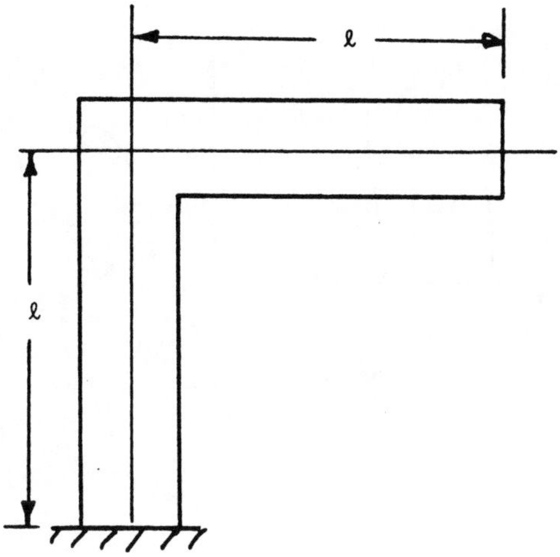

Fig. 4.3.5. An "L" frame

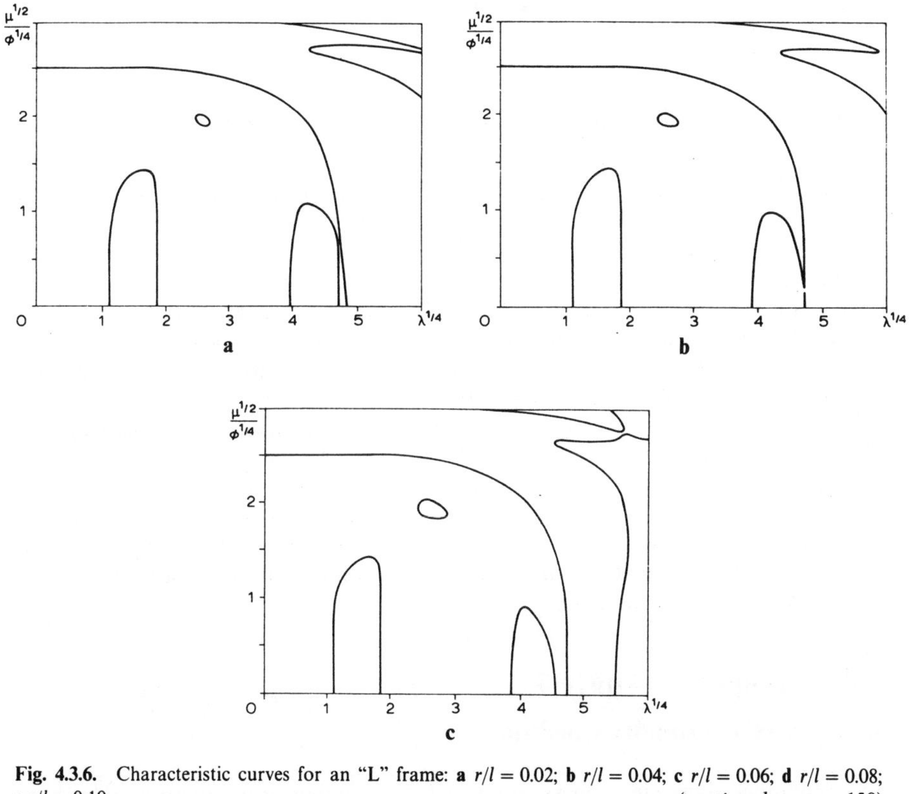

Fig. 4.3.6. Characteristic curves for an "L" frame: **a** $r/l = 0.02$; **b** $r/l = 0.04$; **c** $r/l = 0.06$; **d** $r/l = 0.08$; **e** $r/l = 0.10$ (*continued on page 158*)

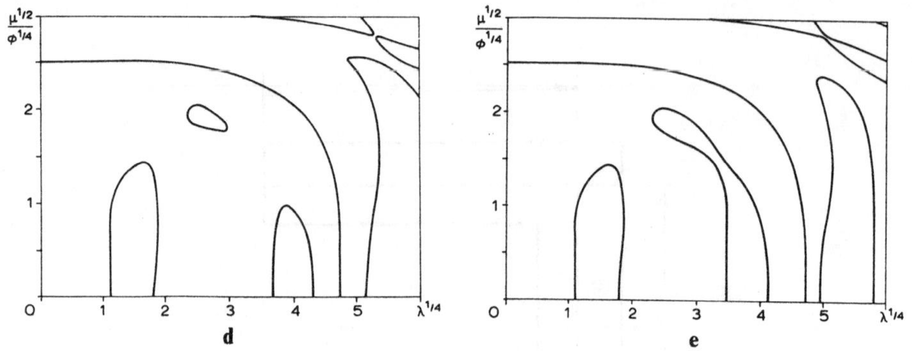

Fig. 4.3.6 (*continued*)

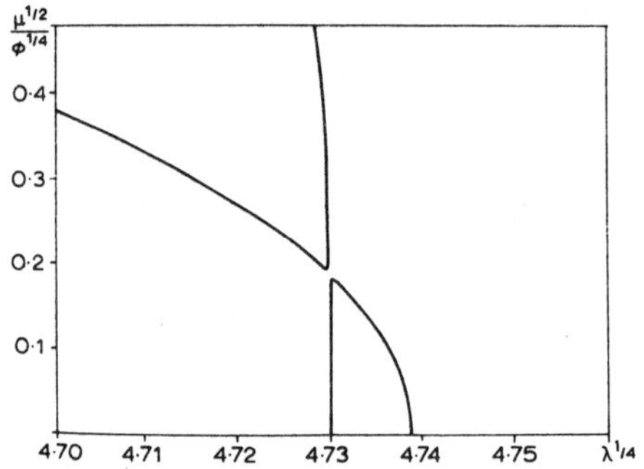

Fig. 4.3.7. An enlarged view of the gap in Fig. 4.3.6b

coupling of flexural and torsional modes is possible. Due to the mixed effects of softening and hardening, the follower moment can produce an isola response curve. It is observed in Fig. 4.3.4a, b that when r/l changes from 0.02 to 0.04 the first unstressed torsional mode changes from $\lambda^{1/4} = 9.8699$ to 6.979, which crosses the third flexural mode $\lambda^{1/4} = 7.854$ and is unaffected by r/l. As a result, an isola loop results. A closer view is presented in Fig. 4.3.4k–n when r/l varies from 0.028 to 0.034 in steps of 0.004. At $r/l = 0.032$, the first torsional mode coincides with the third flexural mode. An isola loop is formed.

The L-shaped frame subject to a follower moment is taken as the last example (Fig. 4.3.5). The physical properties of the two member beams are assumed to be identical. When r/l varies from 0.02 to 0.10 in steps of 0.02, the characteristic curves are plotted in Fig. 4.3.6a–e. The gap in Fig. 4.3.6b at $\lambda^{1/4} = 4.73$ is enlarged in Fig. 4.3.7 just when a separation is formed. Figure 4.3.6d, e shows the disappearance of an isola loop when r/l increases from 0.08 to 0.10, which is in contrast to the formation of an isola loop seen before.

4.4. Response Analysis

4.4.1. Orthonormality Condition

The undamped dynamic stiffness equation for a member or a structure under harmonic excitation can conveniently be written in terms of the symmetrical dynamic stiffness $[\mathbf{D}(\omega)]$

$$[\mathbf{D}(\omega)]\{\boldsymbol{\phi}\} = \{\mathbf{f}\} \tag{4.4.1}$$

where ω is the excitation frequency, $\{\mathbf{f}\}e^{i\omega t}$ and $\{\boldsymbol{\phi}\}e^{i\omega t}$ are the excitation and response vectors respectively. When the forcing vector vanishes, the non-trivial solutions of the response vector $\{\boldsymbol{\phi}\}e^{i\omega t}$ constitute the eigenvalue problem,

$$[\mathbf{D}(\omega_i)]\{\boldsymbol{\phi}_i\} = \{0\} \tag{4.4.2}$$

where ω_i is the natural frequency which is real for undamped systems. The subscript i denotes the ith natural frequency when the natural frequencies are arranged in ascending order $\omega_i \leq \omega_{i+1}$, etc. The natural frequencies are determined by equating the determinant of the dynamic stiffness to zero,

$$\det[\mathbf{D}(\omega_i)] = 0 \tag{4.4.3}$$

There is no obvious orthonormality condition for the modal vectors $\{\phi_i\}$. In order to be in parallel with the classical modal analysis, the following normalization condition is first suggested:

$$\{\boldsymbol{\phi}_i\}^{\mathrm{T}}[\mathbf{M}(\omega_i)]\{\boldsymbol{\phi}_i\} = 1 \tag{4.4.4}$$

where

$$[\mathbf{M}(\omega)] = -\frac{\mathrm{d}}{\mathrm{d}\omega^2}[\mathbf{D}(\omega)] \tag{4.4.5}$$

is the frequency-dependent mass matrix. Then the orthonormality condition is established in the following manner.

Premultiply Eq. (4.4.2) by $\{\boldsymbol{\phi}_j\}^{\mathrm{T}}$, $i \neq j$, $\omega_i \neq \omega_j$:

$$\{\boldsymbol{\phi}_j\}^{\mathrm{T}}[\mathbf{D}(\omega_i)]\{\boldsymbol{\phi}_i\} = 0 \tag{4.4.6}$$

Exchange the subscripts i and j and transpose:

$$\{\boldsymbol{\phi}_j\}^{\mathrm{T}}[\mathbf{D}(\omega_j)]\{\boldsymbol{\phi}_i\} = 0 \tag{4.4.7}$$

Subtract Eq. (4.4.7) from Eq. (4.4.6):

$$\{\boldsymbol{\phi}_j\}^{\mathrm{T}}[\mathbf{D}(\omega_i) - \mathbf{D}(\omega_j)]\{\boldsymbol{\phi}_i\} = 0 \tag{4.4.8}$$

The orthonormality condition (4.4.8) holds for $i \neq j$, $\omega_i \neq \omega_j$. When $i = j$ or $\omega_i = \omega_j$, the condition is identically zero and is not useful. We consider the case of distinct natural frequencies first and suggest a slight modification of Eq. (4.4.8),

$$\{\boldsymbol{\phi}_j\}^{\mathrm{T}}\frac{1}{\omega_j^2 - \omega_i^2}[\mathbf{D}(\omega_i) - \mathbf{D}(\omega_j)]\{\boldsymbol{\phi}_i\} = \delta_{ij} \tag{4.4.9}$$

where δ_{ij} is the Kronecker delta. The orthonormality condition (4.4.9) is equivalent to the orthonormality condition (4.4.8) if $i \neq j$, and to the normalization condition (4.4.4) if $i = j$. We further define the mixed mass matrix

$$[\mathbf{M}(\omega_i, \omega_j)] = \frac{1}{\omega_j^2 - \omega_i^2}[\mathbf{D}(\omega_i) - \mathbf{D}(\omega_j)] \tag{4.4.10}$$

then the orthonormal condition (4.4.9) becomes

$$\{\boldsymbol{\phi}_j\}^{\mathrm{T}}[\mathbf{M}(\omega_i, \omega_j)]\{\boldsymbol{\phi}_i\} = \delta_{ij} \tag{4.4.11}$$

The mixed mass matrix is symmetrical with respect to ω_i and ω_j, that is,

$$[\mathbf{M}(\omega_i, \omega_j)] = [\mathbf{M}(\omega_j, \omega_i)] \tag{4.4.12}$$

and degenerates to the mass matrix if $\omega_i = \omega_j$,

$$[\mathbf{M}(\omega_i, \omega_j)] = \lim_{\omega_i \to \omega_j} \frac{1}{\omega_j^2 - \omega_i^2} [\mathbf{D}(\omega_i) - \mathbf{D}(\omega_j)]$$

$$= -\frac{\mathrm{d}}{\mathrm{d}\omega^2} [\mathbf{D}(\omega_i)] = [\mathbf{M}(\omega_i)] \tag{4.4.13}$$

In particular, when $[\mathbf{D}(\omega)] = [\mathbf{K}] - \omega^2[\mathbf{M}]$, where $[\mathbf{K}]$ and $[\mathbf{M}]$ are constant matrices, as a result of the usual finite element method the orthonormal condition (4.4.11) degenerates to the classical condition

$$\{\boldsymbol{\phi}_i\}^{\mathrm{T}}[\mathbf{M}]\{\boldsymbol{\phi}_j\} = \delta_{ij}$$

Also, if we let

$$[\mathbf{D}(\omega)] = [\mathbf{K}(\omega)] - \omega^2[\mathbf{M}(\omega)] \tag{4.4.14}$$

then Eq. (4.4.2) implies that

$$\{\boldsymbol{\phi}_i\}^{\mathrm{T}}[\mathbf{D}(\omega_i)]\{\boldsymbol{\phi}_i\} = 0$$

or

$$\{\boldsymbol{\phi}_i\}^{\mathrm{T}}[\mathbf{K}(\omega_i)]\{\boldsymbol{\phi}_i\} = \omega_i^2\{\boldsymbol{\phi}_i\}^{\mathrm{T}}[\mathbf{M}(\omega_i)]\{\boldsymbol{\phi}_i\} = \omega_i^2 \tag{4.4.15}$$

4.4.2. Multiple Natural Frequencies

If $\omega_i = \omega_{i+1}$ is a double root of the determinantal equation (4.4.3), then the dynamic stiffness matrix $[\mathbf{D}(\omega_i)]$ has a degeneracy of order two and there are two independent associated eigenvectors $\{\boldsymbol{\phi}_i\}$ and $\{\mathbf{y}_i\}$. These two vectors are not necessarily orthogonal. However, they can be orthonormalized in the following manner.

Let $\{\boldsymbol{\phi}_{i+1}\}$ be undetermined but orthogonal to $\{\boldsymbol{\phi}_i\}$. Since $[\mathbf{D}(\omega_i)]$ is of second-order degeneracy, one of the three vectors $\{\boldsymbol{\phi}_i\}$, $\{\mathbf{y}_i\}$ and $\{\boldsymbol{\phi}_{i+1}\}$ is a linear combination of the other two. Let $\{\mathbf{y}_i\}$ be linearly dependent on the orthogonal vectors $\{\boldsymbol{\phi}_i\}$ and $\{\boldsymbol{\phi}_{i+1}\}$:

$$\{\mathbf{y}_i\} = \alpha\{\boldsymbol{\phi}_i\} + \beta\{\boldsymbol{\phi}_{i+1}\} \tag{4.4.16}$$

then

$$[\boldsymbol{\phi}_i\boldsymbol{\phi}_{i+1}]^{\mathrm{T}}[\mathbf{M}(\omega_i)][\boldsymbol{\phi}_i\boldsymbol{\phi}_{i+1}] = [\mathbf{I}] \tag{4.4.17}$$

where $[\mathbf{I}]$ is the identity matrix of order 2. Premultiply Eq. (4.4.16) by $\{\boldsymbol{\phi}_i\}^{\mathrm{T}}[\mathbf{M}(\omega_i)]$:

$$\{\boldsymbol{\phi}_i\}^{\mathrm{T}}[\mathbf{M}(\omega_i)]\{\mathbf{y}_i\} = \alpha\{\boldsymbol{\phi}_i\}^{\mathrm{T}}[\mathbf{M}(\omega_i)]\{\boldsymbol{\phi}_i\} + \beta\{\boldsymbol{\phi}_i\}^{\mathrm{T}}[\mathbf{M}(\omega_i)]\{\boldsymbol{\phi}_{i+1}\} \tag{4.4.18}$$

Apply Eq. (4.4.17), arriving at

$$\alpha = \{\boldsymbol{\phi}_i\}^{\mathrm{T}}[\mathbf{M}(\omega_i)]\{\mathbf{y}_i\} \tag{4.4.19}$$

and

$$\beta\{\phi_{i+1}\} = \{\mathbf{y}_i\} - \alpha\{\boldsymbol{\phi}_i\} \tag{4.4.20}$$

The normalizing factor β is determined so as to satisfy

$$\{\boldsymbol{\phi}_{i+1}\}^{\mathrm{T}}[\mathbf{M}(\omega_i)]\{\boldsymbol{\phi}_{i+1}\} = 1 \tag{4.4.21}$$

Therefore, for any two distinct vectors $\{\boldsymbol{\phi}_i\}$ and $\{\mathbf{y}_i\}$ associated with the degenerate dynamic stiffness matrix $[\mathbf{D}(\omega_i)]$, a modal vector $\{\boldsymbol{\phi}_{i+1}\}$ orthogonal to $\{\boldsymbol{\phi}_i\}$ with respect to the mass matrix $[\mathbf{M}(\omega_i)]$ can be found. The orthonormality condition (4.4.11) is assured even for double roots.

The extension to multiple natural frequencies is straightforward. We agree in the following development that all natural modes are orthonormalized according to condition (4.4.11), whether the natural frequencies are multiple or not.

4.4.3. Expansion Theorem

Suppose that there are m orthonormal modes $\{\boldsymbol{\phi}_i\}$, $i = 1, 2, \ldots, m$ available for the system (4.4.2) with n degrees of freedom. If $m > n$, the $\{\boldsymbol{\phi}_i\}$ are not linearly independent, so the expansion of an arbitrary vector $\{\boldsymbol{\phi}\}$ in terms of $\{\boldsymbol{\phi}_i\}$ presents difficulties. These can be removed by considering the distributed displacement within any two nodes instead of nodal displacement at the nodes.

In deriving the dynamic stiffness matrix for a harmonically vibrating member with frequency ω, the relation between the distributed displacement $\{\mathbf{u}_e(x)\}e^{i\omega t} = \{uvw\}e^{i\omega t}$ at the Cartesian coordinate $\mathbf{x} = (x, y, z)$ and the nodal displacement $\{\mathbf{q}_e\}e^{i\omega t}$ of a member has been used:

$$\{\mathbf{u}_e(x)\}e^{i\omega t} = [\mathbf{N}_e(\omega)]\{\mathbf{q}_e\}e^{i\omega t} \tag{4.4.22}$$

where $[\mathbf{N}_e(\omega)]$ is the frequency-dependent shape function matrix. Equation (4.4.22) represents the nodal transformation. The corresponding relation for the system is

$$\{\mathbf{u}(\mathbf{X})\}e^{i\omega t} = [\mathbf{N}(\omega)]\{\mathbf{q}\}e^{i\omega t} \tag{4.4.23}$$

When the members are connected at the boundary nodes to form the system, the following transformation is assumed:

$$\{\mathbf{q}_e\} = [\mathbf{T}_e]\{\mathbf{q}\} \quad \text{and} \quad \{\mathbf{u}(x)\} = \sum_e \{\mathbf{u}_e(x)\} \tag{4.4.24}$$

since the distributed displacements of individual members are exclusive. The system shape-function matrix relating the system internal displacements and system nodal displacements is established,

$$[\mathbf{N}(\omega)]\{\mathbf{q}\} = \sum_e [\mathbf{N}_e(\omega)]\{\mathbf{q}_e\} = \left(\sum_e [\mathbf{N}_e(\omega)][\mathbf{T}_e]\right)\{\mathbf{q}\} \tag{4.4.25}$$

or

$$[\mathbf{N}(\omega)] = \sum_e [\mathbf{N}_e(\omega)][\mathbf{T}_e] \tag{4.4.26}$$

If the system is vibrating in the ith natural mode, Eq. (4.4.23) becomes

$$\{\boldsymbol{\psi}_i(x)\}e^{i\omega_i t} = [\mathbf{N}(\omega_i)]\{\boldsymbol{\phi}_i\}e^{i\omega_i t} \tag{4.4.27}$$

where $\{\boldsymbol{\Psi}_i(x)\}$ is a three-component modal vector in terms of distributed displacements. Expanding an arbitrary three-component distributed vector $\{\mathbf{v}(\mathbf{X})\}$ in the distributed modal vectors $\{\boldsymbol{\psi}_i(x)\}$ will present no difficulties:

$$\{\mathbf{v}(x)\} = \sum_i \alpha_i\{\boldsymbol{\psi}_i(x)\} \tag{4.4.28}$$

Let $\{\mathbf{v}(x)\}$ represent an arbitrary distributed displacement of the system. The coefficients α_i can be determined in the following manner. The nodal transformation is

$$\{\mathbf{v}(x)\} = [\mathbf{N}(0)]\{\mathbf{q}\} = [\mathbf{N}]\{\mathbf{q}\} \tag{4.4.29}$$

where $[\mathbf{N}] = [\mathbf{N}\ (\omega = 0)]$ is the classical finite-element shape function. Premultiply Eq. (4.4.28) by $\{\boldsymbol{\psi}_j(x)\}^{\mathrm{T}}[\mathbf{m}]$, where $[\mathbf{m}]$ is the mass density matrix, and integrate over the whole volume:

$$\int \{\boldsymbol{\psi}_j(x)\}^{\mathrm{T}}[\mathbf{m}]\{\mathbf{v}(x)\}\,\mathrm{d}\,vol = \sum_i \alpha_i \int \{\boldsymbol{\psi}_j(x)\}^{\mathrm{T}}[\mathbf{m}]\{\boldsymbol{\psi}_i(x)\}\,\mathrm{d}\,vol \tag{4.4.30}$$

where d vol is an element of volume.

By means of the nodal transformation (4.4.27), the right-hand side of (4.4.30) becomes

$$\sum_i \alpha_i \int \{\boldsymbol{\phi}_i\}^{\mathrm{T}}[\mathbf{N}(\omega_j)]^{\mathrm{T}}[\mathbf{m}][\mathbf{N}(\omega_i)]\{\boldsymbol{\phi}_i\}\,\mathrm{d}\,vol$$

$$= \sum_i \alpha_i \{\boldsymbol{\phi}_j\}^{\mathrm{T}} \int [\mathbf{N}(\omega_j)]^{\mathrm{T}}[\mathbf{m}][\mathbf{N}(\omega_i)]\,\mathrm{d}\,vol\,\{\boldsymbol{\phi}_i\} \tag{4.4.31}$$

However, we shall prove in Sect. 4.4.5 that

$$[\mathbf{M}(\omega_i, \omega_j)] = \int [\mathbf{N}(\omega_i)]^{\mathrm{T}}[\mathbf{m}][\mathbf{N}(\omega_j)]\,\mathrm{d}\,vol \tag{4.4.32}$$

then the right-hand side of Eq. (4.4.30) becomes

$$\sum_i \alpha_i \{\boldsymbol{\phi}_j\}^{\mathrm{T}}[\mathbf{M}(\omega_i, \omega_j)] = \alpha_j \tag{4.4.33}$$

due to the orthonormality condition (4.4.11). By means of the nodal transformation (4.4.29), the left-hand side of Eq. (4.4.30) is simply

$$\int \{\boldsymbol{\phi}_j\}^{\mathrm{T}}[\mathbf{N}(\omega_j)]^{\mathrm{T}}[\mathbf{m}][\mathbf{N}(0)]\{\mathbf{q}\}\,\mathrm{d}\,vol = \{\boldsymbol{\phi}_j\}^{\mathrm{T}}[\mathbf{M}(\omega_j, 0)]\{\mathbf{q}\} \tag{4.4.34}$$

Combining Eqs (4.4.33) and (4.4.34), we have

$$\alpha_j = \{\boldsymbol{\phi}_j\}^{\mathrm{T}}[\mathbf{M}(\omega_j, 0)]\{\mathbf{q}\} \tag{4.4.35}$$

Equations (4.4.28) and (4.4.35) establish the expansion theorem.

4.4.4. Modal Analysis

For an undamped system, the natural modes ω_i, $\{\boldsymbol{\psi}_i(\mathbf{X})\}$ and $\{\boldsymbol{\phi}_i\}$ are real. Suppose that the system is subject to a non-harmonic distributed force $\{\mathbf{f}(\mathbf{X}, t)\}$ and initial conditions

$$\{\mathbf{u}(\mathbf{X}, 0)\} = \{\mathbf{u}_0\} \quad \text{and} \quad \{\dot{\mathbf{u}}(\mathbf{X}, 0)\} = \{\dot{\mathbf{u}}_0\} \tag{4.4.36}$$

The distributed response $\{\mathbf{u}(\mathbf{X}, t)\}$ can be obtained as follows by the above expansion theorem

$$\{\mathbf{u}(\mathbf{X}, t)\} = \sum_i p_i(t)\{\boldsymbol{\psi}_i(\mathbf{X})\} \tag{4.4.37}$$

where $p_i(t)$ is the time-dependent principal coordinate to be determined and $\{\boldsymbol{\psi}_i(\mathbf{x})\}$

is the ith distributed natural mode. When x is taken at the nodes, Eq. (4.4.37) becomes

$$\{\mathbf{q}(t)\} = \sum_i p_i(t)\{\boldsymbol{\phi}_i\} \tag{4.4.38}$$

Lagrange's equations are

$$\frac{d}{dt}\frac{\partial \mathbf{L}}{\partial \dot{p}} - \frac{\partial \mathbf{L}}{\partial p_i} = P_i \tag{4.4.39}$$

where $\mathbf{L} = \mathbf{T} - \mathbf{U}$ is the Lagrangian, and

$$\mathbf{T} = \frac{1}{2}\int \{\dot{\mathbf{u}}(x,t)\}^{\mathrm{T}}[\mathbf{m}]\{\dot{\mathbf{u}}(\mathbf{x},t)\}\,\mathrm{d}\,vol \tag{4.4.40}$$

and

$$\mathbf{U} = \frac{1}{2}\int \{\varepsilon(x,t)\}^{\mathrm{T}}[\mathbf{E}]\{\varepsilon(\mathbf{x},t)\}\,\mathrm{d}\,vol \tag{4.4.41}$$

are the kinetic energy and strain energy of the system respectively. The P_i are the principal forces to be derived later by means of external work, $[\mathbf{m}]$ is the mass density per unit volume and $[\mathbf{E}]$ is the matrix of elastic moduli. The strain is related to the displacement by

$$[\varepsilon(x,t)] = [\mathbf{b}]\{\mathbf{u}(x,t)\} \tag{4.4.42}$$

where $[\mathbf{b}]$ is a given differential operator in spatial coordinates.

Under the modal transformation (4.4.37) and the nodal transformation (4.4.27), the kinetic energy (4.4.40) reduces to

$$\begin{aligned}\mathbf{T} &= \frac{1}{2}\sum_{ij}\dot{p}_i(t)\dot{p}_j(t)\int \{\boldsymbol{\psi}_i(x)\}^{\mathrm{T}}[\mathbf{m}]\{\boldsymbol{\psi}_j(x)\}\,\mathrm{d}\,vol \\ &= \frac{1}{2}\sum_{ij}\dot{p}_i(t)\dot{p}_j(t)\{\boldsymbol{\phi}_i\}^{\mathrm{T}}\int [\mathbf{N}(\omega_i)]^{\mathrm{T}}[\mathbf{m}][\mathbf{N}(\omega_j)]\,\mathrm{d}\,vol\,\{\boldsymbol{\phi}_j\} \\ &= \frac{1}{2}\sum_{ij}\dot{p}_i(t)\dot{p}_j(t)\{\boldsymbol{\phi}_i\}[\mathbf{M}(\omega_i,\omega_j)][\boldsymbol{\phi}_j\} \\ &= \frac{1}{2}\sum_{ij}\dot{p}_i^2(t)\end{aligned} \tag{4.4.43}$$

where the definition of the mixed mass matrix (4.4.32) and the orthonormality condition (4.4.11) have been used. Before processing the strain energy, we note that

$$\begin{aligned}[\varepsilon(x,t)] &= [\mathbf{b}]\{\mathbf{u}(\mathbf{X},t)\} = \sum_i p_i(t)[\mathbf{b}]\{\boldsymbol{\psi}_i(x)\} \\ &= \sum_i p_i(t)[\mathbf{b}][\mathbf{N}(\omega_i)]\{\boldsymbol{\phi}_i\} = \sum_i p_i(t)[\mathbf{B}(\omega_i)]\,\mathrm{d}\,vol\end{aligned} \tag{4.4.44}$$

$$\begin{aligned}\mathbf{U} &= \frac{1}{2}\sum_i p_i^2(t)\int \{\boldsymbol{\psi}_i(x)\}^{\mathrm{T}}[\mathbf{E}]\{\boldsymbol{\psi}_i(x)\}\,\mathrm{d}\,vol \\ &= \frac{1}{2}\sum_i p_i^2(t)\{\boldsymbol{\phi}_i\}\int [\mathbf{B}(\omega_i)]^{\mathrm{T}}[\mathbf{E}][\mathbf{B}(\omega_i)]\,\mathrm{d}\,vol\end{aligned}$$

$$= \frac{1}{2} \sum_i p_i^2(t) \{\boldsymbol{\phi}_i\} [\mathbf{K}(\omega_i)] \{\boldsymbol{\phi}_i\}$$

$$= \frac{1}{2} \sum \omega_i^2 p_i^2(t) \tag{4.4.45}$$

where

$$[\mathbf{K}(\omega_i)] = \int [\mathbf{B}(\omega_i)]^{\mathrm{T}} [\mathbf{E}] [\mathbf{B}(\omega_i)] \, \mathrm{d}\, vol$$

and Eq. (4.4.15) has been used. The equivalent work done due to the external distributed force $\{\mathbf{f}(\mathbf{X}, t)\}$ is given by

$$\mathbf{W} = \int \{\mathbf{f}(x, t)\}^{\mathrm{T}} \{\mathbf{u}(x, t)\} \, \mathrm{d}\, vol \tag{4.4.46}$$

Further reduction is possible if the distributed force $\{\mathbf{f}(\mathbf{X}, t)\}$ is expanded in terms of the natural modes:

$$\{\mathbf{f}(\mathbf{X}, t)\} = \sum_i P_i(t) \{\boldsymbol{\psi}_i(\mathbf{X})\} \tag{4.4.47}$$

By means of the expansion theorem, we premultiply Eq. (4.4.47) by $\{\boldsymbol{\psi}_i(\mathbf{X})\}^{\mathrm{T}}[\mathbf{m}]$ and integrate over the whole volume:

$$\int \{\boldsymbol{\psi}_j(\mathbf{X})\}^{\mathrm{T}} [\mathbf{m}] \{\mathbf{f}(\mathbf{x}, t)\} \, \mathrm{d}\, vol = \sum_i P_i(t) \int \{\boldsymbol{\psi}_j(\mathbf{X})\}^{\mathrm{T}} [\mathbf{m}] \{\boldsymbol{\psi}_i(\mathbf{X})\} \, \mathrm{d}\, vol = P_j(t) \tag{4.4.48}$$

The integration in Eq. (4.4.48) can be simplified if the distributed force vector can be approximated in the finite-element sense:

$$\{\mathbf{f}(x, t)\} = [\mathbf{N}(0)] \{\mathbf{F}(t)\} \tag{4.4.49}$$

where $\{\mathbf{F}(t)\}$ is the nodal value of $\{\mathbf{f}(x, t)\}$. Then Eq. (4.4.48) requires that

$$P_i(t) = \int \{\boldsymbol{\phi}_i\}^{\mathrm{T}} [\mathbf{N}(\omega_i)]^{\mathrm{T}} [\mathbf{m}] [\mathbf{N}(0)] \{\mathbf{F}(t)\} \, \mathrm{d}\, vol$$

$$= \{\boldsymbol{\phi}_i\}^{\mathrm{T}} [\mathbf{M}(\omega_i, 0)] \{\mathbf{F}(t)\} = \{\boldsymbol{\gamma}_i\}^{\mathrm{T}} \{\mathbf{F}(t)\} \tag{4.4.50}$$

and explicit integration is completely eliminated. Here $\{\boldsymbol{\gamma}_i\}^{\mathrm{T}}$ is defined as the distributed modal participation factor. If there is another external nodal force $\{\mathbf{Q}(t)\}$, then the additional work done is $\{\mathbf{Q}(t)\}^{\mathrm{T}}\{\mathbf{q}(t)\}$ and

$$\{\mathbf{Q}(t)\} = \sum_i P_i(t) \{\boldsymbol{\phi}_i\} \tag{4.4.51}$$

After premultiplying by $\{\boldsymbol{\phi}_j\}^{\mathrm{T}}[\mathbf{M}(\omega_i, \omega_j)]$, the additional modal force is given by

$$P_j(t) = \{\boldsymbol{\phi}_j\}^{\mathrm{T}} [\mathbf{M}(\omega_j)] \{\mathbf{Q}(t)\} = \{\boldsymbol{\Gamma}_i\}^{\mathrm{T}} \{\mathbf{Q}(t)\} \tag{4.4.52}$$

where $\{\boldsymbol{\Gamma}_i\}$ is the vector of modal participation factors. When both a distributed force and concentrated force are present,

$$P_i(t) = \{\boldsymbol{\phi}_i\}^{\mathrm{T}} [\mathbf{M}(\omega_i, 0)] \{\mathbf{F}(t)\} + \{\boldsymbol{\phi}_i\}^{\mathrm{T}} [\mathbf{M}(\omega_i)] \{\mathbf{Q}(t)\}$$

$$= \{\boldsymbol{\gamma}_i\}^{\mathrm{T}} \{\mathbf{F}(t)\} + \{\boldsymbol{\Gamma}_i\}^{\mathrm{T}} \{\mathbf{Q}(t)\} \tag{4.4.53}$$

Substituting Eqs (4.4.43), (4.4.45) and (4.4.53) into Eq. (4.4.39) gives

$$\ddot{p}_i(t) + \omega_i^2 p_i(t) = P_i(t) \tag{4.4.54}$$

The initial conditions for the principal coordinates are derived in the following. Assume the nodal and modal transformations

$$\{\mathbf{u}_0\} = \sum_i p_i(0)\{\boldsymbol{\psi}_i(x)\} = [\mathbf{N}(0)]\{\mathbf{q}_0\}$$

and

$$\{\dot{\mathbf{u}}_0\} = \sum_i \dot{p}_i(0)\{\boldsymbol{\psi}_i(x)\} = [\mathbf{N}(0)]\{\dot{\mathbf{q}}_0\}$$

Premultiply by $\{\boldsymbol{\psi}_j(x)\}^{\mathrm{T}}[\mathbf{m}]$ and integrate over the whole volume:

$$\int \{\boldsymbol{\psi}_j(x)\}^{\mathrm{T}}[\mathbf{m}][\mathbf{N}(0)]\{\mathbf{q}_0\}\,\mathrm{d}\,vol = \sum_i p_i(0)\int \{\boldsymbol{\psi}_j(\mathbf{X})\}^{\mathrm{T}}[\mathbf{m}]\{\boldsymbol{\psi}_i(x)\}\,\mathrm{d}\,vol$$

and

$$\left.\begin{aligned} p_i(0) &= \{\boldsymbol{\phi}_i\}^{\mathrm{T}}[\mathbf{M}(\omega_i, 0)]\{\mathbf{q}_0\} = \{\gamma_i\}^{\mathrm{T}}\{\mathbf{q}_0\} \\ p_i(0) &= \{\boldsymbol{\phi}_i\}^{\mathrm{T}}[\mathbf{M}(\omega_i, 0)]\{\dot{\mathbf{q}}_0\} = \{\gamma_i\}^{\mathrm{T}}\{\dot{\mathbf{q}}_0\} \end{aligned}\right\} \tag{4.4.55}$$

The solution of the uncoupled second order differential equations (4.4.54) with the initial conditions (4.4.55) is given by

$$p_i(t) = \int_0^t P_i(\tau)h_i(t-\tau)\,\mathrm{d}\tau + g_i(t)p_i(0) + h_i(t)\dot{p}_i(0) \tag{4.4.56}$$

where

$$h_i(t) = \frac{1}{\omega_i}\sin\omega_i t, \qquad g_i(t) = \cos\omega_i t$$

Example 4.4.1

Consider the system with three degrees of freedom shown in Fig. 4.4.1. Let $EI = EA = L = 1$ where E, A, I, L are the Young's modulus, cross-sectional area, second moment of area and element length respectively. The dynamic and mass matrices are given by Kolousek [17, 18]

$$[\mathbf{D}(\omega)] = \begin{bmatrix} F_6 & -F_4 & F_3 \\ -F_4 & F_2 & F_1 \\ F_3 & F_1 & 2F_2 \end{bmatrix} \quad \text{and} \quad [\mathbf{M}(\omega)] = \begin{bmatrix} G_6 & -G_4 & G_3 \\ -G_4 & G_2 & G_1 \\ G_3 & G_1 & 2G_2 \end{bmatrix}$$

where $G_i = -\partial F_i/\partial\lambda^4$ and

$$\begin{aligned}
F_1 &= -\lambda(\sinh\lambda - \sin\lambda)/\delta & G_1 &= (F_1F_2 - F_3 - F_1)/4\lambda^4 \\
F_2 &= -\lambda(\cosh\lambda\sin\lambda - \sinh\lambda\cos\lambda)/\delta & G_2 &= (F_1^2 - F_2)/4\lambda^4 \\
F_3 &= -\lambda^2(\cosh\lambda - \cos\lambda)/\delta & G_3 &= -(F_1F_4 + 2F_3)/4\lambda^4 \\
F_4 &= \lambda^2(\sinh\lambda\sin\lambda)/\delta & G_4 &= -(F_1F_3 + 2F_4)/4\lambda^4 \\
F_5 &= \lambda^3(\sinh\lambda + \sin\lambda)/\delta & G_5 &= (F_3F_4 - 3F_5)/4\lambda^4 \\
F_6 &= -\lambda^3(\cosh\lambda\sin\lambda + \sinh\lambda\cos\lambda)/\delta & G_6 &= (F_3^2 - 3F_6)/4\lambda^4 \\
\delta &= \cosh\lambda\cos\lambda - 1
\end{aligned}$$

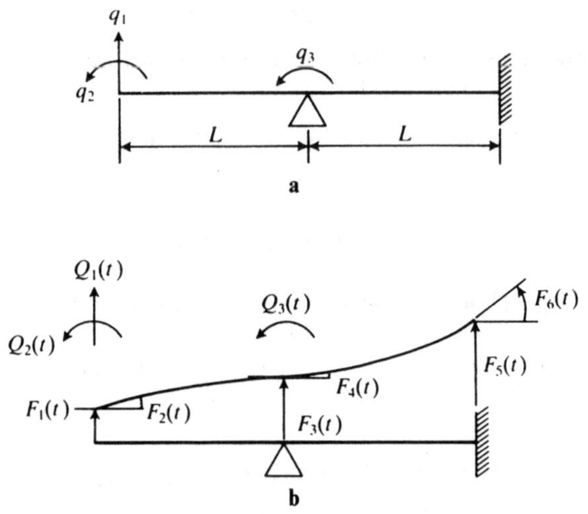

Fig. 4.4.1. **a** An example continuous beam. **b** Generalized forces

The first 50 modes ω_i are computed by letting the determinantal equation $\det[\mathbf{D}(\omega)]$ vanish, and the modal vectors $\{\boldsymbol{\phi}_i\}$ are othonormalized with respect to the mass matrix at resonance, $[\mathbf{M}(\omega_i)]$. If only concentrated nodal forces are of interest, the nodal participation factors $\{\boldsymbol{\Gamma}_i\}^{\mathrm{T}} = \{\boldsymbol{\phi}_i\}^{\mathrm{T}}[\mathbf{M}(\omega_i, 0)]$ are required. The mixed mass matrix is given by

$$[\mathbf{M}(\omega_i, 0)] = \frac{1}{\omega_i}\begin{bmatrix} 12 - F_6 & 6 + F_4 & -12 - F_5 & 6 - F_3 & 0 & 0 \\ 6 + F_4 & 4 - F_2 & -6 + F_3 & 2 - F_1 & 0 & 0 \\ 6 - F_3 & 2 - F_1 & 0 & 8 - 2F_2 & -6 + F_3 & 2 - F_1 \end{bmatrix},$$

where the third, fifth and sixth rows have been deleted due to zero displacements.

Table 4.4.1 shows the mode numbers in the first column, natural frequencies in rad s^{-1} in the second column, normalized modal vectors in the third to fifth columns and participation factors $\{\boldsymbol{\Gamma}_i\}$ in the sixth to eight columns; Table 4.4.2 lists the distributed participation factors $\{\boldsymbol{\gamma}_i\}$.

When $\{\mathbf{Q}(t)\}$ and $\{\mathbf{F}(t)\}$ are given, it is a simple matter to calculate the modal forces using Eq. (4.4.53) and solve the modal coordinates with the help of Eq. (4.4.56). The required response is obtained from Eq. (4.4.38). The initial conditions can also be evaluated using distributed modal participation factors $\{\boldsymbol{\gamma}_i\}$ in Eq. (4.4.55).

4.4.5. Proof that $[\mathbf{M}(\omega_1, \omega_2)] = \int [\mathbf{N}(\omega_1)]^{\mathrm{T}}[\mathbf{m}][\mathbf{N}(\omega_2)]\,\mathrm{d}\,vol$

Let the body of interest be excited by two systems of nodal forces $\{\mathbf{Q}_1\}\mathrm{e}^{\mathrm{i}\omega_1 t}$, $\{Q_2\}\mathrm{e}^{\mathrm{i}\omega_2 t}$. The steady-state responses are $\{\mathbf{u}_1\}\mathrm{e}^{\mathrm{i}\omega_1 t}$ and $\{\mathbf{u}_2\}\mathrm{e}^{\mathrm{i}\omega_2 t}$, and the corresponding nodal responses are $\{\mathbf{q}_1\}\mathrm{e}^{\mathrm{i}\omega_1 t}$ and $\{\mathbf{q}_2\}\mathrm{e}^{\mathrm{i}\omega_2 t}$ respectively. The following relations are obvious:

$$\left.\begin{aligned} \{\mathbf{u}_1\} &= [\mathbf{N}(\omega_1)]\{\mathbf{q}_1\}, & \{\mathbf{u}_2\} &= [\mathbf{N}(\omega_2)]\{\mathbf{q}_2\} \\ [\mathbf{D}(\omega_1)]\{\mathbf{q}_1\} &= \{\mathbf{Q}_1\}, & [\mathbf{D}(\omega_2)]\{\mathbf{q}_2\} &= \{\mathbf{Q}_2\} \end{aligned}\right\} \tag{4.4.57}$$

Table 4.4.1. The first 50 modes and participation factors

Mode	Natural frequency	Natural mode			Participation factor		
		1	2	3	1	2	3
1	2.467	1.861	−2.221	−0.885	0.609242	0.084123	−0.060179
2	15.418	−1.453	5.710	−3.924	0.149813	0.065284	−0.215347
3	49.965	−1.413	10.000	7.071	0.254576	0.042984	0.131527
4	104.248	−1.414	14.439	−10.209	0.284332	0.031242	−0.093152
5	178.270	−1.414	18.881	13.351	0.300664	0.024501	0.072099
6	272.040	−1.427	23.932	−16.386	0.317707	0.020557	−0.058680
7	385.531	1.414	−27.763	−19.631	−0.317684	−0.017096	−0.049643
8	518.920	1.366	−29.233	23.254	−0.296882	−0.013726	0.043191
9	671.751	−1.417	36.863	25.952	0.326001	0.013105	0.037719
10	844.474	1.399	−40.648	28.781	−0.336514	−0.011986	0.034172
11	1036.929	1.408	−45.019	−32.116	−0.332115	−0.010654	−0.030767
12	1249.190	−1.438	52.455	−34.908	0.347286	0.010106	−0.027767
13	1481.079	−1.427	55.875	38.722	0.333854	0.008915	0.025266
14	1732.844	1.385	−55.386	42.189	−0.319910	−0.007892	0.023842
15	2004.150	−1.419	63.936	44.870	0.337284	0.007711	0.021966
16	2296.327	1.326	−56.247	49.608	−0.290363	−0.006219	0.020868
17	2606.194	1.411	−71.676	−50.965	−0.340037	−0.006796	−0.019475
18	2937.055	−1.448	82.234	−53.195	0.360879	0.006778	−0.018148
19	3287.222	1.402	−79.085	−57.004	−0.342313	−0.006077	−0.017529
20	3657.352	1.399	−82.827	60.919	−0.332950	−0.005603	0.016442
21	4047.164	−1.421	91.228	63.825	0.341753	0.005460	0.015473
22	4457.717	1.343	−81.233	68.741	−0.303074	−0.004621	0.015008
23	4886.072	1.413	−98.556	−69.851	−0.343580	−0.004988	−0.014228
24	5335.682	−1.459	113.244	−71.239	0.370766	0.005141	−0.013456
25	5803.969	1.404	−105.547	−75.821	−0.345190	−0.004592	−0.013190
26	6292.492	1.412	−111.556	79.430	−0.343061	−0.004381	0.012533
27	6800.796	−1.423	118.738	82.816	0.344078	0.004225	0.011930
28	7329.602	1.358	−107.558	87.681	−0.313410	−0.003710	0.011708
29	7876.561	−1.415	125.658	88.774	0.345530	0.003938	0.011197
30	8445.122	−1.468	145.459	−89.034	0.379150	0.004167	−0.010677
31	9031.324	1.406	−132.234	−94.675	−0.346850	−0.003689	−0.010562
32	9638.315	−1.423	141.553	−97.713	0.351865	0.003621	−0.010115
33	10265.045	−1.425	146.465	101.843	0.345455	0.003444	0.009698
34	10912.034	1.372	−135.208	106.418	−0.322602	−0.003121	0.009591
35	11577.665	−1.417	152.981	107.734	0.346723	0.003253	0.009223
36	12265.420	−1.478	178.853	−106.570	0.386733	0.003520	−0.008838
37	12969.289	1.408	−159.147	−113.566	−0.347895	−0.003082	−0.008801
38	13694.873	−1.435	172.796	−115.758	0.359936	0.003101	−0.008471
39	14439.913	−1.427	174.406	120.906	0.346328	0.002907	0.008164
40	15205.066	1.386	−164.165	124.940	−0.331125	−0.002709	0.008117
41	15989.385	−1.419	180.523	126.730	0.347495	0.002770	0.007835
42	16796.375	1.330	−152.974	134.021	−0.300472	−0.002350	0.007772
43	17617.867	1.411	−186.281	−132.495	−0.348585	−0.002646	−0.007538
44	18462.215	−1.446	205.261	−133.557	0.367528	0.002724	−0.007280
45	19325.370	1.402	−191.673	−138.197	−0.349593	−0.002533	−0.007269
46	20208.751	1.398	−194.409	143.238	−0.339199	−0.002404	0.007030
47	21111.722	−1.421	208.282	145.763	0.348010	0.002412	0.006806
48	22036.635	1.344	−180.861	152.783	−0.308980	−0.002098	0.006780
49	22977.057	1.413	−213.638	−151.460	−0.349051	−0.002318	−0.006588
50	23940.390	−1.456	238.923	−151.102	0.374771	0.002437	−0.006377

Table 4.4.2. The first 50 distributed modal participation factors

Mode	Distributed modal participation factors					
	1	2	3	4	5	6
1	0.605672278	0.083371874	0.173261727	−0.059378353	−0.027796936	0.006409451
2	−0.028258642	0.026395025	−0.042373572	−0.120672339	−0.251226496	0.056121791
3	0.034327287	0.018279187	−0.016793547	0.027305991	−0.217370165	0.034012039
4	0.000773878	0.002654828	−0.006389586	−0.005638846	−0.132852084	0.011684957
5	0.005551170	0.002949612	−0.003029996	0.004281977	−0.108431254	0.008772522
6	0.000468680	0.000653817	0.000309670	−0.001366610	−0.083879019	0.004723250
7	−0.001799136	−0.000954227	0.001006537	−0.001373114	0.072804345	−0.003931690
8	−0.000012007	−0.000388288	0.005756799	0.000303133	0.063062618	−0.002618391
9	0.000955198	0.000421369	−0.000733319	0.000615549	−0.054982091	0.002223084
10	−0.000146076	−0.000137916	0.000318773	0.000221086	0.047961811	−0.001578066
11	−0.000162684	−0.000221474	−0.000217416	−0.000300965	0.043982975	−0.001420046
12	0.000233410	0.000052305	0.001655192	−0.000172556	−0.039423341	0.001074479
13	0.000790260	0.000130663	−0.001050899	0.000212560	−0.037090234	0.000996316
14	0.000067561	−0.000072023	0.001946323	0.000037195	0.034393357	−0.000800143
15	0.000326481	0.000083113	−0.000378568	0.000125579	−0.031730692	0.000729630
16	0.000031505	−0.000087922	0.004605152	−0.000019852	0.030828034	−0.000625685
17	−0.000000558	−0.000056448	−0.000122517	−0.000076326	0.027700999	−0.000556740
18	0.000236188	0.000005351	0.001746902	−0.000067374	−0.025631282	0.000461299
19	0.000237265	−0.000040444	−0.000511202	−0.000046691	0.024561547	−0.000438390
20	0.000038593	−0.000022231	0.000718522	0.000013814	0.023537630	−0.000380549
21	0.000228812	0.000029050	−0.000322258	0.000046425	−0.022327699	0.000358395
22	0.000064724	−0.000036365	0.002720406	−0.000013700	0.021948225	−0.000322125
23	−0.000008646	−0.000022109	−0.000032373	−0.000030504	0.020235189	−0.000295086
24	0.000258569	−0.000003171	0.001759861	−0.000039514	−0.018937561	0.000254457
25	0.000166241	−0.000017403	−0.000328492	−0.000019581	0.018491177	−0.000247042
26	0.000003568	−0.000007771	0.000109409	0.000009464	0.017839781	−0.000221034
27	0.000201946	0.000013305	−0.000306310	0.000022883	−0.017234418	0.000212437
28	0.000064948	−0.000017705	0.001701797	−0.000007900	0.016987936	−0.000195277
29	0.000030720	0.000010804	−0.000029221	0.000015557	−0.015947760	0.000182458
30	0.000284453	−0.000004983	0.001761415	−0.000027986	−0.014988586	0.000160686
31	0.000111175	−0.000009004	−0.000211444	−0.000010144	0.014833556	−0.000158336
32	0.000029572	0.000002269	0.000255527	−0.000008215	−0.014333057	0.000143956
33	0.000192485	0.000007135	−0.000300331	0.000013390	−0.014039418	0.000140446
34	0.000054746	−0.000009285	0.001063776	−0.000004151	0.013824272	−0.000130603
35	0.000050707	0.000006042	−0.000072294	0.000009262	−0.013164711	0.000123912
36	0.000311432	−0.000005130	0.001761321	−0.000021915	−0.012385571	0.000110456
37	0.000069906	−0.000005233	−0.000130762	−0.000006063	0.012388359	−0.000110096
38	0.000060694	−0.000000137	0.000499096	−0.000007674	−0.011959323	0.000100992
39	0.000188857	0.000004244	−0.000297697	0.000008783	−0.011847945	0.000099727
40	0.000040235	−0.000004954	0.000626657	−0.000001758	0.011632619	−0.000093283
41	0.000067125	0.000003700	−0.000103738	0.000006141	−0.011212298	0.000089638
42	0.000037800	−0.000009798	0.001651495	−0.000007453	0.011397476	−0.000087015
43	0.000038330	−0.000003295	−0.000071979	−0.000004024	0.010638372	−0.000080984
44	0.000090430	−0.000001261	0.000673591	−0.000007334	−0.010247143	0.000074649
45	0.000129700	−0.000002995	−0.000231985	−0.000002323	0.010117826	−0.000073506
46	0.000023510	−0.000002531	0.000308450	−0.000000192	0.010025925	−0.000069840
47	0.000080512	0.000002418	−0.000127586	0.000004405	−0.009766907	0.000067860
48	0.000041064	−0.000006284	0.001211386	−0.000005009	0.009874027	−0.000065899
49	0.000013505	−0.000002199	−0.000027320	−0.000002893	0.009323990	−0.000062074
50	0.000119329	−0.000001788	0.000805089	−0.000007064	−0.008954713	0.000057357

The reciprocal theorem states that the work done by the first set of forces (including inertia) acting on the second set of displacements is equal to the work done by the second set of forces acting on the first set of displacements, that is,

$$(\{\mathbf{q}_1\}^{\mathrm{T}}\{\mathbf{Q}_2\} + \omega_2^2 \int \{\mathbf{u}_1\}^{\mathrm{T}}[\mathbf{m}]\{\mathbf{u}_2\}\,\mathrm{d}\,vol)\,\mathrm{e}^{\mathrm{i}(\omega_1+\omega_2)t}$$

$$= (\{\mathbf{q}_2\}^{\mathrm{T}}\{\mathbf{Q}_1\} + \omega_1^2 \int \{\mathbf{u}_2\}^{\mathrm{T}}[\mathbf{m}]\{\mathbf{u}_1\}\,\mathrm{d}\,vol)\,\mathrm{e}^{\mathrm{i}(\omega_1+\omega_2)t}$$

where the inertia forces are

$$-\frac{\mathrm{d}^2}{\mathrm{d}t^2}[\mathbf{m}]\{\mathbf{u}_1\}\,\mathrm{e}^{\mathrm{i}\omega_1 t} = \omega_1^2[\mathbf{m}]\{\mathbf{u}_1\}\,\mathrm{e}^{\mathrm{i}\omega_1 t}$$

and

$$\omega_2^2[\mathbf{m}]\{\mathbf{u}_2\}\,\mathrm{e}^{\mathrm{i}\omega_2 t}$$

for the two systems respectively. By means of the relations (4.4.57),

$$\{\mathbf{q}_1\}^{\mathrm{T}}[\mathbf{D}(\omega_2)]\{\mathbf{q}_2\} + \omega_2^2\{\mathbf{q}_1\}^{\mathrm{T}} \int [\mathbf{N}(\omega_1)]^{\mathrm{T}}[\mathbf{m}][\mathbf{N}(\omega_2)]\,\mathrm{d}\,vol\,\{\mathbf{q}_2\}$$

$$= \{\mathbf{q}_2\}^{\mathrm{T}}[\mathbf{D}(\omega_1)]\{\mathbf{q}_1\} + \omega_1^2\{\mathbf{q}_2\}^{\mathrm{T}} \int [\mathbf{N}(\omega_2)]^{\mathrm{T}}[\mathbf{m}][\mathbf{N}(\omega_1)]\,\mathrm{d}\,vol\,\{\mathbf{q}_1\}$$

Since $\{\mathbf{q}_1\}$ and $\{\mathbf{q}_2\}$ are not identically zero,

$$\int [\mathbf{N}(\omega_1)]^{\mathrm{T}}[\mathbf{m}][\mathbf{N}(\omega_1)]\,\mathrm{d}\,vol = \frac{[\mathbf{D}(\omega_1) - \mathbf{D}(\omega_2)]}{\omega_2^2 - \omega_1^2}$$

By the definition of the mixed mass matrix,

$$[\mathbf{M}(\omega_1, \omega_2)] = \int [\mathbf{N}(\omega_1)]^{\mathrm{T}}[\mathbf{m}][\mathbf{N}(\omega_2)]\,\mathrm{d}\,vol$$

as required.

4.5. Non-conservative Modal Analysis

The modal analysis of non-conservative linear systems is usually studied by first order formulation including velocities as generalized coordinates [19, 20] or by second order formulation using displacements alone [21, 22]. Most authors disregard defective matrices when the number of eigenvectors is less than the number of eigenvalues. Recently, Newland [20] presented a general modal analysis including defective matrices by first order formulation. It is necessary that the coefficient matrices are constant and that all modes are available in order to find $[W]^{-1}$.

If the non-conservative problem, however, is formulated in terms of dynamic stiffness, the coefficient matrices are no longer constant, but dependent on the frequency of oscillation. It is well known that the governing equation for the response can be obtained simply by replacing $\lambda = \mathrm{i}\omega$ by $\mathrm{d}/\mathrm{d}t$ in the dynamic stiffness equation, where ω is the complex frequency and t is the time variable. The inverse of the dynamic stiffness matrix (the flexibility) is expressed in spectral form so that the partially uncoupled modal equations can be integrated.

Only the essential displacement coordinates are employed to reduce the problem size and the band form of the matrices is preserved. However, for conciseness, the extended formulation is used in the derivation so that the existing algebraic eigenvalue theorems can be directly assumed. The unwanted degrees of freedom (generalized velocities) are treated as slave coordinates and eliminated. The corresponding orthonormality conditions, the spectral decomposition of the dynamic flexibility and the required expansion theorem are derived. A new theorem on the mixed-frequency derivative matrix of the dynamic stiffness is introduced to make the solution of the initial value problem possible. The response is conveniently calculated in two parts, one depending on the forcing function alone (steady state), and the other depending on the initial conditions. For a wide class of problems in which the forcing function is expressed as a product of exponentials, sinusoidals and polynomials, the steady state can be obtained by purely algebraic manipulation without integration. Since, for continuous systems, the number of modes is infinite, convergence criteria for computing the response are also recommended.

4.5.1. Discrete System

By means of substructure or stiffness methods, the governing equation for harmonic response has the form

$$[\mathbf{D}(\lambda)]\{\bar{\mathbf{u}}\} = \{\bar{\mathbf{f}}\} \tag{4.5.1}$$

where λ is a complex frequency parameter occurring when a solution of the form $\{\bar{\mathbf{u}}\}e^{\lambda t}$ is of interest. $\{\bar{\mathbf{u}}\}$ and $\{\bar{\mathbf{f}}\}$ are the response and force amplitude vectors respectively. The corresponding governing equation for time response is [23]

$$\left[\mathbf{D}\left(\frac{\mathrm{d}}{\mathrm{d}t}\right)\right]\{\mathbf{u}(t)\} = \{\mathbf{f}(t)\} \tag{4.5.2}$$

with the initial conditions $\{\mathbf{u}(0)\}$ and $\{\dot{\mathbf{u}}(0)\}$ given, where a dot denotes derivative with respect to time t. The vectors $\{\bar{\mathbf{u}}\}$ and $\{\mathbf{u}(t)\}$, $\{\bar{\mathbf{f}}\}$ and $\{\mathbf{f}(t)\}$ are Fourier transform pairs respectively.

To make use of the existing theory of linear algebraic eigenvalue problems, we assume that the system is formulated initially by a discrete model, such as the finite element method, and by extended state variables, resulting in the vibration problem

$$[\mathbf{A}]\{\boldsymbol{\phi}\} = \lambda[\mathbf{B}]\{\boldsymbol{\phi}\} \quad \text{and} \quad \{\boldsymbol{\psi}\}^{\mathrm{T}}[\mathbf{A}] = \lambda\{\boldsymbol{\psi}\}^{\mathrm{T}}[\mathbf{B}] \tag{4.5.3}$$

All unwanted coordinates (including all generalized velocities) will be eliminated later. In Eqs (4.5.3), λ is the eigenvalue extracted from the constant matrices $[\mathbf{A}]$ and $[\mathbf{B}]$, and $\{\boldsymbol{\phi}\}$ and $\{\boldsymbol{\psi}\}$ are the corresponding eigenvector and eigenrow respectively. The frequency-dependent matrix $[\mathbf{D}(\lambda)] = [\mathbf{A} - \lambda\mathbf{B}]$ is called the (extended) dynamic stiffness matrix and $[\mathbf{A} - \lambda\mathbf{B}]^{-1}$ the dynamic flexibility.

It is well known from the theory of linear algebraic eigenvalue problems [24] for any constant matrix $[\mathbf{C}]$, there exists a non-singular matrix $[\mathbf{P}]$, such that

$$[\mathbf{P}]^{-1}[\mathbf{C}][\mathbf{P}] = [\mathbf{J}] \tag{4.5.4}$$

has a canonical form,

$$[\mathbf{J}] = \mathrm{diag}[\mathbf{J}_1, \mathbf{J}_2, \ldots, \mathbf{J}_p] \tag{4.5.5}$$

where

$$[\mathbf{J}_i] = \begin{bmatrix} \lambda_i & 1 & & 0 \\ & \lambda_i & \ddots & \\ & & \ddots & 1 \\ 0 & & & \lambda_i \end{bmatrix}_{p_i \times p_i}$$

in which $\sum_i p_i$ is the order of matrix $[\mathbf{C}]$. The eigenvalues are arranged in absolute ascending order, $|\lambda_1| \leq |\lambda_2| \leq \cdots \leq |\lambda_p|$. Let $[\mathbf{C}] = [\mathbf{B}]^{-1}[\mathbf{A}]$. Then from Eq. (4.5.4),

$$[\mathbf{BP}]^{-1}[\mathbf{A}][\mathbf{P}] = [\mathbf{J}], \qquad [\mathbf{A}][\mathbf{P}] = [\mathbf{B}][\mathbf{P}][\mathbf{J}] \tag{4.5.6}$$

or

$$[\mathbf{Q}]^{\mathrm{T}}[\mathbf{A}][\mathbf{P}] = [\mathbf{J}] \tag{4.5.7}$$

where $[\mathbf{Q}] = [\mathbf{BP}]^{-1}$. In practice, $[\mathbf{Q}]$ is generated along with $[\mathbf{P}]$ and matrix inversion is not required. When compared with Eqs (4.5.3), it is obvious that, for non-defective matrices, $[\mathbf{P}]$ and $[\mathbf{Q}]$ are the collections of eigenvectors and eigenrows respectively and $[\mathbf{J}]$ is a diagonal matrix containing the λ_i. The orthonormality condition is

$$[\mathbf{Q}]^{\mathrm{T}}[\mathbf{B}][\mathbf{P}] = [\mathbf{P}]^{-1}[\mathbf{B}]^{-1}[\mathbf{B}][\mathbf{P}] = [\mathbf{I}] \tag{4.5.8}$$

In component form,

$$[\mathbf{P}] = [\mathbf{P}_1, \mathbf{P}_2, \ldots, \mathbf{P}_p] \quad \text{and} \quad [\mathbf{Q}] = [\mathbf{Q}_1, \mathbf{Q}_2, \ldots, \mathbf{Q}_p] \tag{4.5.9}$$

When $p_i > 1$, the defective eigenvalue λ_i of multiplicity p_i has only one eigenvector $\{\boldsymbol{\phi}_1\}$ satisfying Eqs (4.5.3). The other $p_i - 1$ vectors, called the principal vectors, are obtained by writing Eq. (4.5.6) in component form,

$$[\mathbf{A}][\boldsymbol{\phi}_1, \boldsymbol{\phi}_2, \ldots, \boldsymbol{\phi}_{p_i}] = [\mathbf{B}][\boldsymbol{\phi}_1, \boldsymbol{\phi}_2, \ldots, \boldsymbol{\phi}_{p_i}][\mathbf{J}_i]$$

or

$$\begin{aligned} [\mathbf{A}]\{\boldsymbol{\phi}_1\} &= \lambda_i[\mathbf{B}]\{\boldsymbol{\phi}_1\} \\ [\mathbf{A} - \lambda_i\mathbf{B}]\{\boldsymbol{\phi}_2\} &= [\mathbf{B}]\{\boldsymbol{\phi}_1\} \\ [\mathbf{A} - \lambda_i\mathbf{B}]\{\boldsymbol{\phi}_3\} &= [\mathbf{B}]\{\boldsymbol{\phi}_2\}, \text{ etc.} \end{aligned} \tag{4.5.10}$$

The eigenvector $\{\boldsymbol{\phi}_1\}$ is called the leading vector. We call the leading vector and the principal vectors collectively as principal vectors. The principal rows are similarly defined.

The spectral decomposition of the dynamic flexibility $[\mathbf{A} - \lambda\mathbf{B}]^{-1}$ in terms of the principal vectors and rows is achieved using the orthonormality conditions (4.5.7) and (4.5.8),

$$[\mathbf{Q}]^{\mathrm{T}}[\mathbf{A} - \lambda\mathbf{B}][\mathbf{P}] = [\mathbf{J} - \lambda\mathbf{I}] \tag{4.5.11}$$

Taking reciprocals, we have

$$[\mathbf{A} - \lambda\mathbf{B}]^{-1} = [\mathbf{P}][\mathbf{J} - \lambda\mathbf{I}]^{-1}[\mathbf{Q}]^{\mathrm{T}} = [\mathbf{Z}(\lambda)] = [\mathbf{D}(\lambda)]^{-1} \tag{4.5.12}$$

or in component form,

$$[\mathbf{Z}(\lambda)] = \sum_{i=1}^{p} [\mathbf{P}_i][\mathbf{J}_i - \lambda\mathbf{I}]^{-1}[\mathbf{Q}_i]^{\mathrm{T}} \tag{4.5.13}$$

where

$$[\mathbf{J}_i - \lambda\mathbf{I}]^{-1} = \begin{bmatrix} (\lambda_i - \lambda)^{-1} & -(\lambda_i - \lambda)^{-2} & \cdots & (-1)^{p_i+1}(\lambda_i - \lambda)^{-p_i} \\ & (\lambda_i - \lambda)^{-1} & \cdots & (-1)^{p_i}(\lambda_i - \lambda)^{-p_i+1} \\ & & \cdots & \cdots \\ & & & (\lambda_i - \lambda)^{-1} \end{bmatrix}$$

4.5.2. Condensation

It is commonly known that if the number of finite elements in an analysis increases, the physically continuous model is represented with increasing accuracy. However, it is not always practical to include all coordinates and condensation as necessary. If the coordinates are partitioned according to the masters (subscript m) which are the retained coordinates and the slaves (subscript s) which are unwanted coordinates,

$$[\mathbf{D}(\lambda)]\{\bar{\mathbf{u}}\} = \begin{bmatrix} \mathbf{D}_{mm}(\lambda) & \mathbf{D}_{ms}(\lambda) \\ \mathbf{D}_{sm}(\lambda) & \mathbf{D}_{ss}(\lambda) \end{bmatrix} \begin{Bmatrix} \bar{\mathbf{u}}_m \\ \bar{\mathbf{u}}_s \end{Bmatrix} = \begin{Bmatrix} \bar{\mathbf{f}}_m \\ \mathbf{0} \end{Bmatrix} \tag{4.5.14}$$

where without loss of generality, $\{\bar{\mathbf{f}}_s\}$ is assumed to be zero, if otherwise $\{\bar{\mathbf{f}}_m\}$ is replaced by $\{\bar{\mathbf{f}}_m\} - \{\mathbf{D}_{ms}(\lambda)\mathbf{D}_{ss}^{-1}(\lambda)\bar{\mathbf{f}}_s\}$ in the subsequent study. From the second equation of (4.5.14), we can establish the condensation transformation

$$\{\bar{\mathbf{u}}\} = \begin{Bmatrix} \bar{\mathbf{u}}_s \\ \bar{\mathbf{u}}_m \end{Bmatrix} = \begin{bmatrix} \mathbf{I} \\ -\mathbf{D}_{ss}^{-1}\mathbf{D}_{sm} \end{bmatrix} \{\bar{\mathbf{u}}_{sm}\} = [\mathbf{S}(\lambda)]\{\bar{\mathbf{u}}_m\} \tag{4.5.15}$$

The adjoint transformation is

$$\{\bar{\mathbf{v}}\}^{\mathrm{T}} = \{\bar{\mathbf{v}}_m\}^{\mathrm{T}}[\mathbf{T}(\lambda)]^{\mathrm{T}} \tag{4.5.16}$$

where

$$[\mathbf{S}(\lambda)] = \begin{bmatrix} \mathbf{I} \\ -\mathbf{D}_{ss}^{-1}\mathbf{D}_{sm} \end{bmatrix} \quad \text{and} \quad [\mathbf{T}(\lambda)]^{\mathrm{T}} = [\mathbf{I} - \mathbf{D}_{ms}\mathbf{D}_{ss}^{-1}] \tag{4.5.17}$$

After eliminating $\{\bar{\mathbf{u}}_m\}$ from Eq. (4.5.14), we have

$$[\mathbf{D}^*(\lambda)]\{\bar{\mathbf{u}}_m\} = \{\bar{\mathbf{f}}_m\} \tag{4.5.18}$$

where

$$[\mathbf{D}^*(\lambda)] = [\mathbf{D}_{mm} - \mathbf{D}_{ms}\mathbf{D}_{ss}^{-1}\mathbf{D}_{sm}] = [\mathbf{T}(\lambda)]^{\mathrm{T}}[\mathbf{S}(\lambda)] \tag{4.5.19}$$

is the condensed dynamic stiffness associated with the essential displacement coordinates only.

We shall establish the orthonormality conditions, Eqs (4.5.7)–(4.5.10), and the spectral decomposition of $[\mathbf{D}^*(\lambda)]^{-1}$, Eq. (4.5.13) in the following sections.

4.5.3. Orthonormal Condition

If λ_i is an eigenvalue of degeneracy p_i, then from the definition, Eq. (4.5.11) gives

$$[\mathbf{Q}_i]^{\mathrm{T}}[\mathbf{D}(\lambda_i)][\mathbf{P}_i] = [\mathbf{Q}_i]^{\mathrm{T}}[\mathbf{A} - \lambda_i\mathbf{B}][\mathbf{P}_i] = [\mathbf{J}_i - \lambda_i\mathbf{I}] \tag{4.5.20}$$

By means of the condensation transformation and its adjoint, Eqs (4.5.15)–(4.5.17), the extended orthonormality condition (4.5.20) is condensed to

$$[\mathbf{Q}_{mi}]^{\mathrm{T}}[\mathbf{D}^*(\lambda_i)][\mathbf{P}_{mi}] = [\mathbf{J}_i - \lambda_i\mathbf{I}) \tag{4.5.21}$$

where

$$\left.\begin{aligned} &[\mathbf{P}_i] = [\boldsymbol{\phi}_1, \boldsymbol{\phi}_2, \ldots, \boldsymbol{\phi}_{p_i}] = [\mathbf{S}(\lambda_i)][\mathbf{P}_{mi}] = [\mathbf{S}(\lambda_i)][\boldsymbol{\phi}_{m1}, \boldsymbol{\phi}_{m2}, \ldots, \boldsymbol{\phi}_{mp_i}] \\ &[\mathbf{Q}_i] = [\boldsymbol{\psi}_1, \boldsymbol{\psi}_2, \ldots, \boldsymbol{\psi}_{p_i}] = [\mathbf{T}(\lambda_i)][\mathbf{Q}_{mi}] = [\mathbf{T}(\lambda_i)][\boldsymbol{\psi}_{m1}, \boldsymbol{\psi}_{m2}, \ldots, \boldsymbol{\psi}_{mp_i}] \\ &[\mathbf{D}^*(\lambda_i)] = [\mathbf{T}(\lambda_i)]^{\mathrm{T}}[\mathbf{D}(\lambda_i)][\mathbf{S}(\lambda_i)] \end{aligned}\right\} \tag{4.5.22}$$

The relations between the principal and leading vectors and rows, Eqs (4.5.10), become

$$\left.\begin{aligned} &[\mathbf{D}^*(\lambda_i)]\{\boldsymbol{\phi}_{m1}\} = \{\mathbf{0}\}, && \{\boldsymbol{\psi}_{m1}\}^{\mathrm{T}}[\mathbf{D}^*(\lambda_i)] = [\mathbf{0}] \\ &[\mathbf{D}^*(\lambda_i)]\{\boldsymbol{\phi}_{mj}\} = -[\mathbf{D}'(\lambda_i)]\{\boldsymbol{\phi}_{mj-1}\}, && \{\boldsymbol{\psi}_{mj}\}^{\mathrm{T}}[\mathbf{D}^*(\lambda_i)] = -[\boldsymbol{\psi}_{mj-1}]^{\mathrm{T}}[\mathbf{D}'(\lambda_i)] \\ &\{\boldsymbol{\psi}_{mj}\}^{\mathrm{T}}[\mathbf{D}^*(\lambda_i)] = -\{\boldsymbol{\psi}_{mj-1}\}^{\mathrm{T}}[\mathbf{D}'(\lambda_i)], && j = 2, 3, \ldots, p_i \end{aligned}\right\} \tag{4.5.23}$$

where

$$[\mathbf{D}'(\lambda_i)] = \frac{\mathrm{d}}{\mathrm{d}\lambda}[\mathbf{D}^*(\lambda_i)] \tag{4.5.24}$$

The vectors $\{\boldsymbol{\phi}_{m1}\}$ and $\{\boldsymbol{\psi}_{m1}\}$ determined by the first two equations of (4.5.23) are arbitrary up to a constant factor. To agree with the first of (4.5.21), we assume

$$\{\boldsymbol{\psi}_{m1}\}^{\mathrm{T}}[\mathbf{D}^*(\lambda_i)]\{\boldsymbol{\phi}_{m1}\} = \lambda_i \tag{4.5.25}$$

Subsequently, corresponding to Eq. (4.5.8), we implicitly assume also that

$$\{\boldsymbol{\psi}_{m1}\}^{\mathrm{T}}[\mathbf{D}'(\lambda_i)]\{\boldsymbol{\phi}_{m1}\} = -1 \tag{4.5.26}$$

Equation (4.5.21) is the collective form of the orthonormality condition and Eqs (4.5.23)–(4.5.26); we have made use of the fact that

$$[\mathbf{D}'(\lambda_i)] = \frac{\mathrm{d}}{\mathrm{d}\lambda}[\mathbf{D}^*(\lambda_i)] = -[\mathbf{T}(\lambda_i)][\mathbf{B}][\mathbf{S}(\lambda_i)] \tag{4.5.27}$$

which can be proved easily by matrix algebra.

4.5.4. Mixed-Frequency Dynamic Derivative

If we write Eq. (4.5.8) in component form, then

$$[\mathbf{Q}_i]^{\mathrm{T}}[\mathbf{B}][\mathbf{P}_j] = \begin{cases} [\mathbf{0}], & i \neq j \\ [\mathbf{I}], & i = j \end{cases}$$

We simplify the presentation by writing

$$[\mathbf{Q}_i]^{\mathrm{T}}[\mathbf{B}][\mathbf{P}_j] = \delta_{ij}[\mathbf{I}] \tag{4.5.28}$$

although the left-hand side is not always square. Applying the condensation transformation to Eq. (4.5.28) gives

$$\delta_{ij}[\mathbf{I}] = [\mathbf{Q}_{mi}]^{\mathrm{T}}[\mathbf{T}(\lambda_i)]^{\mathrm{T}}[\mathbf{B}][\mathbf{S}(\lambda_j)][\mathbf{P}_{mj}] = [\mathbf{Q}_{mi}]^{\mathrm{T}}[\mathbf{M}^*(\lambda_i, \lambda_j)][\mathbf{P}_{mj}] \tag{4.5.29}$$

where

$$[\mathbf{M}(\lambda_i, \lambda_j)] = [\mathbf{T}(\lambda_i)]^{\mathrm{T}}[\mathbf{B}][\mathbf{S}(\lambda_j)] \tag{4.5.30}$$

It is obvious in the extended system that $[\mathbf{T}(\lambda_i)] = [\mathbf{S}(\lambda_j)] = [\mathbf{I}]$ and

$$[\mathbf{M}(\lambda_i, \lambda_j)] = \frac{1}{\lambda_j - \lambda_i}[\mathbf{D}(\lambda_i) - \mathbf{D}(\lambda_j)] = [\mathbf{B}] \tag{4.5.31}$$

We shall prove below that for the condensed system,

$$[\mathbf{M}^*(\lambda_i, \lambda_j)] = \frac{1}{\lambda_j - \lambda_i}[\mathbf{D}^*(\lambda_i) - \mathbf{D}^*(\lambda_j)] \tag{4.5.32}$$

which is useful in dealing with initial conditions. The mixed frequency matrix has the symmetrical property that $[\mathbf{M}^*(\lambda_i, \lambda_j)] = [\mathbf{M}^*(\lambda_j, \lambda_i)]$. Substituting Eq. (4.5.32) into Eq. (4.5.28) and applying the condensation transformation gives

$$[\mathbf{Q}_i]^{\mathrm{T}}[\mathbf{D}(\lambda_i)] - \mathbf{D}(\lambda_j)][\mathbf{P}_j] = [\mathbf{Q}_{mi}]^{\mathrm{T}}[\mathbf{I} - \mathbf{D}_{ms}(\lambda_i)]\mathbf{D}_{ss}^{-1}(\lambda_i)]$$

$$\begin{bmatrix} [\mathbf{D}_{mm}(\lambda_i) - \mathbf{D}_{mm}(\lambda_j)] & [\mathbf{D}_{ms}(\lambda_i) - \mathbf{D}_{ms}(\lambda_j)] \\ [\mathbf{D}_{sm}(\lambda_i) - \mathbf{D}_{sm}(\lambda_j)] & [\mathbf{D}_{ss}(\lambda_i) - \mathbf{D}_{ss}(\lambda_j)] \end{bmatrix} \begin{bmatrix} \mathbf{I} \\ -\mathbf{D}_{ss}^{-1}(\lambda_j)\mathbf{D}_{sm}(\lambda_j) \end{bmatrix} [\mathbf{P}_{mj}]$$

$$= [\mathbf{Q}_{mi}]^{\mathrm{T}}[\mathbf{D}_{mm}(\lambda_i) - \mathbf{D}_{ms}(\lambda_i)\mathbf{D}_{ss}^{-1}(\lambda_i)\mathbf{D}_{sm}(\lambda_i) - \mathbf{D}_{mm}(\lambda_j) + \mathbf{D}_{ms}(\lambda_j)\mathbf{D}_{ss}^{-1}(\lambda_j)\mathbf{D}_{sm}(\lambda_j)][\mathbf{P}_{mj}]$$

$$= [\mathbf{Q}_{mi}]^{\mathrm{T}}[\mathbf{D}^*(\lambda_i) - \mathbf{D}^*(\lambda_j)][\mathbf{P}_{mj}] = (\lambda_j - \lambda_i)\delta_{ij}[\mathbf{I}]$$

or

$$[\mathbf{M}^*(\lambda_i, \lambda_j)] = [\mathbf{T}(\lambda_i)]^{\mathrm{T}}[\mathbf{B}][\mathbf{S}(\lambda_j)] = \frac{1}{\lambda_j - \lambda_i}[\mathbf{D}^*(\lambda_i) - \mathbf{D}^*(\lambda_j)] \tag{4.5.33}$$

If we let $\lambda_i \to \lambda_j$, then

$$[\mathbf{M}^*(\lambda_i, \lambda_j)] = -\frac{\mathrm{d}}{\mathrm{d}\lambda}[\mathbf{D}^*(\lambda_i)] = -[\mathbf{D}'(\lambda_i)] \tag{4.5.34}$$

The matrix $[\mathbf{M}^*]$ is called the dynamic derivative. When $\lambda_i = \lambda_j$ in the conservative system, $[\mathbf{M}^*]$ is the equivalent mass matrix. Equation (4.5.29) will be considered again in the expansion theorem.

4.5.5. Spectral Decomposition of Dynamic Flexibility

The spectral decomposition of the dynamic flexibility, Eq. (4.5.12), will be given in terms of master coordinates. The partition of Eq. (4.5.12), according to masters and slaves, is

$$\begin{bmatrix} \mathbf{Z}_{mm}(\lambda) & \mathbf{Z}_{ms}(\lambda) \\ \mathbf{Z}_{sm}(\lambda) & \mathbf{Z}_{ss}(\lambda) \end{bmatrix} = \sum_{i=1}^{p} \begin{bmatrix} \mathbf{P}_{mi} \\ \mathbf{P}_{si} \end{bmatrix} [\mathbf{J}_i - \lambda\mathbf{I}]^{-1} \begin{bmatrix} \mathbf{Q}_{mi} \\ \mathbf{Q}_{si} \end{bmatrix}^{\mathrm{T}} \tag{4.5.35}$$

Comparing terms, we have

$$[\mathbf{Z}_{mm}(\lambda)] = \sum_{i=1}^{p} [\mathbf{P}_{mi}][\mathbf{J}_i - \lambda\mathbf{I}]^{-1}[\mathbf{Q}_{mi}]^{\mathrm{T}} = [\mathbf{D}^*(\lambda)]^{-1} \tag{4.5.36}$$

which is the spectral decomposition of the dynamic flexibility in terms of master coordinates only.

4.5.6. Expansion Theorem

A substructure dynamic stiffness contains a greater number of modes than masters retained, and expanding a vector by more modes than the order is difficult. This problem can be overcome by considering the extended system initially and then condensing.

To expand a constant vector $\{\mathbf{v}\}$ in the extended system using the n principal vectors $[\mathbf{P}]$, let

$$\{\mathbf{v}\} = [\mathbf{P}]\{\boldsymbol{\alpha}\} = \sum_j [\mathbf{P}_j]\{\boldsymbol{\alpha}_j\} \tag{4.5.37}$$

where $\{\boldsymbol{\alpha}\}$ is an n-vector to be determined and $\{\boldsymbol{\alpha}_j\}$ is a p_j-vector to be determined. Partition $[\mathbf{v}]$ according to masters and slaves $\{\mathbf{v}_m, \mathbf{v}_s\}$. Usually $\{\mathbf{v}_m\}$ is given and $\{\mathbf{v}_s\}$ is ignored. The modal expansion is required when considering initial conditions. If no vibration is assumed initially, for a consistent analysis,

$$\begin{bmatrix} \mathbf{D}_{mm}(0) & \mathbf{D}_{ms}(0) \\ \mathbf{D}_{sm}(0) & \mathbf{D}_{ss}(0) \end{bmatrix} \begin{Bmatrix} \mathbf{v}_m \\ \mathbf{v}_s \end{Bmatrix} = \begin{Bmatrix} \mathbf{V}_m \\ \mathbf{0} \end{Bmatrix} \tag{4.5.38}$$

where $\{\mathbf{V}_m\}$ is the generalized force required to produce $\{\mathbf{v}_m\}$ initially. From the second of Eq. (4.5.38), we have

$$\{\mathbf{v}\} = [\mathbf{S}(0)]\{\mathbf{v}_m\} \tag{4.5.39}$$

which is a form of the condensation transformation Eq. (4.5.17). Premultiply Eq. (4.5.37) by $[\mathbf{Q}_i]^{\mathrm{T}}[\mathbf{B}]$ and making use of Eq. (4.5.28) gives

$$[\mathbf{Q}_{mi}]^{\mathrm{T}}[\mathbf{T}(\lambda_i)]^{\mathrm{T}}[\mathbf{B}][\mathbf{S}(0)]\{\mathbf{v}_m\} = \sum_j [\mathbf{Q}_i]^{\mathrm{T}}[\mathbf{B}][\mathbf{P}_j]\{\boldsymbol{\alpha}_j\}$$

or

$$[\mathbf{Q}_{mi}]^{\mathrm{T}}[\mathbf{M}^*(\lambda_i, 0)]\{\mathbf{v}_m\} = \{\boldsymbol{\alpha}_i\} \tag{4.5.40}$$

or, in view of Eq. (4.5.32),

$$\{\boldsymbol{\alpha}_i\} = \frac{1}{\lambda_i}[\mathbf{Q}_{mi}]^{\mathrm{T}}[\mathbf{D}^*(\lambda_i) - \mathbf{D}^*(0)]\{\mathbf{v}_m\} \tag{4.5.41}$$

Therefore, an arbitrary constant vector $\{\mathbf{v}_m\}$ in master coordinates can be expanded in modal form, Eq. (4.5.37), by means of Eq. (4.5.40).

However, if the system under steady-state vibration has the forcing history $\{\mathbf{V}_m\}e^{\lambda t}$, $\lambda = \sigma + i\omega$, then Eq. (4.5.38) is more appropriately written as

$$\begin{bmatrix} \mathbf{D}_{mm}(\lambda) & \mathbf{D}_{ms}(\lambda) \\ \mathbf{D}_{sm}(\lambda) & \mathbf{D}_{ss}(\lambda) \end{bmatrix} \begin{Bmatrix} \mathbf{v}_m \\ \mathbf{v}_s \end{Bmatrix} = \begin{Bmatrix} \mathbf{V}_m \\ \mathbf{0} \end{Bmatrix} \tag{4.5.42}$$

and Eq. (4.5.40) becomes

$$\{\boldsymbol{\alpha}_i\} = [\mathbf{Q}_{mi}]^{\mathrm{T}}[\mathbf{M}^*(\lambda_i, \lambda)]\{\mathbf{v}_m\} \tag{4.5.43}$$

4.5.7. Forced Response

The solution of the forced system

$$\left[\mathbf{D}\left(\frac{\mathrm{d}}{\mathrm{d}t}\right)\right]\{\mathbf{u}(t)\} = \{\mathbf{f}(t)\} \tag{4.5.44}$$

is separated into two parts: steady state (forced vibration) and transient. The steady-state solution is in fact the particular integral of Eq. (4.5.44) and the transient solution depends on the initial conditions, being known as the complementary function in ordinary differential equation theory. The transient solution will be considered in the next section. Since condensed systems are of interest, the asterisk and subscript m are omitted.

From Eq. (4.5.36), the particular integral of Eq. (4.5.44) is given by

$$\begin{aligned}[\mathbf{u}_f(t)\} &= \left[\mathbf{D}\left(\frac{\mathrm{d}}{\mathrm{d}t}\right)\right]^{-1}\{\mathbf{f}(t)\} = \left[\mathbf{Z}\left(\frac{\mathrm{d}}{\mathrm{d}t}\right)\right]\{\mathbf{f}(t)\} \\ &= \sum_j [\mathbf{P}_j]\left[\mathbf{J}_j - \mathbf{I}\frac{\mathrm{d}}{\mathrm{d}t}\right]^{-1}[\mathbf{Q}_j]^{\mathrm{T}}\{\mathbf{f}(t)\} \end{aligned} \tag{4.5.45}$$

the subscript f denoting forced response.

Let us consider a special solution first and a general solution afterwards. Suppose $\{\mathbf{f}(t)\}$ is an exponentially varying harmonic [2], then

$$\{\mathbf{f}(t)\} = \mathrm{e}^{\lambda t}\{\mathbf{F}\} \tag{4.5.46}$$

where $\lambda = \sigma + \mathrm{i}\omega$ for given real σ and ω and $\{\mathbf{F}\}$ is a given complex constant vector. Therefore, by substituting Eq. (4.5.46) into Eq. (4.5.45),

$$\{\mathbf{u}_f(t)\} = \mathrm{e}^{\lambda t}[\mathbf{Z}(\lambda)]\{\mathbf{F}\} \tag{4.5.47}$$

More generally, let

$$\{\mathbf{f}(t)\} = t^q\mathrm{e}^{\lambda t}\{\mathbf{F}\} \tag{4.5.48}$$

where q is a positive integer, then

$$\{\mathbf{u}_f(0)\} = \mathrm{e}^{\lambda t}\left(\frac{\partial}{\partial\lambda} + t\right)^q[\mathbf{Z}(\lambda)]\{\mathbf{F}\} \tag{4.5.49}$$

When $[\mathbf{Z}]$ is expressed in its spectral form the differentiation is straightforward. The initial conditions inherent in Eq. (4.5.47) are

$$\{\mathbf{u}_f(0)\} = [\mathbf{Z}(\lambda)]\{\mathbf{F}\} \quad \text{and} \quad \{\dot{\mathbf{u}}_f(0)\} = \lambda[\mathbf{Z}(\lambda)]\{\mathbf{F}\} \tag{4.5.50}$$

Since the vibration is steady, we must use Eq. (4.5.43) instead of Eq. (4.5.40) when expanding $\{\mathbf{u}_f(0)\}$ and $\{\dot{\mathbf{u}}_f(0)\}$ in modal components. Similarly, for Eq. (4.5.49),

$$\begin{aligned}\{\mathbf{u}_f(0)\} &= \left(\frac{\partial}{\partial\lambda}\right)^q[\mathbf{Z}(\lambda)]\{\mathbf{F}\} \\ \{\dot{\mathbf{u}}_f(0)\} &= \left[\lambda\left(\frac{\partial}{\partial\lambda}\right)^q + q\left(\frac{\partial}{\partial\lambda}\right)^{q-1}\right][\mathbf{Z}(\lambda)]\{\mathbf{F}\} \end{aligned} \tag{4.5.51}$$

It is seen that an explicit integral is not required if the excitation is expressible in a product form of elementary functions of exponentials, sinusoidals and polynomials.

Consider the case when $\{\mathbf{f}(t)\}$ is not given explicitly. Taking the Laplace transform of Eq. (4.5.45) and disregarding the initial conditions,

$$\{\bar{\mathbf{u}}_f(s)\} = \sum_j [\mathbf{P}_j][\mathbf{J}_j - s\mathbf{I}]^{-1}[\mathbf{Q}_j]^{\mathrm{T}}\{\mathbf{f}(s)\} \tag{4.5.52}$$

Since the inverse transformation of $[\mathbf{J}_j - s\mathbf{I}]^{-1}$ is given by $\exp[\mathbf{J}_j t]$, from the convolution theorem,

$$\{\mathbf{u}_f(t)\} = \sum_j [\mathbf{P}_j] \int_0^t \exp[\mathbf{J}_j(t-\tau)][\mathbf{Q}_j]^{\mathrm{T}}\{\mathbf{f}(\tau)\}\,\mathrm{d}\tau \tag{4.5.53}$$

It is seen from Eqs (4.5.47), (4.5.49) and (4.5.53) that the initial conditions of $\{\mathbf{u}_f(t)\}$ and $\{\dot{\mathbf{u}}_f(t)\}$ do not always vanish identically.

4.5.8. Modal Solution

Equation (4.5.44) can equivalently be solved by the modal expansion

$$\{\mathbf{u}(t)\} = \sum_k [P_k]\{\boldsymbol{\alpha}_k(t)\} \tag{4.5.54}$$

where $\{\boldsymbol{\alpha}_k(t)\}$ are vectors to be determined and p_k is the order of degeneracy of the kth mode. Let the modal expansion of the excitation be

$$\{\mathbf{F}_k(t)\} = [\mathbf{Q}_k]^{\mathrm{T}}\{\mathbf{f}(t)\} \tag{4.5.55}$$

From Eq. (4.5.45),

$$\left[\mathbf{J}_j - \mathbf{I}\frac{\mathrm{d}}{\mathrm{d}t}\right]\{\boldsymbol{\alpha}_j(t)\} = \{\mathbf{F}_j\} \tag{4.5.56}$$

When the initial conditions $\{\mathbf{u}(0)\}$ and $\{\dot{\mathbf{u}}(0)\}$ are given, then from the expansion theorem, Eq. (4.5.40),

$$\{\boldsymbol{\alpha}_j(0)\} = [\mathbf{Q}_j]^{\mathrm{T}}[\mathbf{M}(\lambda_j, 0)]\{\mathbf{u}(0)\} \tag{4.5.57}$$

and

$$\{\dot{\boldsymbol{\alpha}}_j(0)\} = [\mathbf{Q}_j]^{\mathrm{T}}[\mathbf{M}(\lambda_j, 0)]\{\dot{\mathbf{u}}(0)\} \tag{4.5.58}$$

Since $\{\boldsymbol{\alpha}_j(t)\}$ is complex in general, let $\{\boldsymbol{\alpha}_j(t)\} = \{\boldsymbol{\beta}_j(t)\} + \mathrm{i}\{\boldsymbol{\gamma}_j(t)\}$, $\lambda_j = \sigma_j + \mathrm{i}\omega_j$, $\{\mathbf{F}_j\} = \{\mathbf{G}_j\} + \mathrm{i}\{\mathbf{H}_j\}$, where the quantities on the right-hand sides are real. Comparing the real and imaginary parts of Eq. (4.5.56), we have

$$\left[\begin{bmatrix} \sigma_j & -\omega_j \\ \omega_j & \sigma_j \end{bmatrix} - \mathbf{I}\frac{\mathrm{d}}{\mathrm{d}t}\right]\begin{Bmatrix} \beta_j \\ \gamma_j \end{Bmatrix} = \begin{Bmatrix} \mathbf{G}_j \\ \mathbf{H}_j \end{Bmatrix} \tag{4.5.59}$$

where

$$[\boldsymbol{\sigma}_j] = \begin{bmatrix} \sigma_j & 1 & & 0 \\ & \sigma_j & & \\ & & \ddots & 1 \\ 0 & & & \sigma_j \end{bmatrix}_{p_j \times p_j} \quad \text{and} \quad [\boldsymbol{\omega}_j] = \omega_j[\mathbf{I}]$$

The uncoupled form of Eq. (4.5.59) is given by

$$\left[\frac{\mathrm{d}^2}{\mathrm{d}t^2} - 2\sigma_j\frac{\mathrm{d}}{\mathrm{d}t} + \sigma_j^2 + \omega_j^2\right]\begin{Bmatrix} \beta_j \\ \gamma_j \end{Bmatrix} = \left[\begin{bmatrix} \sigma_j & \omega_j \\ -\omega_j & \sigma_j \end{bmatrix} - \mathbf{I}\frac{\mathrm{d}}{\mathrm{d}t}\right]\begin{Bmatrix} \mathbf{G}_j \\ \mathbf{H}_j \end{Bmatrix} \tag{4.5.60}$$

The initial conditions for $\{\boldsymbol{\beta}_j\}$ and $\{\boldsymbol{\gamma}_j\}$ are given by the real and imaginary parts of Eqs (4.5.57) and (4.5.58) respectively,

$$\{\boldsymbol{\beta}_j(0) + \mathrm{i}\boldsymbol{\gamma}_j(0)\} = [\mathbf{Q}_j]^{\mathrm{T}}[\mathbf{M}(\lambda_j, 0)]\{\mathbf{u}(0)\} \tag{4.5.61}$$

and

$$\{\boldsymbol{\beta}_j(0) + i\dot{\boldsymbol{\gamma}}_j(0)\} = [\mathbf{Q}_j]^T[\mathbf{M}(\lambda_j, 0)]\{\dot{\mathbf{u}}(0)\} \tag{4.5.62}$$

Since complex arithmetic is no problem for modern computer languages, computation of the forced response using the method presented in the previous section is simple. However, the initial conditions are difficult to satisfy. Now, Eq. (4.5.60) is solved with

$$\{\mathbf{f}_i(t)\} = [\boldsymbol{\sigma}_j]\{\mathbf{G}_j\} - \{\mathbf{G}_j\} + [\boldsymbol{\omega}_j]\{\mathbf{H}_j\} \tag{4.5.63}$$

subject to initial conditions (4.5.61) and (4.5.62).

When the degeneracy $p_j = 1$

$$\ddot{\beta}_j - 2\sigma_j\dot{\beta}_j + (\sigma_j^2 + \omega_j^2)\beta_j = f_j(t) \tag{4.5.64}$$

and the solution is

$$\beta_j(t) = g_j(t)\beta_j(0) + h_j(t)\dot{\beta}_j(0) + \int_0^t h_j(t-\tau)f_j(\tau)\,d\tau \tag{4.5.65}$$

where $g_j(t) = e^{\sigma_j t}\cos\gamma_j t - \sigma_j h_j(t)$ and $h_j(t) = \gamma_j^{-1}e^{\alpha_j t}\sin\gamma_j t$ when $\gamma_j^2 = (1 - \sigma_j^4\omega_j^{-4})\omega_j^2$ is positive. If γ_j^2 is negative, the mode is overdamped.

When $p_j > 1$, let $\{\beta_j(t)\} = \{b_k(t): k = 1, 2, \ldots, p_j\}$; then the first of Eq. (4.5.60) becomes

$$\begin{gathered} \ddot{b}_k - 2\sigma_k\ddot{b}_k + (\sigma_j^2 + \omega_j^2)b_k = 2\dot{b}_{k+1} - 2\sigma_j b_{k+1} - b_{k+2} \\ k = p_j, \quad p_j - 1, \ldots, 1 \qquad b_{p_j+2} = 0 \end{gathered} \tag{4.5.66}$$

where, without loss of generality, $\{\mathbf{f}_j(t)\}$ is assumed to be zero. The solutions of Eqs (4.5.66), in the order of $k = p_j, p_j - 1, \ldots, 1$, are

$$b_k(t) = g_j(t)b_k(0) + h_j(t)\dot{b}_k(0) + \int_0^t h_j(t-\tau)B_k(\tau)\,d\tau \tag{4.5.67}$$

where

$$B_k(t) = 2\dot{b}_{k+1} - 2\sigma_j b_{k+1} - b_{k+2} \tag{4.5.68}$$

Similar results hold for $\{\boldsymbol{\gamma}_j(t)\}$.

4.5.9. Second Order System with Constant Matrices

Discussion is now specialized to the second order system

$$[\mathbf{M}]\{\ddot{\mathbf{u}}\} + [\mathbf{C}]\{\dot{\mathbf{u}}\} + [\mathbf{K}]\{\mathbf{u}\} = \{\mathbf{f}(t)\} \tag{4.5.69}$$

subject to the given initial conditions $\{\mathbf{u}(0)\}$ and $\{\dot{\mathbf{u}}(0)\}$, where the real constant matrices $[\mathbf{M}]$, $[\mathbf{C}]$ and $[\mathbf{K}]$ need not be symmetrical. The corresponding eigenvalue problem,

$$[\mathbf{D}(\lambda)]\{\boldsymbol{\phi}\} = [\lambda^2\mathbf{M} + \lambda\mathbf{C} + \mathbf{K}]\{\boldsymbol{\phi}\} = \{0\} \tag{4.5.70}$$

is obtained by letting $\{\mathbf{u}\} = e^{\lambda t}\{\boldsymbol{\phi}\}$ and $\{\mathbf{f}\} = \{0\}$. Suppose the jth eigensolution λ_j, $[\mathbf{P}_j]$ and $[\mathbf{Q}_j]$ are of degeneracy p_j, the orthonormal conditions are specialized to

$$[\mathbf{Q}_j]^T[\mathbf{D}(\lambda_j)][\mathbf{P}_i] = [\mathbf{J}_j - \lambda_j\mathbf{I}]\delta_{ij} \tag{4.5.71}$$

and

$$[\mathbf{Q}_j]^{\mathrm{T}}[\mathbf{D}'(\lambda_j)][\mathbf{P}_i] = -[\mathbf{I}]\delta_{ij} \tag{4.5.72}$$

where

$$[\mathbf{D}'(\lambda_j)] = 2\lambda_j[\mathbf{M}] + [\mathbf{C}] \tag{4.5.73}$$

If all eigensolutions are of unit degeneracy, the collective orthonormal condition is the same as that given in reference [11], namely

$$[\Lambda][\mathbf{Q}]^{\mathrm{T}}[\mathbf{M}][\mathbf{P}] + [\mathbf{Q}]^{\mathrm{T}}[\mathbf{M}][\mathbf{P}][\Lambda] + [\mathbf{Q}]^{\mathrm{T}}[\mathbf{C}][\mathbf{P}] = [\mathbf{I}] \tag{4.5.74}$$

where $[\Lambda] = \mathrm{diag}[\lambda_1, \lambda_2, \ldots]$. When all coefficient matrices are symmetrical, $[\mathbf{P}] = [\mathbf{Q}]$; and when only $[\mathbf{C}]$ is skew-symmetrical, while $[\mathbf{M}]$ and $[\mathbf{K}]$ are symmetrical, $[\mathbf{P}]^{\mathrm{H}} = [\mathbf{Q}]^{\mathrm{T}}$, where H denotes Hermitian and λ_i are purely imaginary. The expansion theorem, Eqs (4.5.40) and (4.5.43), becomes

$$\{\boldsymbol{\alpha}_j\} = \frac{1}{\lambda_j - \lambda}[\mathbf{Q}]^{\mathrm{T}}[\mathbf{D}(\lambda) - \mathbf{D}(\lambda_j)][\mathbf{P}] = -[\mathbf{Q}]^{\mathrm{T}}\{(\lambda + \lambda_j)\mathbf{M} + \mathbf{C}][\mathbf{P}] \tag{4.5.75}$$

where $\lambda = 0$ for non-oscillatory initial conditions. When all eigensolutions are simply degenerated, $\{\boldsymbol{\alpha}_j\}$ are scalars. If the system is conservative, $[\mathbf{Q}]^{\mathrm{T}} = [\mathbf{P}]^{\mathrm{H}}$ and the λ_j are purely imaginary. The unit step and impulsive responses in Eq. (4.5.65) are given respectively by

$$g_j(t) = \cos\omega_j \quad \text{and} \quad h_j(t) = \omega_j^{-1}\sin\omega_j t \tag{4.5.76}$$

and the integral

$$\int_0^t h_j(t)(t - \tau)f_j(\tau)\,\mathrm{d}\tau$$

becomes the Duhamel integral.

4.6. Exponentially Varying Harmonic Excitations

The dynamic stiffness method has been applied almost exclusively to harmonic, or periodic, oscillations. This is due mainly to the rather misleading intuition that only harmonic vibrations can be described by solutions with separate time- and space-dependent factors. It is shown here that a much wider class of problems involving exponentially varying harmonic excitations can also be analysed by the dynamic stiffness method. The extension is achieved simply by using complex frequency parameters. The forced response (that is, the part of the response which is independent of the initial conditions) can be obtained directly by the solution of linear equations. A single d.o.f. system has been considered in Sect. 1.1. It is shown that the present method is equivalent to the usual Duhamel integral method except that integration is completely avoided and the transient effects due to the initial conditions can be considered separately. The method is applied to undamped straight beam members and is modified so that damped vibration can be covered as well. Distributed loads are then considered and explicit formulae are introduced. Finally, for completeness of presentation, the responses are compared with those obtained by using modal analysis. The method is proved to be equivalent to modal analysis and has the advantages over the latter in that:

1. Integrations in the time variable are completely avoided.
2. The forced response can be obtained directly.
3. Decomposition into generalized forces is not required.
4. The force–response relation is easily visualized.

A frame consists of beam members. Straight beams are the most widely used structural elements in civil engineering. The construction of the dynamic stiffness matrix for a uniform straight beam member under exponentially varying harmonic excitation is considered. The extensions to arches, folded plates and similar members are straightforward, but will not be included here.

The governing equation for the flexural vibration of an undamped beam member subjected to boundary (nodal) forces only is

$$EI\frac{\partial^4 v}{\partial x^4} + \rho A\frac{\partial^2 v}{\partial t^2} = 0 \tag{4.6.1}$$

where EI and ρA are the flexural rigidity and the mass per unit length respectively. For a variables-separable form of solution,

$$v(x,t) = V(x)T(t) \tag{4.6.2}$$

we have, from Eq. (4.6.1),

$$-\frac{EIV^{\text{iv}}}{\rho AV} = \frac{\ddot{T}}{T} = \alpha^2 \tag{4.6.3}$$

where α^2 is an arbitrary constant, independent of both x and t. Roman superscripts and dots denote derivatives with respect to x and t, respectively. Thus, from Eq. (4.6.17),

$$EIV^{\text{iv}} + \alpha^2 \rho AV = 0 \tag{4.6.4}$$

$$\ddot{T} - \alpha^2 T = 0 \tag{4.6.5}$$

From Eq. (4.6.5), according to the value of α^2, one has the following types of excitation possible for the variables-separable form of solution:

$$\left.\begin{array}{ll}\alpha^2 > 0, & T = A_1 e^{\alpha t} + A_2 e^{-\alpha t}\\ \alpha^2 = 0, & T = A_1 + A_2 t\\ \alpha^2 < 0, & T = A_1 \cos vt + A_2 \sin vt,\ v^2 = -\alpha^2\\ \alpha^2 \text{ complex}, & T = \cos vt(A_1 e^{\beta t} + A_2 e^{-\beta t}) \text{ or } T = \cos vt(A_1 e^{\beta t} + A_2 e^{-\beta t})\end{array}\right\} \tag{4.6.6a–d}$$

The assumption

$$\alpha = -\beta + \mathrm{i}v \tag{4.6.7}$$

is made in the last expression and the solution depends on whether real or imaginary parts are under consideration. Equation (4.6.1) as it stands is, of course a homogeneous equation which also governs the free vibrations of the beam, and for this problem the constant α^2 is to be determined as a frequency of free vibration, and the constants A_1 and A_2 by the initial conditions. However, the forms of t in Eqs (4.6.6) are valid irrespective of the values of α^2, A_1 and A_2. Hence, for forced motion of the beam, with a variables-separable form of the forcing term $F(x)f(t)$ replacing zero on the right-hand side of Eq. (4.6.1), permissible forms of $f(t)$ corresponding to Eqs (4.6.6) are

$$e^{-\beta t}, \quad \sin \nu t, \quad \cos \nu t, \quad A_1 + A_2 t, \quad e^{-\beta t}\sin \nu t, \quad e^{-\beta t}\cos \nu t \qquad (4.6.8)$$

and their linear combinations, where β, ν, A_1 and A_2 are any real constants. Except for Eq. (4.6.6b), all the functions can be expressed concisely by $e^{\alpha t}$ where α is a complex number, and the results for (4.6.6b) can be obtained by letting the complex parameter α tend to zero.

The forms (4.6.8) are valid, in particular, when the beam is forced only at its ends, $x = 0, l$, say. The homogeneous Eq. (4.6.1) then applies in the region $0 < x < l$, and the forces at $x = 0$ and $x = l$ can be expressed in terms of spatial derivatives of the displacements there. Hence, after solving for $v(x, t)$, subject to the boundary conditions $v(0, t) = q_1 e^{\alpha t}$, $v'(0, t) = q_2 e^{\alpha t}$, $v(l, t) = q_3 e^{\alpha t}$ and $v'(l, t) = q_4 e^{\alpha t}$, the boundary generalized forces are obtained as

$$Q_1 e^{\alpha t} = EI \frac{\partial^3 v(0)}{\partial x^3}, \qquad Q_2 e^{\alpha t} = -EI \frac{\partial^2 v(0)}{\partial x^2}$$

$$Q_3 e^{\alpha t} = -EI \frac{\partial^3 v(l)}{\partial x^3}, \qquad Q_4 e^{\alpha t} = EI \frac{\partial^2 v(l)}{\partial x^2}$$

and the beam dynamic stiffness matrix is immediately found to be

$$[\mathbf{D}] = \frac{EI}{l^3}\begin{bmatrix} F_1 & -F_4 l & F_5 & F_3 l \\ -F_4 l & F_2 l & -F_3 l & F_1 l \\ F_5 & -F_3 l & F_6 & F_4 l \\ F_3 l & F_1 l & F_4 l & F_2 l \end{bmatrix} \qquad (4.6.9)$$

Thus

$$[\mathbf{D}]\{\mathbf{q}\} = \{\mathbf{Q}\}$$

where the complex frequency functions F_i are given by

$$\left.\begin{aligned} F_1 &= -\lambda(\sinh\lambda + \sin\lambda)/\Delta, & F_2 &= -\lambda(\cosh\lambda \sin\lambda - \sinh\lambda\cos\lambda)/\Delta \\ F_3 &= -\lambda^2(\cosh\lambda - \cos\lambda)/\Delta, & F_4 &= \lambda^2 \sinh\lambda \sin\lambda/\Delta \\ F_5 &= \lambda^3(\sinh\lambda - \sin\lambda)/\Delta, & F_6 &= -\lambda^3(\cosh\lambda\sin\lambda + \sinh\lambda\cos\lambda)/\Delta \\ \Delta &= \cosh\lambda\cos\lambda - 1, & \lambda^4 &= -\alpha^2\rho A l^4/EI \end{aligned}\right\} \qquad (4.6.10)$$

Here λ is a complex frequency parameter, with $\alpha = -\beta + i\nu$. The functions take on the following limiting values when $\alpha^2 = 0$:

$$F_1 = 2, \quad F_2 = 4, \quad F_3 = 6, \quad F_4 = -6, \quad F_5 = -12, \quad F_6 = 12 \qquad (4.6.11)$$

Similar expressions can be obtained for flexural vibration in the xz plane, axial vibration and torsional vibration.

Thus, with time dependence of this general form $e^{\alpha t}$, the dynamic stiffness matrix for a straight uniform beam member in space can be established by following standard finite element procedures. The system dynamic stiffness matrix is assembled according to the equilibrium and compatibility conditions at the nodes. The result is a dynamic stiffness relation for the nodal forcing function $\{\overline{\mathbf{Q}}\}e^{\alpha t}$ and the nodal displacement response $\{\overline{\mathbf{q}}\}e^{\alpha t}$ of the structural system of the form

$$\{\overline{\mathbf{D}}(\alpha)\}\{\overline{\mathbf{q}}\}e^{\alpha t} = \{\overline{\mathbf{Q}}\}e^{\alpha t}, \qquad \alpha = -\beta + \mathrm{i}\nu \tag{4.6.12}$$

The real and imaginary parts of Eq. (4.6.12) correspond to excitations of the forms $\{\overline{\mathbf{Q}}\}e^{-\alpha t}\cos \nu t$ and $\{\overline{\mathbf{Q}}\}e^{-\alpha t}\sin \nu t$, respectively.

4.6.1. Damped Systems

The governing equation for flexural vibration of a straight uniform beam member with material damping and subject to nodal forces only is

$$EI\frac{\partial^4 v}{\partial x^4} + 2\gamma EI\frac{\partial^5 v}{\partial x^4 \partial t} + \rho A\frac{\partial^2 v}{\partial^2 t} + 2\delta\rho A\frac{\partial v}{\partial t} = 0 \tag{4.6.13}$$

where the additional damping constants γ and δ are due to stress rate and inertia, respectively. The method of separation of variables cannot be employed in the usual manner. However, if the nodal boundary conditions are

$$v(0,t) = q_1 e^{\alpha t}, \quad v'(0,t) = q_2 e^{\alpha t}, \quad v(l,t) = q_3 e^{\alpha t}, \quad v'(l,t) = q_4 e^{\alpha t} \tag{4.6.14}$$

then we may assume

$$v(x,t) = V(x)e^{\alpha t} \tag{4.6.15}$$

where $\alpha = -\beta + \mathrm{i}\nu$. Equation (4.6.13) then becomes

$$EI(1 + 2\alpha\gamma)\frac{\mathrm{d}^4 V}{\mathrm{d}x^4} + \rho A(\alpha^2 + 2\alpha\delta)V = 0 \tag{4.6.16}$$

Comparing Eq. (4.6.16) with Eq. (4.6.3) shows that the dynamic stiffness is exactly the same as given in Eq. (4.6.9), but now with

$$\lambda^4 = -\frac{(\alpha^2 + 2\alpha\delta)\rho A l^4}{EI(1 + 2\alpha\gamma)} \tag{4.6.17}$$

The method can therefore be applied to a damped system in a straightforward manner. Viscous dampers can be considered by direct superposition at the appropriate diagonal positions of the system dynamic stiffness matrix. The formulae presented in reference [25] for a damped Rayleigh–Timoshenko beam in harmonic vibration can also be extended to the case of exponentially varying harmonic excitation in a similar manner.

4.6.2. Distributed Loads

It is well known from the principle of stationary total potential energy that the equivalent nodal force vector due to a distributed load $\{\mathbf{f}(\mathbf{x},t)\}$ is

$$\{\mathbf{r}(\mathbf{x},t)\} = \int_{vol} [\mathbf{N}(\mathbf{x},t)]^{\mathrm{T}}\{\mathbf{f}(\mathbf{x},t)\}\,\mathrm{d}\,vol \tag{4.6.18}$$

where $\mathbf{x} = (x, y, z)$ and $[\mathbf{N}(x,t)]$ represents the shape functions, so that

$$\{\mathbf{u}(\mathbf{x},t)\} = [\mathbf{N}(\mathbf{x},t)]\{\mathbf{q}(t)\} \tag{4.6.19}$$

Generally speaking, the form (4.6.19) is not always possible. However, if $\{\mathbf{u}(\mathbf{x}, t)\} = \{\mathbf{u}(\mathbf{x})\}e^{\alpha t}$, $\alpha = -\beta + i\nu$, then

$$\{\mathbf{u}(\mathbf{x})\}e^{\alpha t} = [\mathbf{N}(\mathbf{x}, \alpha)]\{\mathbf{q}\}e^{\alpha t} \tag{4.6.20}$$

and $\{\mathbf{q}\}$ is a vector of generalized nodal displacement amplitudes. Under condition (4.6.20), Eq. (4.6.18) becomes

$$\{\mathbf{r}(\mathbf{x})\}e^{\alpha t} = \int_{vol} [\mathbf{N}(\mathbf{x}, \alpha)]^{\mathrm{T}}\{\mathbf{f}(x)\}\, \mathrm{d}\,vol\, e^{\alpha t} \tag{4.6.21}$$

Now, let $\{\mathbf{f}(\mathbf{x})\}$ be approximated as

$$\{\mathbf{f}(\mathbf{x})\} = [\mathbf{N}(\mathbf{x}, 0)]\{\mathbf{F}\} \tag{4.6.22}$$

where $[\mathbf{N}(\mathbf{x}, 0)]$ represents the shape functions when $\alpha = 0$. In the case of a flexural beam member

$$[\mathbf{N}(\mathbf{x}, 0)] = [1 - 3\xi^2 + 2\xi^3 \quad (\xi - 2\xi^2 + \xi^3)l \quad 3\xi^2 - 2\xi^3 \quad (-\xi^2 + \xi^3)l] \tag{4.6.23}$$

where $\xi = x/l$ and $\{\mathbf{F}\}$ is a given coefficient vector representing the values (and/or their derivatives) of $\{\mathbf{f}(\mathbf{x})\}$ at the nodes. The integral in Eq. (4.6.21) can be obtained explicitly by the application of the reciprocal theorem below.

The relations for the system under dynamic and static forces are

$$[\mathbf{D}(\alpha)]\{\mathbf{q}\}e^{\alpha t} = \{\mathbf{Q}\}e^{\alpha t}, \qquad \{\mathbf{u}(\mathbf{x}, \alpha)\}e^{\alpha t} = [\mathbf{N}(\mathbf{x}, \alpha)]\{\mathbf{q}\}e^{\alpha t}$$

$$[\mathbf{D}(0)]\{\mathbf{q}_0\} = \{\mathbf{Q}_0\}, \qquad \{\mathbf{u}(\mathbf{x}, 0)\} = [\mathbf{N}(\mathbf{x}, 0)]\{\mathbf{q}_0\} \text{ or } \{\mathbf{u}_0\} = [\mathbf{N}_0]\{\mathbf{q}_0\}$$

respectively. The reciprocal theorem states that

$$\{\mathbf{q}\}^{\mathrm{T}}\{\mathbf{Q}_0\}e^{\alpha t} = \left(\{\mathbf{Q}\}^{\mathrm{T}}\{\mathbf{q}_0\} - \alpha^2 \int \{\mathbf{u}\}^{\mathrm{T}}[\rho]\{\mathbf{u}_0\}\, \mathrm{d}\,vol\right)e^{\alpha t}$$

After simplification, we have

$$\int [\mathbf{N}]^{\mathrm{T}}[\rho][\mathbf{N}_0]\, \mathrm{d}\,vol = (1/\alpha^2)\{[\mathbf{D}(\alpha)] - [\mathbf{D}(0)]\} = [\mathbf{M}(0, \alpha)] \tag{4.6.24}$$

or, more generally,

$$\int [\mathbf{N}(\alpha_1)]^{\mathrm{T}}[\rho][\mathbf{N}(\alpha_2)]\, \mathrm{d}\,vol = [1/(\alpha_1^2 - \alpha_2^2)]\{[\mathbf{D}(\alpha_1) - \mathbf{D}(\alpha_2)]\} = [\mathbf{M}(\alpha_1, \alpha_2)]$$

However, from Eq. (4.6.21),

$$\{\mathbf{r}(\mathbf{x})\} = \int_{vol} [\mathbf{N}(\mathbf{x}, \alpha)]^{\mathrm{T}}[\mathbf{N}(\mathbf{x}, 0)]\, \mathrm{d}\,vol\{\mathbf{F}\} \tag{4.6.25}$$

Comparing this with Eq. (4.6.24), we have

$$\{\mathbf{r}(\mathbf{x})\} = \int_{vol} [\mathbf{N}]^{\mathrm{T}}[\mathbf{N}_0]\, \mathrm{d}\,vol\{\mathbf{F}\} = (1/\alpha^2)\{[\tilde{\mathbf{D}}(\alpha) - \tilde{\mathbf{D}}(0)]\} \tag{4.6.26}$$

where $[\tilde{\mathbf{D}}]$ is the dynamic stiffness matrix obtained by setting the mass density equal to unity.

For a flexural beam member, $[\tilde{\mathbf{D}}(\alpha)]$ has the same form as Eq. (4.6.9) but with $\lambda^4 = -\alpha^2 l^4/EI$, with the frequency functions given by Eq. (4.6.11).

4.6.3. Comparison with Modal Analysis

For purposes of comparing the method to the modal analysis method it is convenient to consider the undamped vibration of frames.

Let $\{\mathbf{u}(x,t)\}$ be the total response of a frame with coordinate x, and $\{\phi_i(x)\}$ and ω_i the ith mode and modal frequency, respectively. Then, applying the usual modal method, we have

$$\{\mathbf{u}(x,t)\} = \sum_i \{\phi_i(x)\}\left[\omega_i^{-1}\int_0^t \sin\omega_i(t-\tau)\int\{\phi_i(x)\}^{\mathrm{T}}\{\mathbf{f}(x,\tau)\}\,\mathrm{d}x\,\mathrm{d}\tau \right.$$
$$\left. + \cos\omega_i t\int\{\phi_i(x)\}^{\mathrm{T}}[\rho]\{\mathbf{u}_0(x)\}\,\mathrm{d}x + \sin\omega_i t\int\{\phi_i(x)\}^{\mathrm{T}}\,\mathrm{d}x\,\omega_i^{-1}\right] \tag{4.6.27}$$

where $\{\mathbf{f}(x,t)\} = [\mathbf{N}_0]\{\mathbf{F}\}\mathrm{e}^{\alpha t}$ as in Eq. (4.6.8), and

$$\{\mathbf{u}_0(x)\} = \{\mathbf{u}(x,0)\} = [\mathbf{N}_0]\{\mathbf{q}_0\} \quad \text{and} \quad \{\dot{\mathbf{u}}(x)\} = \{\dot{\mathbf{u}}(x,0)\} = [\mathbf{N}_0]\{\dot{\mathbf{q}}_0\} \tag{4.6.28}$$

are the initial conditions. The corresponding solution according to the present method is

$$\{\mathbf{u}_0(x)\} = \{\mathbf{u}_s(x)\}\mathrm{e}^{\alpha t} + \sum_i \{\phi_i(x)\}\left(\cos\omega_i t\int\{\phi_i(x)\}^{\mathrm{T}}[\rho]\{\mathbf{u}_0(x)-\mathbf{u}_s(x)\}\,\mathrm{d}x\right.$$
$$\left. + \sin\omega_i t\int\{\phi_i(x)\}^{\mathrm{T}}[\rho]\{\dot{\mathbf{u}}_0(x)-\alpha\mathbf{u}_s(x)\}\,\mathrm{d}x\right) \tag{4.6.29}$$

where $\{\mathbf{u}_s(x)\}$ is the forced response. The real and imaginary parts of Eq. (4.6.29) have to be considered separately according to the forms of the forcing functions.

Now, if

$$\{\mathbf{u}(x,t)\} = [\mathbf{N}(x,\alpha)]\{\mathbf{q}\}\mathrm{e}^{\alpha t}, \qquad \{\phi_i(x)\} = [\mathbf{N}(x,\mathrm{i}\omega_i)]\{\mathbf{\Phi}_i\}$$
$$\{\mathbf{u}_0(x)\} = [\mathbf{N}(x,0)]\{\mathbf{q}_0\} \quad \{\dot{\mathbf{u}}_0(x)\} = [\mathbf{N}(x,0)]\{\dot{\mathbf{q}}_0\} \quad \{\mathbf{u}_s(x)\} = [\mathbf{N}(x,\alpha)]\{\mathbf{q}_s\} \tag{4.6.30}$$

then Eq. (4.6.14) becomes

$$\{\mathbf{u}(x,t)\} = \{\mathbf{u}_s(x)\}\mathrm{e}^{\alpha t} + \sum_i \{\phi_i(x)\}\cos\omega_i t\{\mathbf{\Phi}_i\}^{\mathrm{T}}([\mathbf{M}(i\omega_i,0)]\{\mathbf{q}_0\} - [\mathbf{M}(\alpha,0)]\{\mathbf{q}_s\})$$
$$+ \sum_i \{\phi_i(x)\}\sin\omega_i t\{\phi_i\}^{\mathrm{T}}([\mathbf{M}(i\omega_i,0)]\{\dot{\mathbf{q}}_0\} - [\mathbf{M}(\alpha,0)]\{\mathbf{q}_s\}) \tag{4.6.31}$$

or in terms of nodal coordinates,

$$\{\mathbf{q}\}\mathrm{e}^{\alpha t} = \{\mathbf{q}_s\}\mathrm{e}^{\alpha t} + \sum_i \{\mathbf{\Phi}_i\}\{\mathbf{\Phi}_i\}^{\mathrm{T}}([\mathbf{M}(\mathrm{i}\omega_i,0)]\{\mathbf{q}_0\} - [\mathbf{M}(\alpha,0)]\cos\omega_i t$$
$$+ \sum_i \{\mathbf{\Phi}_i\}\{\mathbf{\Phi}_i\}^{\mathrm{T}}([\mathbf{M}(\mathrm{i}\omega_i,0)]\{\dot{\mathbf{q}}_0\} - \alpha[\mathbf{M}(\alpha,0)]\sin\omega_i t \tag{4.6.32}$$

Comparing the solutions (4.6.27) and (4.6.31), one can conclude that, for exponentially varying excitations, the distinct advantage of the present method is that no time integrations are required. In the case of the consistent formulation represented by Eqs (4.6.30), the modal decomposition of the initial conditions is achieved by means of the mixed mass matrix as defined in Eq. (4.6.24). If the forced response only is of interest, $\{\mathbf{q}_s\}$ in Eq. (4.6.32) is obtained by the solution of the linear dynamic stiffness equation directly, without using the natural modes.

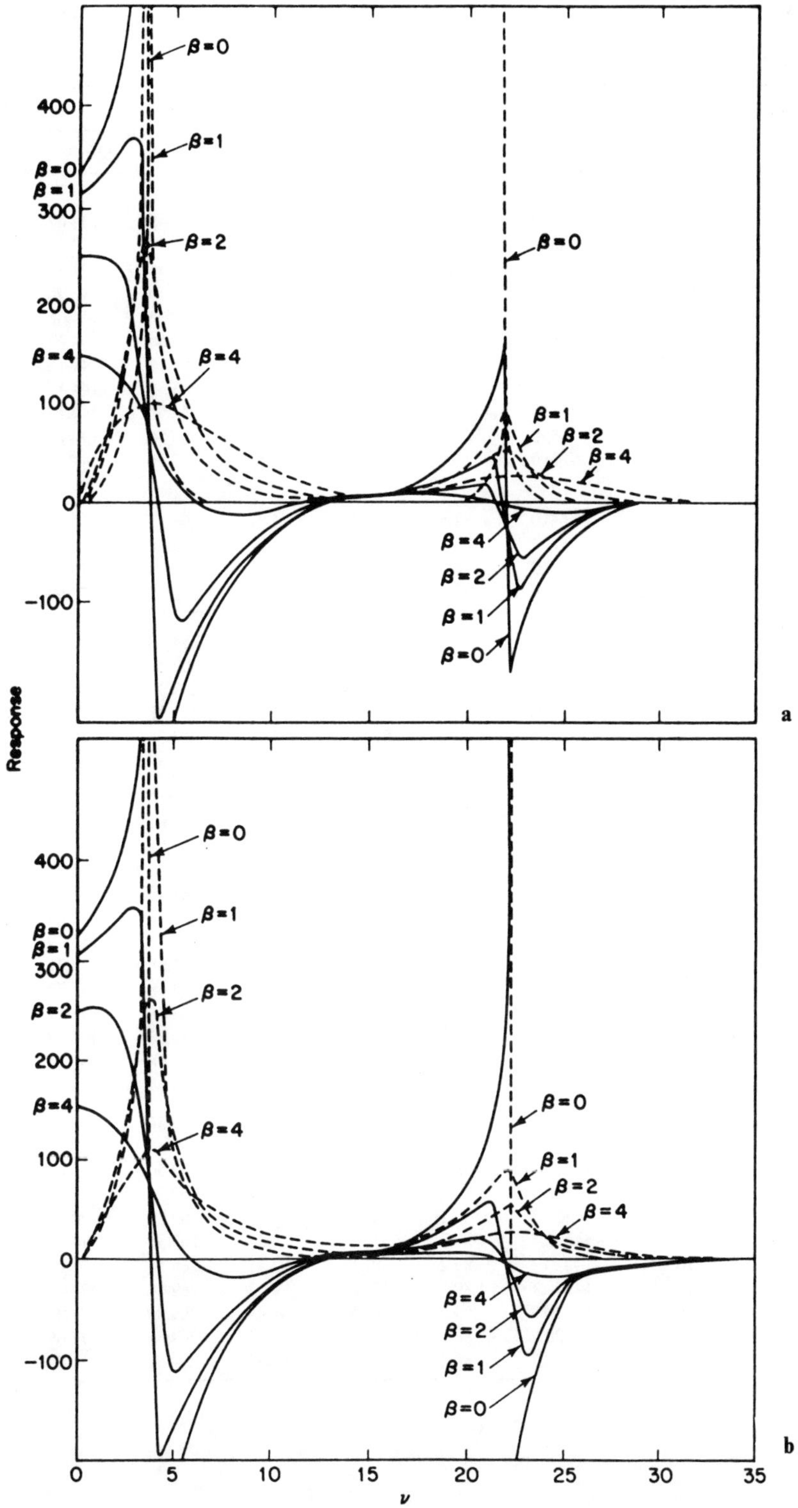

Fig. 4.6.1. Response amplitudes for **a** $\gamma = \delta = 0$; **b** $\gamma = 0$, $\delta = 0.04$

Table 4.6.1. Comparison with modal analysis results (PM: present method)

Mode	$\beta=0$		$\beta=1$		$\beta=2$		$\beta=4$	
	cos	sin	cos	sin	cos	sin	cos	sin
(a) $\delta=0.000$, $\eta=3.5160$								
1	352.0	000.0	315.3	51.0	243.85	63.49	134.67	39.38
2	360.3	000.0	323.6	51.0	252.03	63.56	142.66	39.50
3	361.3	000.0	324.6	51.0	253.08	63.56	143.71	39.51
4	361.6	000.0	324.9	51.0	253.36	63.56	143.98	39.51
5	361.7	000.0	325.0	51.0	253.46	63.56	144.08	39.51
6	361.8	000.0	325.0	51.0	253.50	63.56	144.13	39.51
7	361.8	000.0	325.0	51.0	253.53	63.56	144.15	39.51
8	361.8	000.0	325.1	51.0	253.54	63.56	144.17	39.51
9	361.8	000.0	325.1	51.0	253.55	63.56	144.17	39.51
10	361.8	000.0	325.1	51.0	253.55	63.56	144.18	39.51
PM	361.8	000.0	325.1	51.0	253.55	63.56	144.18	39.51
(b) $\delta=0.000$, $\eta=1.0000$								
1	000.0	000.0	79.3	557.6	74.84	263.13	61.12	107.44
2	8.5	000.0	87.7	557.7	83.21	263.38	69.27	107.91
3	9.5	000.0	88.8	557.7	84.27	263.38	70.32	107.92
4	9.8	000.0	89.0	557.7	84.54	263.38	70.59	107.92
5	9.9	000.0	89.2	557.7	84.64	263.38	70.69	107.92
6	9.9	000.0	89.2	557.7	84.69	263.38	70.74	107.92
7	10.0	000.0	89.2	557.7	84.71	263.38	70.76	107.92
8	10.0	000.0	89.2	557.7	84.72	263.38	70.77	107.92
9	10.0	000.0	89.2	557.7	84.73	263.38	70.78	107.92
10	10.0	000.0	89.2	557.7	84.73	263.38	70.78	107.92
PM	10.0	000.0	89.2	557.7	84.73	263.38	70.78	107.92
(c) $\delta=0.040$, $\eta=1.0000$								
1	352.0	−2.5	317.9	49.7	246.71	63.62	136.23	39.90
2	360.3	−2.5	326.1	49.7	254.90	63.68	144.23	40.03
3	361.3	−2.5	327.2	49.7	255.96	63.68	145.27	40.03
4	361.6	−2.5	327.5	49.7	256.23	63.68	145.55	40.03
5	361.7	−2.5	327.6	49.7	256.33	63.68	145.65	40.03
6	361.8	−2.5	327.6	49.7	256.37	63.68	145.69	40.03
7	361.8	−2.5	327.6	49.7	256.40	63.68	145.71	40.03
8	361.8	−2.5	327.6	49.7	256.41	63.68	145.73	40.03
9	361.8	−2.5	327.7	49.7	256.42	63.68	145.74	40.03
10	361.8	−2.5	327.7	49.7	256.42	63.68	145.74	40.03
PM	361.8	−2.5	327.7	49.7	256.42	63.68	145.74	40.03
(d) $\delta=0.040$, $\eta=3.5160$								
1	000.0	−14220.7	79.3	581.7	75.03	269.31	61.41	109.06
2	8.5	−14220.7	87.7	581.8	83.41	269.56	69.57	109.53
3	9.5	−14220.7	88.8	581.8	84.47	269.56	70.62	109.54
4	9.8	−14220.7	89.0	581.8	84.74	269.56	70.89	109.54
5	9.9	−14220.7	89.1	581.8	84.84	269.56	70.99	109.54
6	9.9	−14220.7	89.2	581.8	84.88	269.56	71.04	109.54
7	10.0	−14220.7	89.2	581.8	84.91	269.56	71.06	109.54
8	10.0	−14220.7	89.2	581.8	84.92	269.56	71.07	109.54
9	10.0	−14220.7	89.2	581.8	84.93	269.56	71.08	109.54
10	10.0	−14220.7	89.2	581.8	84.93	269.56	71.09	109.54
PM	10.0	−14220.7	89.2	581.8	84.93	269.56	71.09	109.54

The formulation can be extended to damped vibration by means of complex property constants (cf. Eq. (4.6.17)). In all cases, both real and imaginary parts of $\{\mathbf{q}_s\}$ are required, because, although the forcing function may be of the type $e^{-\alpha t}\cos \nu t$, the forced response will generally be of the form $e^{-\beta t}(R\cos \nu t - S\sin \nu t)$.

4.6.4. Numerical Examples

To illustrate the power of the method, without too much computational involvement, we can consider a vibrating cantilever subjected to an exponentially varying harmonic lateral force $1000e^{\alpha t}$, $\alpha = -\beta + i\nu$, acting at the free end. We are interested in the forced response only. Let $\rho A = EI = l = 1$. The forced response amplitudes corresponding to $e^{-\alpha t}\cos \nu t$ (solid lines) and $e^{-\alpha t}\sin \nu t$ (dashed lines) at the free end are plotted in Fig. 4.6.1 for damping (a) $\delta = 0$ and (b) $\delta = 0.04$. It is interesting to note the phase difference phenomena, similar to those of pure harmonic vibration.

The same problem has also been solved by the usual modal analysis, with the exception that, for the forced response, the uncoupled equations of motion in the principal coordinates were solved by the complex number method, as used in Sect. 1.1 on single d.o.f. systems. The results, for $\beta = 0, 1, 2, 4$, $\nu = 1$ and 3.516 (at resonance), and $\delta = 0$ and 0.04, are shown in Table 4.6.1.

Numerical experiments have been performed for frame problems with distributed loads. All the results are coincident with those obtained by modal analysis up to the computer accuracy.

References

1. VV Bolotin 1963. Nonconservative problems of the theory of elastic stability. Pergamon Press, London
2. CJF Ridders 1985. Accurate computation of $F'(x)$ and $F''(x)$. In: Software for engineering problems (ed. RA Adey), pp. 83–84. CML Publications
3. WP Howson, JR Banerjee, FW Williams 1983. Concise equations and program for exact eigensolutions of plane frames including member shear. Engng Soft 3, 443–452
4. D Pearson, WH Wittrick 1986. An exact solution for the vibration of helical springs using a Bernoulli–Euler model. Int J Mech Sci 28, 83–96
5. J Henrych 1981. The Dynamics of Arches and Frames. Elsevier, Amsterdam
6. AYT Leung 1978. An accurate method of dynamic condensation in structural analysis. Int J Num Meth Engng 12, 1705–1716
7. AYT Leung 1979. An accurate method of dynamic substructing with simplified computation. Int J Num Meth Engng 14, 1241–1256
8. VV Bolotin 1964. The dynamic stability of elastic systems. Holden-Day, San Francisco
9. BAH Abbas, J Thomas 1978. Dynamic stability of Timoshenko beams resting on an elastic foundation. J Sound Vib 60, 33–44
10. BAH Abbas 1986. Dynamic stability of a rotating Timoshenko beam with a flexible root. J Sound Vib 108, 25–32
11. JR Banerjee, FW Williams 1985. Exact Bernoulli–Euler dynamic stiffness matrix for a range of tapered beams. Int J Num Meth Engng 21, 2289–2302
12. RA Horn, CA Johnson 1985. Matrix analysis. Cambridge University Press, Cambridge
13. B Parlett 1980. The symmetric eigenvalue problem. Prentice-Hall, Englewood Cliffs, NJ
14. AH Chilver (ed.) 1967. Thin-walled structures. Chatto and Windus, London
15. K Huseyin 1989. The convexity of the stability boundary of symmetric structural systems. Acta Mech 8, 205–211
16. WF Chen, T Atsuta, 1977. Theory of beam-columns, vol 2. McGraw-Hill, New York, p 87
17. V Kolousek 1973. Dynamics in engineering structures. Butterworth, London

18. DC Johnson, RED Bishop 1960. The mechanics of vibrations. Cambridge University Press, Cambridge
19. DL Woodcock 1963. On the interpretation of the vector plots of forced vibrations of a linear system with viscous damping. Aero Q 14, 45–62
20. DE Newland 1987. On the modal analysis of nonconservative linear systems. J Sound Vib 112, 69–96
21. I Fawzy, RED Bishop 1976. On the dynamics of linear non-conservative systems. Proc R Soc Lond A352, 25–40
22. IFA Wahed, RED Bishop 1976. On the equations governing the free and forced vibrations of general nonconservative systems. J Mech Engng Sci 18, 6–10
23. RA Frazer, WJ Duncan, AR Collar 1947. Elementary matrices. Cambridge University Press, Cambridge
24. JH Wilkinson 1965. The algebraic eigenvalue problem. Clarendon Press, Oxford
25. R Lundén, B. Åkesson 1983. Damped second-order Rayleigh–Timoshenko beam vibration in space: an exact complex dynamic member stiffness matrix. Int J Num Meth Engng 19, 431–449

Chapter 5

General Formulation

"Exact" solutions are difficult to obtain due to the fact that it is not easy to replace the governing partial differential equations by ordinary differential equations, and that the number and order of the resulting ordinary differential equations are large, so that manual solutions are generally impossible. In this chapter a generalized Kantorovich method is presented which will produce the governing ordinary differential equations automatically in a similar manner to the finite element method which produces algebraic governing equations. The mth order and nth degree governing ordinary differential equations will be solved in a computer-oriented manner so that the dynamic stiffness matrix can be formed automatically.

5.1. Initial Stress Formulation

The governing equations including the effects of initial stress for a vibrating structure are of interest. The equations will determine the critical load when there exists at least one additional distinct equilibrium configuration in a neighbourhood very close to the original configuration. If such an adjacent equilibrium configuration exists, the body may change suddenly from one equilibrium configuration to the other under the stimulus of small external disturbances. This is the most complicated problem which can be solved by linear differential equations. The stability problem will be formulated by this approach, which is sometimes called the Euler method, for a body under non-conservative external forces.

Consider an elastic body which is in static equilibrium and in a reference state with initial stresses $\sigma^{(0)\lambda\mu}$. Initial stresses refers to those stresses which exist in a body in the initial state, that is, before the start of a deformation of interest. Choose the initial state as the reference state of an initial stress problem.

Let a rectangular Cartesian coordinate system (x^1, x^2, x^3) be fixed in space. Form an infinitesimal rectangular parallelepiped enclosed by the six surfaces: $x^\lambda =$ constant and $x^\lambda + \mathrm{d}x^\lambda =$ constant ($\lambda = 1, 2, 3$). Denoting the initial internal forces per unit area acting on the surface $x^\lambda =$ constant by $-\sigma^{(0)\lambda}$ and using the summation convention for a repeated index, we define components of the initial stress as

$$\sigma^{(0)\lambda} = \sigma^{(0)\lambda\mu}\mathbf{i}_\mu \tag{5.1.1}$$

where $\mathbf{i}_\mu$ is the unit vector in the direction of the x^μ-axis. Initial body forces and surface tractions are denoted by $\overline{\mathbf{P}}^{(0)}$ and $\overline{\mathbf{F}}^{(0)}$, respectively, and their components by

$$\overline{\mathbf{P}}^{(0)} = \overline{P}^{(0)\lambda}\mathbf{i}_\lambda, \qquad \overline{\mathbf{F}}^{(0)} = \overline{F}^{(0)\lambda}\mathbf{i}_\lambda \qquad (5.1.2, 5.1.3)$$

For the sake of simplicity, assume that these initial stresses and forces form a self-equilibrating system, i.e.

$$\sigma^{(0)\lambda\mu}_{,\mu} + \overline{P}^{(0)\lambda} = 0 \qquad (5.1.4)$$

in the interior of the body and

$$\sigma^{(0)\lambda\mu}\mathbf{n}_\mu = \overline{F}^{(0)\lambda} \qquad (5.1.5)$$

on the surface of the body, where $(\ )_{,\mu} = \partial(\ )/\partial x^\mu$.

Define a dynamic problem for an elastic body with initial stresses by prescribing additional body forces $\overline{P}^\lambda$, additional surface forces $\overline{F}^\lambda$ on S_1 and surface displacements $\overline{u}^\lambda$ on S_2, where the displacements are measured from the reference state. It is noted that $\overline{P}^\lambda$, $\overline{F}^\lambda$ and $\overline{u}^\lambda$ are prescribed functions of time as well as the space coordinates. Then, the principle of virtual work for the present problem is given by

$$\int_{t_1}^{t_2}\left\{\iiint_V (\sigma^{(0)\lambda\mu} + \sigma^{\lambda\mu})\,\delta e_{\lambda\mu}\,\mathrm{d}V - \delta\iiint_V \frac{1}{2}\rho\left(\frac{\mathrm{d}\mathbf{r}}{\mathrm{d}t}\right)^2\mathrm{d}V - \iiint_V (\overline{\mathbf{P}}^{(0)} + \overline{\mathbf{P}})\,\delta\mathbf{r}\,\mathrm{d}V\right.$$
$$\left. - \iint_{S_1} (\overline{\mathbf{F}}^{(0)} + \overline{\mathbf{F}})\,\delta\mathbf{r}\,\mathrm{d}S\right\}\mathrm{d}t = 0 \qquad (5.1.6)$$

where

$$\mathbf{r} = \mathbf{r}^{(0)} + \mathbf{u} = \mathbf{r}^{(0)} + u^k\mathbf{i}_k \qquad (5.1.7)$$

$$e_{\lambda\mu} = \tfrac{1}{2}(u^\lambda_{,\mu} + u^\mu_{,\lambda} + u^k_{,\lambda}u^k_{,\mu}) \qquad (5.1.8)$$

$$\begin{aligned}\overline{\mathbf{P}}^{(0)} &= \overline{P}^{(0)\lambda}\mathbf{i}_\lambda, & \overline{\mathbf{P}} &= \overline{P}^\lambda\mathbf{i}_\lambda\\ \overline{\mathbf{F}}^{(0)} &= \overline{F}^{(0)\lambda}\mathbf{i}_\lambda, & \overline{\mathbf{F}} &= \overline{F}^\lambda\mathbf{i}_\lambda\end{aligned} \qquad (5.1.9a\text{–}d)$$

and

$$\delta\mathbf{u} = 0 \quad \text{on} \quad S_2 \qquad (5.1.10)$$

Subsequent interest will be confined to deriving governing equations for small motion only, assuming that $u^\lambda = \mathbf{0}(\varepsilon)$ and $\sigma^{(0)\lambda\mu} = \mathbf{0}(1)$. Then, after some manipulation, the principle of virtual work (5.1.6) is reduced to

$$\int_{t_1}^{t_2}\left\{\iiint_V [\sigma^{\lambda\mu}\delta\varepsilon_{\lambda\mu} + \tfrac{1}{2}\sigma^{(0)\lambda\mu}\delta(u^k_{,\lambda}u^k_{,\mu})]\,\mathrm{d}V - \delta\iiint_V \tfrac{1}{2}\rho\dot{u}^k\dot{u}^k\,\mathrm{d}V\right.$$
$$\left. - \iiint_V \overline{\mathbf{P}}\delta\mathbf{r}\,\mathrm{d}V - \iint_{S_1} \overline{\mathbf{F}}\,\delta\mathbf{r}\,\mathrm{d}S\right\}\mathrm{d}t = 0 \qquad (5.1.11)$$

where

$$\varepsilon_{\lambda\mu} = \tfrac{1}{2}(u^\lambda_{,\mu} + u^\mu_{,\lambda}) \qquad (5.1.12)$$

and

$$u^k = \overline{u}^k \quad \text{on} \quad S_2 \qquad (5.1.13)$$

and where Eqs (5.1.4) and (5.1.5) have been used in deriving Eq. (5.1.11).

If the stress–strain relations

$$\sigma^{\lambda\mu} = a^{\lambda\mu\alpha\beta}\varepsilon_{\alpha\beta} \tag{5.1.14}$$

assure the existence of the strain energy function defined by

$$A = \tfrac{1}{2}a^{\lambda\mu\alpha\beta}\varepsilon_{\lambda\mu}\varepsilon_{\alpha\beta} \tag{5.1.15}$$

the principle of virtual work is transformed into

$$\delta \int_{t_1}^{t_2} \left\{ T - U - \iiint_V \tfrac{1}{2}\sigma^{(0)\lambda\mu} u^k_{,\lambda} u^k_{,\mu} \, \mathrm{d}V \right\} \mathrm{d}t + \int_{t_1}^{t_2} \left\{ \iiint_V \overline{\mathbf{P}}\,\delta\mathbf{r}\,\mathrm{d}V + \iint_{S_1} \overline{\mathbf{F}}\,\delta\mathbf{r}\,\mathrm{d}S \right\} \mathrm{d}t = 0 \tag{5.1.16}$$

where

$$T = \iiint_V \tfrac{1}{2}\rho \dot{u}^k \dot{u}^k \, \mathrm{d}V \tag{5.1.17}$$

$$U = \iiint_V A(u^k)\,\mathrm{d}V \tag{5.1.18}$$

and Eq. (5.1.13) are taken as subsidiary conditions.

If the existence of two potential functions Φ and Ψ defined by

$$\delta\Phi = -\overline{\mathbf{P}}^{\lambda}\delta u^{\lambda}, \qquad \delta\Psi = -\overline{\mathbf{F}}^{\lambda}\delta u^{\lambda} \tag{5.1.19}$$

is also assumed, the principle (5.1.16) reduces to

$$\delta \int_{t_1}^{t_2} H\,\mathrm{d}t = \delta \int_{t_1}^{t_2} \left\{ T - U - \iiint_V \tfrac{1}{2}\sigma^{(0)\lambda\mu} u^k_{,\lambda} u^k_{,\mu}\,\mathrm{d}V - \iiint_V \Phi\,\mathrm{d}V - \iint_{S_1} \Psi\,\mathrm{d}S \right\} \mathrm{d}t = 0 \tag{5.1.20}$$

where Eqs (5.1.4) are taken as subsidiary conditions and the variation is taken with respect to u^k. Equation (5.1.20) is Hamilton's principle applied to the dynamic problem with initial stresses. The Hamiltonian H is defined by

$$H = T - U - \iiint_V \tfrac{1}{2}\sigma^{(0)\lambda\mu} u^k_{,\lambda} u^k_{,\mu}\,\mathrm{d}V - \iiint_V \phi\,\mathrm{d}V - \iint_{S_1} \Psi\,\mathrm{d}S \tag{5.1.21}$$

5.2. Finite Element Method

The finite element method uses the fact that the system Hamiltonian H is equal to the summation of the element Hamiltonians H_e,

$$H = \sum_e H_e \tag{5.2.1}$$

or

$$\delta \int_{t_1}^{t_2} H\,\mathrm{d}t = \sum_e \int_{t_1}^{t_2} H_e\,\mathrm{d}t = 0 \tag{5.2.2}$$

Since the elements do not overlap, Hamilton's principle requires that

$$\delta \int_{t_1}^{t_2} H_e \,\mathrm{d}t = 0 \tag{5.2.3}$$

where the element Hamiltonian is written in vector form in free vibration,

$$H_e = T_e - U_e - \frac{1}{2}\iiint_V \{\boldsymbol{\theta}\}^{\mathrm{T}}[\boldsymbol{\sigma}^0]\{\boldsymbol{\theta}\}\,\mathrm{d}V \tag{5.2.4}$$

$$T_e = \frac{1}{2}\iiint_V \{\dot{\mathbf{u}}\}^{\mathrm{T}}[\mathbf{m}]\{\dot{\mathbf{u}}\}\,\mathrm{d}V \tag{5.2.5}$$

$$U_e = \frac{1}{2}\iiint_V \{\boldsymbol{\varepsilon}\}^{\mathrm{T}}[\mathbf{E}]\{\boldsymbol{\varepsilon}\}\,\mathrm{d}V \tag{5.2.6}$$

where $[\mathbf{m}]$ is the inertia density matrix, $[\mathbf{E}]$ the elastic modulus matrix, $[\mathbf{u}]$ the displacement field vector, $\{\boldsymbol{\varepsilon}\}$ the engineering strain vector, $\{\boldsymbol{\theta}\}$ the displacement gradient vector and $[\boldsymbol{\sigma}^0]$ the initial stress matrix defined in Cartesian coordinates by

$$\{\mathbf{u}(x, y, z, t)\} = \{\mathbf{u}(\mathbf{x}, t)\} = [u_x, u_y, u_z]^{\mathrm{T}} \tag{5.2.7}$$

$$\{\boldsymbol{\varepsilon}(\mathbf{x}, t)\} = [\varepsilon_x, \varepsilon_y, \varepsilon_z, 2\varepsilon_{xy}, 2\varepsilon_{yz}, 2\varepsilon_{xz}]^{\mathrm{T}} \tag{5.2.8}$$

$$\{\boldsymbol{\theta}(\mathbf{x}, t)\} = \left[\frac{\partial \mathbf{u}}{\partial x}, \frac{\partial \mathbf{u}}{\partial y}, \frac{\partial \mathbf{u}}{\partial z}\right]^{\mathrm{T}} \tag{5.2.9}$$

$$[\boldsymbol{\sigma}^0] = \begin{bmatrix} \sigma_x^0\mathbf{I} & \sigma_{xy}^0\mathbf{I} & \sigma_{xz}^0\mathbf{I} \\ \sigma_{yx}^0\mathbf{I} & \sigma_y^0\mathbf{I} & \sigma_{yz}^0\mathbf{I} \\ \sigma_{zx}^0\mathbf{I} & \sigma_{zy}^0\mathbf{I} & \sigma_z^0\mathbf{I} \end{bmatrix} \tag{5.2.10}$$

and $[\mathbf{I}]$ is a 3×3 identity matrix. In Eqs (5.2.4)–(5.2.6), all matrix quantities are given and all vector quantities are to be determined. The finite element method assumes

$$\{\mathbf{u}(\mathbf{x}, t)\} = [\mathbf{N}(\mathbf{x})]\{\mathbf{q}_e(t)\} \tag{5.2.11}$$

where $[\mathbf{N}(\mathbf{x})]$ is a matrix of predetermined shape functions and $\{\mathbf{q}_e(t)\}$ is a vector of nodal displacements.

Substituting Eq. (5.2.11) into (5.1.12), we have

$$\{\boldsymbol{\varepsilon}(\mathbf{x}, t)\} = [\mathbf{B}(\mathbf{x})]\{\mathbf{q}_e(t)\} \tag{5.2.12}$$

where $[\mathbf{B}(\mathbf{x})]$ is called the strain matrix obtained by spatial differentiation of the shape function $[\mathbf{N}(\mathbf{x})]$ according to the strain–displacement equation (5.1.12). Similarly, substituting Eq. (5.2.11) into (5.2.9), we have

$$\{\boldsymbol{\theta}(\mathbf{x}, t)\} = [\mathbf{G}(\mathbf{x})]\{\mathbf{q}_e(t)\} \tag{5.2.13}$$

where $[\mathbf{G}(\mathbf{x})]$ is called the gradient matrix obtained by spatial differentiation of $[\mathbf{N}(\mathbf{x})]$ according to Eq. (5.2.9). Finally, substituting Eqs (5.2.11–5.2.13) into (5.2.3–5.2.4), we have

$$[\mathbf{M}_e]\{\ddot{\mathbf{q}}_e\} + [\mathbf{K}_e]\{\mathbf{q}_e\} + [\mathbf{K}_e^\sigma]\{\mathbf{q}_e\} = \{\mathbf{0}\} \tag{5.2.14}$$

where the mass matrix, stiffness matrix and initial stress matrix are given by, respectively,

$$[\mathbf{M}_e] = \iiint_V [\mathbf{N}(\mathbf{x})]^{\mathrm{T}}[\mathbf{m}][\mathbf{N}(\mathbf{x})]\,\mathrm{d}V \tag{5.2.15}$$

$$[\mathbf{K}_e] = \iiint_V [\mathbf{B}(\mathbf{x})]^{\mathrm{T}}[\mathbf{E}][\mathbf{B}(\mathbf{x})]\,\mathrm{d}V \tag{5.2.16}$$

$$[\mathbf{K}_e^\sigma] = \iiint_V [\mathbf{G}(\mathbf{x})]^{\mathrm{T}}[\boldsymbol{\sigma}^0][\mathbf{G}(\mathbf{x})]\,\mathrm{d}V \tag{5.2.17}$$

If the element coordinate $\{\mathbf{q}_e\}$ relates to the global coordinate $\{\mathbf{q}\}$ by

$$\{\mathbf{q}_e\} = [\mathbf{T}_e]\{\mathbf{q}\} \tag{5.2.18}$$

where $[\mathbf{T}_e]$ is a transformation matrix, then Hamilton's principle, Eq. (5.2.2), gives

$$[\mathbf{M}]\{\ddot{\mathbf{q}}\} + [\mathbf{K}]\{\mathbf{q}\} + [\mathbf{K}^\sigma]\{\mathbf{q}\} = \{\mathbf{0}\} \tag{5.2.19}$$

where

$$[\mathbf{M}] = \sum_e [\mathbf{T}_e]^{\mathrm{T}}[\mathbf{M}_e][\mathbf{T}_e]$$

$$[\mathbf{K}] = \sum_e [\mathbf{T}_e]^{\mathrm{T}}[\mathbf{K}_e][\mathbf{T}_e]$$

$$[\mathbf{K}^\sigma] = \sum_e [\mathbf{T}_e]^{\mathrm{T}}[\mathbf{K}_e^\sigma][\mathbf{T}_e]$$

For free vibration problems

$$\{\mathbf{q}(t)\} = \mathrm{e}^{\mathrm{i}\omega t}\{\bar{\mathbf{q}}\} \tag{5.2.20}$$

where ω is the vibration frequency, and, from Eq. (5.2.19),

$$[\mathbf{K} + \mathbf{K}^\sigma - \omega^2\mathbf{M}]\{\bar{\mathbf{q}}\} = \{\mathbf{0}\} \tag{5.2.21}$$

is an eigenvalue problem to determine the natural frequency ω and the corresponding mode $\{\bar{\mathbf{q}}\}$.

5.3. Dynamic Stiffness Method

In a free vibration analysis, the finite element method results in a set of algebraic equations in the form of Eq. (5.2.21), whose solution method is fairly standard. Alternatively, we may use the Kantorovich method, Eq. (5.3.1) below and obtain a set of ordinary differential equations (when the time variable is eliminated for harmonic oscillation). In conjunction with the natural boundary conditions, the solutions of the ordinary differential equations give the dynamic stiffness matrix. The latter method, which will be presented in detail, involves fewer assumptions and therefore is more accurate.

To apply the Kantorovich method, the three-dimensional displacement field is approximated by, instead of Eq. (5.2.11),

$$\{\mathbf{u}(x, y, z)\} = [\mathbf{N}(x, y)]\{\boldsymbol{\alpha}(z)\} \tag{5.3.1}$$

where the time variable is implicit and $[\mathbf{N}(x, y)]$ is given.

Analogous to the finite element method, call $[\mathbf{N}(x, y)]$ the generalized shape function matrix and $\{\boldsymbol{\alpha}(z)\}$ the generalized displacement vector. By means of the compatibility relation between the linear strain $\{\boldsymbol{\varepsilon}_0(x, y, z)\}$ and the displacement $\{\mathbf{u}(x, y, z)\}$, we can write

$$\{\boldsymbol{\varepsilon}_0(x, y, z)\} = [\mathbf{B}(x, y)]\{\boldsymbol{\alpha}(z)\} \tag{5.3.2}$$

where $[\mathbf{B}(x, y)]$ is the generalized strain matrix. Finally, the displacement gradient vector, $\{\boldsymbol{\theta}(x, y, z)\} = [\boldsymbol{\theta}_x^{\mathrm{T}}, \boldsymbol{\theta}_y^{\mathrm{T}}, \boldsymbol{\theta}_z^{\mathrm{T}}]^{\mathrm{T}} = [\partial u/\partial x, \partial u/\partial y, \partial u/\partial z, \partial v/\partial x, \ldots, \partial w/\partial z]^{\mathrm{T}}$, can be written as

$$\{\boldsymbol{\theta}(x, y, z)\} = [\mathbf{G}(x, y)]\{\boldsymbol{\alpha}(z)\} \tag{5.3.3}$$

where $[\mathbf{G}(x, y)]$ is the generalized gradient matrix. In general, the matrices $[\mathbf{N}(x, y)]$, $[\mathbf{B}(x, y)]$, $[\mathbf{G}(x, y)]$ will involve the partial differential operator $D \equiv \partial/\partial z$ and therefore,

$$[\mathbf{N}(x, y)] = [\mathbf{N}_0] + [\mathbf{N}_1]D + \cdots = \sum_{j=0}^{p} [\mathbf{N}_j]D^j \tag{5.3.4}$$

$$[\mathbf{B}(x, y)] = [\mathbf{B}_0] + [\mathbf{B}_1]D + \cdots = \sum_{j=0}^{p+1} [\mathbf{B}_j]D^j \tag{5.3.5}$$

$$[\mathbf{G}(x, y)] = [\mathbf{G}_0] + [\mathbf{G}_1]D + \cdots = \sum_{j=0}^{p+1} [\mathbf{G}_j]D^j \tag{5.3.6}$$

The strain energy of the elastic member just before buckling (first bifurcation) is given by

$$\begin{aligned} U &= \tfrac{1}{2}\int \{\boldsymbol{\varepsilon}_0 + \boldsymbol{\varepsilon}_n\}^{\mathrm{T}}\{\boldsymbol{\sigma}_0 + \boldsymbol{\sigma}_n\}\,\mathrm{d}\,vol \\ &= \int (\tfrac{1}{2}\{\boldsymbol{\varepsilon}_0\}^{\mathrm{T}}\{\boldsymbol{\sigma}_0\} + \{\boldsymbol{\varepsilon}_n\}^{\mathrm{T}}\{\boldsymbol{\sigma}_0\} + \tfrac{1}{2}\{\boldsymbol{\varepsilon}_n\}^{\mathrm{T}}\{\boldsymbol{\sigma}_n\})\,\mathrm{d}\,vol \\ &= \tfrac{1}{2}\int (\{\boldsymbol{\varepsilon}_0\}^{\mathrm{T}}\{\boldsymbol{\sigma}_0\} + \{\boldsymbol{\theta}\}^{\mathrm{T}}[\boldsymbol{\sigma}^0]\{\boldsymbol{\theta}\} + \{\boldsymbol{\varepsilon}_n\}^{\mathrm{T}}\{\boldsymbol{\sigma}_n\})\,\mathrm{d}\,vol \end{aligned} \tag{5.3.7}$$

where $\{\boldsymbol{\varepsilon}_0\}$ and $\{\boldsymbol{\sigma}_0\}$ are the linear (initial) strain and stress, $\{\boldsymbol{\varepsilon}_n\}$ and $\{\boldsymbol{\sigma}_n\}$ are the non-linear (incremental) strain and stress, $[\boldsymbol{\sigma}_0] = [\sigma_x^0, \sigma_y^0, \sigma_z^0, \sigma_{xy}^0, \sigma_{yz}^0, \sigma_{zx}^0]^{\mathrm{T}}$ are the first order stresses and

$$[\boldsymbol{\sigma}^0] = \begin{bmatrix} \boldsymbol{\sigma}_x^0\mathbf{I} & & \text{sym.} \\ \boldsymbol{\sigma}_{xy}^0\mathbf{I} & \boldsymbol{\sigma}_y^0\mathbf{I} & \\ \boldsymbol{\sigma}_{zx}^0\mathbf{I} & \boldsymbol{\sigma}_{yz}^0\mathbf{I} & \boldsymbol{\sigma}_z^0\mathbf{I} \end{bmatrix} \tag{5.3.8}$$

The term $\{\boldsymbol{\varepsilon}_n\}^{\mathrm{T}}\{\boldsymbol{\sigma}_0\} = \tfrac{1}{2}\{\boldsymbol{\theta}\}^{\mathrm{T}}[\boldsymbol{\sigma}^0]\{\boldsymbol{\theta}\}$ appears in Eq. (5.3.7) due to the particular structure between $\{\boldsymbol{\varepsilon}_n\}$ and $\{\boldsymbol{\theta}\}$. For theories valid up to the initial bifurcation, the last term in Eq. (5.3.7), of second order, is small enough to be neglected. For branch switching problems, this term is essential for finding the secondary solution path when the primary path becomes unstable. If the deformation is within the elastic limit,

$$\{\boldsymbol{\sigma}_0\} = [\mathbf{E}]\{\boldsymbol{\varepsilon}_0\}, \quad \text{and} \quad \{\boldsymbol{\sigma}_0 + \boldsymbol{\sigma}_n\} = [\mathbf{E}]\{\boldsymbol{\varepsilon}_0 + \boldsymbol{\varepsilon}_n\} \tag{5.3.9}$$

where $[\mathbf{E}] = \mathrm{diag}[E, E, E, G, G, G]$ in which E and G are the Young's modulus and shear modulus respectively. Without loss of generality, take $p = 1$ in Eqs (5.3.4–5.3.6). The first variation of the strain energy, Eq. (5.3.7) with the last term omitted, is given by, with the help of Eqs (5.3.2) and (5.3.3),

$$\delta U = \int (\{\boldsymbol{\varepsilon}_0\}^{\mathrm{T}}[\mathbf{E}]\{\delta\boldsymbol{\varepsilon}_0\} + \{\boldsymbol{\theta}\}^{\mathrm{T}}[\boldsymbol{\sigma}_0]\{\delta\boldsymbol{\theta}\})\,\mathrm{d}\,vol$$

$$= \int \{\boldsymbol{\alpha}\}^{\mathrm{T}} \int ([\mathbf{B}]^{\mathrm{T}}[\mathbf{E}][\mathbf{B}] + [\mathbf{G}]^{\mathrm{T}}[\boldsymbol{\sigma}_0][\mathbf{G}])J\,\mathrm{d}A\{\delta\boldsymbol{\alpha}\}\,\mathrm{d}z \qquad (5.3.10)$$

where J is the Jacobian and A the cross-sectional area, so that $\mathrm{d}\,vol = J\,\mathrm{d}A\,\mathrm{d}z$. When Eqs (5.3.5) and (5.3.6) are substituted into Eq. (5.3.10), we have

$$\delta U = \int \begin{Bmatrix} \boldsymbol{\alpha} \\ \boldsymbol{\alpha}' \\ \boldsymbol{\alpha}'' \end{Bmatrix}^{\mathrm{T}} \begin{bmatrix} \mathbf{A}_{00} & \mathbf{A}_{01} & \mathbf{A}_{02} \\ \mathbf{A}_{10} & \mathbf{A}_{11} & \mathbf{A}_{12} \\ \mathbf{A}_{20} & \mathbf{A}_{21} & \mathbf{A}_{22} \end{bmatrix} \delta \begin{Bmatrix} \boldsymbol{\alpha} \\ \boldsymbol{\alpha}' \\ \boldsymbol{\alpha}'' \end{Bmatrix} \mathrm{d}z \qquad (5.3.11)$$

where

$$[\mathbf{A}_{ji}]^{\mathrm{T}} = [\mathbf{A}_{ij}] = \int ([\mathbf{B}_i]^{\mathrm{T}}[\mathbf{E}][\mathbf{B}_j] + [\mathbf{G}_i]^{\mathrm{T}}[\boldsymbol{\sigma}_0][\mathbf{G}_j])J\,\mathrm{d}A \qquad i, j = 0, 1, 2 \qquad (5.3.12)$$

Carrying out integration by parts gives

$$\delta U = \int \left[\begin{Bmatrix} \boldsymbol{\alpha} \\ \boldsymbol{\alpha}' \\ \boldsymbol{\alpha}'' \end{Bmatrix}^{\mathrm{T}} \begin{bmatrix} \mathbf{A}_{00} \\ \mathbf{A}_{10} \\ \mathbf{A}_{20} \end{bmatrix} - \left(\begin{Bmatrix} \boldsymbol{\alpha} \\ \boldsymbol{\alpha}' \\ \boldsymbol{\alpha}'' \end{Bmatrix}^{\mathrm{T}} \begin{bmatrix} \mathbf{A}_{01} \\ \mathbf{A}_{11} \\ \mathbf{A}_{21} \end{bmatrix} \right)' + \left(\begin{Bmatrix} \boldsymbol{\alpha} \\ \boldsymbol{\alpha}' \\ \boldsymbol{\alpha}'' \end{Bmatrix}^{\mathrm{T}} \begin{bmatrix} \mathbf{A}_{02} \\ \mathbf{A}_{12} \\ \mathbf{A}_{22} \end{bmatrix} \right)'' \right] \delta\{\boldsymbol{\alpha}\}\,\mathrm{d}z$$

$$+ \left\{ \left[\begin{Bmatrix} \boldsymbol{\alpha} \\ \boldsymbol{\alpha}' \\ \boldsymbol{\alpha}'' \end{Bmatrix}^{\mathrm{T}} \begin{bmatrix} \mathbf{A}_{01} \\ \mathbf{A}_{11} \\ \mathbf{A}_{21} \end{bmatrix} - \left(\begin{Bmatrix} \boldsymbol{\alpha} \\ \boldsymbol{\alpha}' \\ \boldsymbol{\alpha}'' \end{Bmatrix}^{\mathrm{T}} \begin{bmatrix} \mathbf{A}_{02} \\ \mathbf{A}_{12} \\ \mathbf{A}_{22} \end{bmatrix} \right)' \right] \delta\{\boldsymbol{\alpha}\} \right.$$

$$\left. + \begin{Bmatrix} \boldsymbol{\alpha} \\ \boldsymbol{\alpha}' \\ \boldsymbol{\alpha}'' \end{Bmatrix}^{\mathrm{T}} \begin{bmatrix} \mathbf{A}_{02} \\ \mathbf{A}_{12} \\ \mathbf{A}_{22} \end{bmatrix} \delta\{\boldsymbol{\alpha}'\} \right\} \qquad (5.3.13)$$

on the boundaries.

If the coefficient matrices $[\mathbf{A}_{ij}]$ are constant with respect to x, then

$$\delta U = \int \{\delta\boldsymbol{\alpha}\}^{\mathrm{T}}[\mathbf{A}_{00} - (\mathbf{A}_{10} - \mathbf{A}_{01})D + (\mathbf{A}_{20} - \mathbf{A}_{11} + \mathbf{A}_{02})D^2$$
$$- (\mathbf{A}_{12} - \mathbf{A}_{21})D^3 + \mathbf{A}_{22}D^4]\{\boldsymbol{\alpha}\}\,\mathrm{d}z + \{\delta\boldsymbol{\alpha}\}^{\mathrm{T}}[\mathbf{A}_{10} + (\mathbf{A}_{11} - \mathbf{A}_{20})D$$
$$+ (\mathbf{A}_{12} - \mathbf{A}_{21})D^2 - \mathbf{A}_{22}D^3]\{\boldsymbol{\alpha}\}_{\mathrm{b}} + \{\delta\boldsymbol{\alpha}'\}^{\mathrm{T}}[\mathbf{A}_{20} + \mathbf{A}_{21}D + \mathbf{A}_{22}D^2]\{\boldsymbol{\alpha}\}_{\mathrm{b}} \qquad (5.3.14)$$

where the subscript b indicates that the terms are to be evaluated at the boundaries $z = z_1, z_2$, and that the terms should take negative values at $z = z_1$.

The kinetic energy is given by

$$T = \tfrac{1}{2} \int \{\dot{\mathbf{u}}\}^{\mathrm{T}}\{\dot{\mathbf{u}}\}\rho\,\mathrm{d}\,vol \qquad (5.3.15)$$

where ρ is the mass density. When Eq. (5.3.1) is substituted into Eq. (5.3.15) and the first variation is taken, we have

$$\delta T = \int \{\dot{\boldsymbol{\alpha}}\}^{\mathrm{T}} \int \rho[\mathbf{N}]^{\mathrm{T}}[\mathbf{N}] J\,\mathrm{d}A\{\delta\dot{\boldsymbol{\alpha}}\}\,\mathrm{d}z$$

$$= \int \begin{Bmatrix} \dot{\boldsymbol{\alpha}} \\ \dot{\boldsymbol{\alpha}}' \end{Bmatrix}^{\mathrm{T}} \begin{bmatrix} \mathbf{T}_{00} & \mathbf{T}_{01} \\ \mathbf{T}_{10} & \mathbf{T}_{11} \end{bmatrix} \begin{Bmatrix} \delta\dot{\boldsymbol{\alpha}} \\ \delta\dot{\boldsymbol{\alpha}}' \end{Bmatrix} \mathrm{d}z$$

$$= \int \left[\begin{Bmatrix} \dot{\boldsymbol{\alpha}} \\ \dot{\boldsymbol{\alpha}}' \end{Bmatrix}^{\mathrm{T}} \begin{bmatrix} \mathbf{T}_{00} \\ \mathbf{T}_{10} \end{bmatrix} - \left(\begin{Bmatrix} \dot{\boldsymbol{\alpha}} \\ \dot{\boldsymbol{\alpha}}' \end{Bmatrix}^{\mathrm{T}} \begin{bmatrix} \mathbf{T}_{01} \\ \mathbf{T}_{11} \end{bmatrix} \right)' \right] \delta\{\dot{\boldsymbol{\alpha}}\}\,\mathrm{d}z + \begin{Bmatrix} \dot{\boldsymbol{\alpha}} \\ \dot{\boldsymbol{\alpha}}' \end{Bmatrix}^{\mathrm{T}} \begin{bmatrix} \mathbf{T}_{01} \\ \mathbf{T}_{11} \end{bmatrix} \{\delta\dot{\boldsymbol{\alpha}}\}_{\mathrm{b}} \quad (5.3.16)$$

where

$$[\mathbf{T}_{ji}]^{\mathrm{T}} = [\mathbf{T}_{ij}] = \int [\mathbf{N}_i]^{\mathrm{T}}[\mathbf{N}_j]\rho J\,\mathrm{d}A, \qquad i, j = 0, 1 \quad (5.3.17)$$

If $[T_{ij}]$ are constant with respect to z, then

$$\delta T = \int \{\delta\dot{\boldsymbol{\alpha}}\}^{\mathrm{T}}[\mathbf{T}_{00} - (\mathbf{T}_{10} - \mathbf{T}_{01})D - \mathbf{T}_{11}D^2]\{\dot{\boldsymbol{\alpha}}\}\,\mathrm{d}z$$
$$+ \{\delta\dot{\boldsymbol{\alpha}}\}^{\mathrm{T}}[\mathbf{T}_{10} + \mathbf{T}_{11}D]\{\dot{\boldsymbol{\alpha}}\}_{\mathrm{b}} \quad (5.3.18)$$

For free vibration with frequency ω, Hamilton's principle requires that

$$\delta \int_{t_1}^{t_2} (T - U)\,\mathrm{d}t = 0 \quad (5.3.19)$$

where U includes the strain energy due to initial stress, giving the following equation, for constant coefficient matrices,

$$[\mathbf{A}_{00} - (\mathbf{A}_{10} - \mathbf{A}_{01})D + (\mathbf{A}_{20} - \mathbf{A}_{11} + \mathbf{A}_{02})D^2 - (\mathbf{A}_{12} - \mathbf{A}_{21})D^3 + \mathbf{A}_{22}D^4]\{\boldsymbol{\alpha}\}$$
$$- \omega^2[\mathbf{T}_{00} - (\mathbf{T}_{10} - \mathbf{T}_{01})D - \mathbf{T}_{11}D^2]\{\boldsymbol{\alpha}\} = \{\mathbf{0}\} \quad (5.3.20)$$

with the natural boundary conditions,

$$\{\delta\boldsymbol{\alpha}\}^{\mathrm{T}}\{\mathbf{S}_0\} + \{\delta\boldsymbol{\alpha}'\}^{\mathrm{T}}\{\mathbf{S}_1\} = 0 \quad (5.3.21)$$

The generalized forces $\{\mathbf{S}_j\}$ are given by

$$\left. \begin{aligned} \{\mathbf{S}_0\} &= [\mathbf{A}_{10} - \omega^2\mathbf{T}_{10} + (\mathbf{A}_{11} - \mathbf{A}_{20} - \omega^2\mathbf{T}_{11})D + (\mathbf{A}_{12} - \mathbf{A}_{01})D^2 - \mathbf{A}_{22}D^3]\{\boldsymbol{\alpha}\} \\ \{\mathbf{S}_1\} &= [\mathbf{A}_{20} + \mathbf{A}_{21}D + \mathbf{A}_{22}D^2]\{\boldsymbol{\alpha}\} \end{aligned} \right\} \quad (5.3.22)$$

It is noted that only the amplitude of the generalized displacements and forces are of interest in Eqs (5.3.20) and (5.3.21). If the member is resting on an elastic foundation with elastic constants $[\mathbf{k}] = \mathrm{diag}\,[k_x, k_y, k_z]$ in the three orthogonal directions, the strain energy must be modified to include

$$U_f = \tfrac{1}{2} \int \{\mathbf{u}\}^{\mathrm{T}}[\mathbf{k}]\{\mathbf{u}\}\,\mathrm{d}\,vol$$
$$= \tfrac{1}{2} \int \{\boldsymbol{\alpha}\}^{\mathrm{T}} \int [\mathbf{N}]^{\mathrm{T}}[\mathbf{k}][\mathbf{N}] J\,\mathrm{d}A\{\boldsymbol{\alpha}\}\,\mathrm{d}z \quad (5.3.23)$$

The total work done W by the distributed force amplitude $\{\mathbf{q}(x, y, z)\}$ and the concentrated force amplitude $\{\mathbf{Q}(x, y)\}$ at $z = z_0$, $z_0 = z_1, z_2$, is given by

$$W = \int \{\mathbf{q}\}^{\mathrm{T}}\{\mathbf{u}\}\,\mathrm{d}\,vol + \int \{\mathbf{Q}\}^{\mathrm{T}}\{\mathbf{u}(x, y, z_0)\}J\,\mathrm{d}A$$

$$= \iint \{\mathbf{q}\}^{\mathrm{T}}[\mathbf{N}]J\,\mathrm{d}A[\boldsymbol{\alpha}\}\,\mathrm{d}z + \int \{\mathbf{Q}\}^{\mathrm{T}}[\mathbf{N}(x, y, z_0)]J\,\mathrm{d}A$$

$$= \int \{\mathbf{q}_0 + \mathbf{q}_1 D\}^{\mathrm{T}}\{\boldsymbol{\alpha}\}\,\mathrm{d}z + \{\mathbf{Q}_0 + \mathbf{Q}_1 D\}^{\mathrm{T}}\{\boldsymbol{\alpha}(z_0)\} \tag{5.3.24}$$

where

$$\{\mathbf{q}_j\}^{\mathrm{T}} = \int \{\mathbf{q}\}^{\mathrm{T}}[\mathbf{N}_j]J\,\mathrm{d}A, \quad \{\mathbf{Q}_j\}^{\mathrm{T}} = \int \{\mathbf{Q}\}^{\mathrm{T}}[\mathbf{N}_j(x, y, z_0)]J\,\mathrm{d}A \qquad j = 0, 1 \tag{5.3.25}$$

are the generalized distributed forces and the generalized concentrated forces respectively. Taking variation, we have

$$\delta W = \{\mathbf{q}_1\}^{\mathrm{T}}\{\delta\boldsymbol{\alpha}_{\mathrm{b}}\} + \int \{\mathbf{q}_0 - \mathbf{q}_1'\}^{\mathrm{T}}\{\delta\boldsymbol{\alpha}\}\,\mathrm{d}z + \{\mathbf{Q}_0\}^{\mathrm{T}}\{\delta\boldsymbol{\alpha}\}_{\mathrm{b}} + \{\mathbf{Q}_1\}^{\mathrm{T}}\{\delta\boldsymbol{\alpha}'\}_{\mathrm{b}} \tag{5.3.26}$$

When applying Hamilton's principle,

$$\delta \int_{t_1}^{t_2} (T - U + W)\,\mathrm{d}t = 0 \tag{5.3.27}$$

the right-hand side of Eq. (5.3.20) becomes $\{\mathbf{q}_0 - \mathbf{q}_1'\}$, instead of $\{\mathbf{0}\}$. The last two terms in Eq. (5.3.20) do not appear in the governing equation and are absorbed by the natural boundary conditions. However, if $\{\mathbf{Q}\}$ is a follower which changes direction during deformation, then the corresponding concentrated force amplitude becomes $[\mathbf{R}]\{\mathbf{Q}\}$, where $[\mathbf{R}]$ is a rotation matrix, and the natural boundary conditions are modified by $[\mathbf{R} - \mathbf{I}]\{\mathbf{Q}\}$. The following examples are illustrative.

Example 5.3.1. Euler Beam

For a straight Euler beam in bending, as in Fig. 5.3.1,

$$\{\mathbf{u}(x, y, z)\} = \begin{Bmatrix} u(z) \\ w(x, z) \end{Bmatrix} = \begin{Bmatrix} 1 \\ -xD \end{Bmatrix} u(z)$$

$$\{\boldsymbol{\varepsilon}_0\} = \begin{Bmatrix} \varepsilon_z \\ \gamma_{zx} \end{Bmatrix} = \begin{Bmatrix} \partial w/\partial z \\ \partial w/\partial x + \partial u/\partial z \end{Bmatrix} = \begin{Bmatrix} -xD^2 \\ 0 \end{Bmatrix} u(z)$$

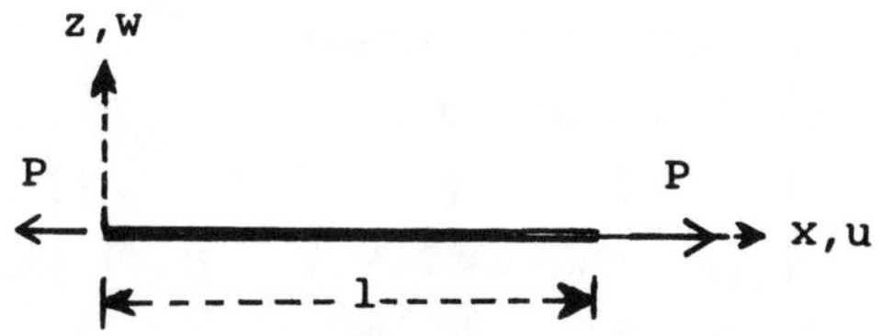

Fig. 5.3.1. The coordinates of a straight beam

Since the non-vanishing stress is

$$\sigma_z^0 = \frac{P}{A}$$

the relevant displacement gradient is

$$\{\boldsymbol{\theta}_z\} = \begin{Bmatrix} \partial u/\partial z \\ \partial w/\partial z \end{Bmatrix} = \begin{Bmatrix} D \\ -xD^2 \end{Bmatrix} u$$

$$\delta U = \int \begin{Bmatrix} u' \\ u'' \end{Bmatrix}^{\mathrm{T}} \begin{bmatrix} P & 0 \\ 0 & EI + Pr^2 \end{bmatrix} \delta \begin{Bmatrix} u' \\ u'' \end{Bmatrix} \mathrm{d}z, \qquad r^2 = \frac{I}{A}$$

$$\delta^T = \int \begin{Bmatrix} \dot{u} \\ \dot{u}' \end{Bmatrix}^{\mathrm{T}} \begin{bmatrix} \rho A & 0 \\ 0 & \rho I \end{bmatrix} \delta \begin{Bmatrix} \dot{u} \\ \dot{u}' \end{Bmatrix} \mathrm{d}z$$

Therefore, for free vibration of a uniform beam,

$$(Pr^2 + EI)u^{\mathrm{iv}} - Pu'' - \omega^2(\rho A u - \rho I u'') = 0 \tag{5.3.28}$$

with the associated natural boundary conditions

$$[(P - \omega^2 \rho I)u' - (Pr^2 + EI)u''']\delta u + (Pr^2 + EI)u''\delta u' = 0 \tag{5.3.29}$$

It is noted that P is positive in tension and negative in compression. The term Pr^2 is usually neglected when small compared with EI and the term $\omega^2 \rho I$ corresponds to the rotatory inertia. If the terms Pr^2 and $\omega^2 \rho I$ are negligible, then, the usual simplified theory for the Euler beam results,

$$EIu^{\mathrm{iv}} - Pu'' - \omega^2 \rho A u = 0 \tag{5.3.30}$$

with the associated natural boundary conditions

$$Q\delta u + M\delta u' = 0 \tag{5.3.31}$$

where the shear force and bending moment are given by

$$Q = Pu' - EIu''' \quad \text{and} \quad M = EIu'' \tag{5.3.32}$$

If P is a follower at $z = l$ then the additional work done by P on the virtual displacement δu is $Pu'\delta u$, therefore, the shear force at $z = l$ must be modified to

$$\bar{Q} = Q - Pu' = -EIu''' \tag{5.3.33}$$

Example 5.3.2. Timoshenko Column

The displacement assumption for a Timoshenko column is

$$\{\mathbf{u}\} = \begin{Bmatrix} u(z) \\ w(x,z) \end{Bmatrix} = \begin{bmatrix} 1 & 0 \\ 0 & x \end{bmatrix} \begin{Bmatrix} u(z) \\ \phi(z) \end{Bmatrix}, \qquad \{\boldsymbol{\alpha}\} = \begin{Bmatrix} u(z) \\ \phi(z) \end{Bmatrix}$$

where $\phi(z)$ is the rotation of the cross-section.

$$\{\boldsymbol{\varepsilon}_0\} = \begin{Bmatrix} \varepsilon_z \\ \gamma_{zx} \end{Bmatrix} = \begin{Bmatrix} \partial w/\partial z \\ \partial w/\partial x + \partial u/\partial z \end{Bmatrix} = \begin{bmatrix} 0 & xD \\ D & 1 \end{bmatrix} \begin{Bmatrix} u(z) \\ \phi(z) \end{Bmatrix}$$

$$\sigma_z^0 = \frac{P}{A}$$

$$\{\boldsymbol{\theta}_z\} = \begin{Bmatrix} \partial u/\partial z \\ \partial w/\partial z \end{Bmatrix} = \begin{bmatrix} D & 0 \\ 0 & xD \end{bmatrix} \begin{Bmatrix} u(z) \\ \phi(z) \end{Bmatrix}$$

$$\delta U = \int \begin{Bmatrix} \boldsymbol{\alpha} \\ \boldsymbol{\alpha}' \end{Bmatrix}^{\mathrm{T}} \begin{bmatrix} 0 & 0 & 0 & 0 \\ 0 & GA & GA & 0 \\ 0 & GA & GA + P & 0 \\ 0 & 0 & 0 & EI + Pr^2 \end{bmatrix} \delta \begin{Bmatrix} \boldsymbol{\alpha} \\ \boldsymbol{\alpha}' \end{Bmatrix} \mathrm{d}z$$

$$\delta T = \int \{\dot{\boldsymbol{\alpha}}\}^{\mathrm{T}} \begin{bmatrix} \rho A & 0 \\ 0 & \rho I \end{bmatrix} \delta\{\dot{\boldsymbol{\alpha}}\}\, \mathrm{d}z$$

Therefore, for the uniform Timoshenko column,

$$\left(\begin{bmatrix} 0 & 0 \\ 0 & GA \end{bmatrix} + \begin{bmatrix} 0 & -GA \\ GA & 0 \end{bmatrix} D - \begin{bmatrix} GA + P & 0 \\ 0 & EI + Pr^2 \end{bmatrix} D^2 - \omega^2 \begin{bmatrix} \rho A & 0 \\ 0 & \rho I \end{bmatrix} \right) \begin{Bmatrix} u \\ \phi \end{Bmatrix} = \{0\} \tag{5.3.34}$$

with the natural boundary conditions

$$\{\delta\boldsymbol{\alpha}\}^{\mathrm{T}}\{\mathbf{S}_0\} = Q\,\delta u + M\,\delta\phi = 0 \tag{5.3.35}$$

where

$$\{\mathbf{S}_0\} = \begin{Bmatrix} Q \\ M \end{Bmatrix} = \left(\begin{bmatrix} 0 & GA \\ 0 & 0 \end{bmatrix} + \begin{bmatrix} GA + P & 0 \\ 0 & EI + Pr^2 \end{bmatrix} D \right) \begin{Bmatrix} u \\ \phi \end{Bmatrix}$$

$$= \begin{Bmatrix} (GA + P)u' + GA\phi \\ (EI + Pr^2)\phi' \end{Bmatrix} \tag{5.3.36}$$

Since the shear stress is assumed to be constant over the cross-section which contradicts the realistic parabolic distribution, a shear factor k is added to all terms involving GA giving kGA for more accurate results. For classical theory, Pr^2 is assumed negligible when compared to EI before buckling.

5.4. Thin-Walled Beam

Consider the thin-walled member shown in Fig. 5.4.1. The origin is at the centroid G and the shear centre is at coordinates (x_s, y_s) with respect to G. The displacement field along the coordinate axes (x, y, z) is given by

$$\{\mathbf{u}\} = \begin{Bmatrix} u_1 \\ u_2 \\ u_3 \end{Bmatrix} = \begin{bmatrix} 1 & 0 & -\bar{y} & 0 \\ 0 & 1 & \bar{x} & 0 \\ -xD & -yD & -\Omega D & 1 \end{bmatrix} \begin{Bmatrix} u \\ v \\ \phi \\ w \end{Bmatrix} = [\mathbf{N}_0 + \mathbf{N}_1 D]\{\alpha\} \tag{5.4.1}$$

where the generalized displacements $\{\boldsymbol{\alpha}\}^{\mathrm{T}} = [u, v, \phi, w]$ are the respective displacements of the centre line, ϕ the rotation about the z axis, Ω the normalized principal warping function, $\bar{x} = x - x_s$, $\bar{y} = y - y_s$ and $D \equiv \partial/\partial z$. The linear strains are given by

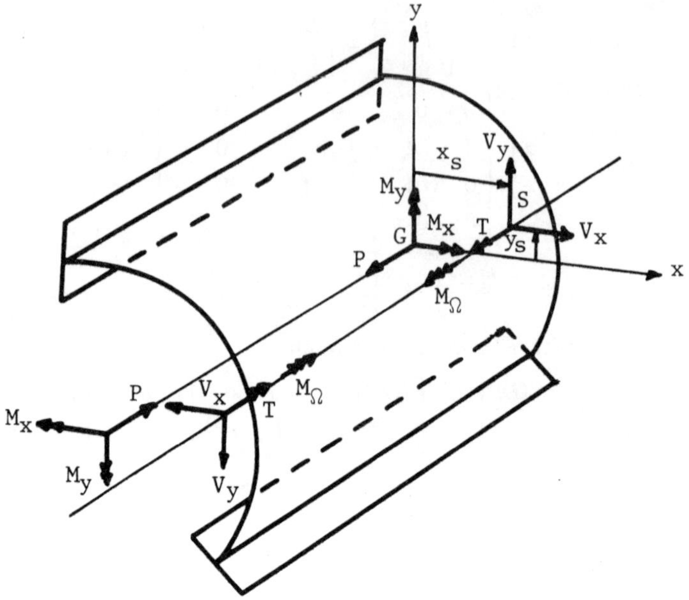

Fig. 5.4.1

$$\{\varepsilon\} = \begin{Bmatrix} \varepsilon_z \\ \gamma_{zx} \\ \gamma_{zy} \end{Bmatrix} = \begin{Bmatrix} \partial u_3/\partial z \\ \partial u_3/\partial x + \partial u_1/\partial z \\ \partial u_3/\partial y + \partial u_2/\partial z \end{Bmatrix} = \begin{bmatrix} -xD^2 & -yD^2 & -\Omega D^2 & D \\ 0 & 0 & -(\bar{y} + \Omega_{,x})D & 0 \\ 0 & 0 & (\bar{x} + \Omega_{,y})D & 0 \end{bmatrix} \{\boldsymbol{\alpha}\}$$

$$= [\mathbf{B}_1 D + \mathbf{B}_2 D^2]\{\boldsymbol{\alpha}\} \tag{5.4.2}$$

where $\Omega_{,x} = \partial\Omega/\partial x$, $\Omega_{,y} = \partial\Omega/\partial y$.

When shear buckling is excluded, the relevant displacement gradient vector is

$$\{\boldsymbol{\theta}\} = \begin{Bmatrix} \partial u_1/\partial z \\ \partial u_2/\partial z \\ \partial u_3/\partial z \end{Bmatrix} = \begin{bmatrix} D & 0 & -\bar{y}D & 0 \\ 0 & D & -\bar{x}D & 0 \\ -xD^2 & -yD^2 & -\Omega D^2 & D \end{bmatrix} \{\boldsymbol{\alpha}\} = [\mathbf{G}_1 D + \mathbf{G}_2 D^2]\{\boldsymbol{\alpha}\} \tag{5.4.3}$$

which corresponds to the initial axial stress

$$[\boldsymbol{\sigma}_0] = \sigma_z[\mathbf{I}]$$
$$\sigma_z = \frac{P}{A} + \frac{M_x}{I_{yy}}y - \frac{M_y}{I_{xx}}x + \frac{M_\Omega}{I_{\Omega\Omega}}\Omega \tag{5.4.4}$$

where P, M_x, M_y, M_Ω are the axial force, the two bending moments and the bi-moment respectively, and

$$I_{xx} = \int x^2\,dA, \qquad I_{yy} = \int y^2\,dA, \qquad I_{\Omega\Omega} = \int \Omega^2\,dA \tag{5.4.5}$$

in which, the integration is over the cross-sectional area A. Therefore, corresponding to Eq. (5.3.11),

$$[\mathbf{A}_{11}] = \begin{bmatrix} P & 0 & -M_x & 0 \\ & P & -M_y & 0 \\ & & \left(GJ + \frac{PI_p}{A} + \frac{M_x\beta_y}{I_{xx}} - \frac{M_y\beta_x}{I_{yy}} + \frac{M_\Omega}{I_{\Omega\Omega}}\beta_\Omega\right) & 0 \\ \text{sym.} & & & EA + P \end{bmatrix}$$

$$[\mathbf{A}_{22}] = \left(E + \frac{P}{A}\right)\mathrm{diag}[I_{xx}, I_{yy}, I_{\Omega\Omega}, 0] + \begin{bmatrix} K_{xx} & K_{xy} & K_{x\Omega} & 0 \\ & K_{yy} & K_{y\Omega} & 0 \\ & & K_{\Omega\Omega} & 0 \\ \text{sym.} & & & 0 \end{bmatrix}$$

$$[\mathbf{A}_{21}]^{\mathrm{T}} = [\mathbf{A}_{12}] = \begin{bmatrix} 0 & 0 & 0 & 0 \\ 0 & 0 & 0 & 0 \\ 0 & 0 & 0 & 0 \\ M_y & -M_x & -M_\Omega & 0 \end{bmatrix} \tag{5.4.6}$$

where

$$K_{pq} = \frac{M_x}{I_{yy}} I_{ypq} - \frac{M_y}{I_{xx}} I_{xpq} + \frac{M_\Omega}{I_{\Omega\Omega}} I_{\Omega_{pq}}, \qquad p, q = x, y, \Omega$$

$$I_{xpq} = \int xpq\,\mathrm{d}A, \qquad I_p = \int (\bar{x}^2 + \bar{y}^2)\,\mathrm{d}A, \qquad J = \int (\bar{x} + \Omega_{,y})^2 + (\bar{y} + \Omega_{,x})^2\,\mathrm{d}A$$

$$\beta_q = \int (\bar{x}^2 + \bar{y}^2) q\,\mathrm{d}A$$

Also, corresponding to Eq. (5.3.16),

$$[\mathbf{T}_{00}] = \rho\,\mathrm{diag}[A, A, I_p, A], \qquad [\mathbf{T}_{01}] = [\mathbf{T}_{10}]^{\mathrm{T}} = [\mathbf{0}]$$
$$[\mathbf{T}_{11}] = \rho\,\mathrm{diag}[I_{xx}, I_{yy}, I_{\Omega\Omega}, 0] \tag{5.4.7}$$

Therefore, the governing equation and the associated natural boundary conditions are given by Eqs (5.1.20)–(5.1.22), for constant conservative end forces P, M_x, M_y and M_Ω.

Suppose P is a follower force acting at the shear centre. The additional shear forces induced by P as the member deforms are given by

$$\begin{Bmatrix} Q_x \\ Q_y \\ 0 \end{Bmatrix} = [\mathbf{R} - \mathbf{I}]\{\mathbf{P}\} = \begin{bmatrix} 0 & -\phi & u' \\ \phi & 0 & v' \\ -u' & -v' & 0 \end{bmatrix} \begin{Bmatrix} 0 \\ 0 \\ P \end{Bmatrix} = \begin{Bmatrix} Pu' \\ Pv' \\ 0 \end{Bmatrix} \tag{5.4.8}$$

which must be subtracted from the appropriated natural boundary conditions. Similarly, if M_x and M_y are followers, then, the additional moments

$$\begin{Bmatrix} \overline{M}_x \\ \overline{M}_y \\ \overline{M}_z \end{Bmatrix} = \begin{bmatrix} 0 & -\phi & u' \\ \phi & 0 & v' \\ -u' & -v' & 0 \end{bmatrix} \begin{Bmatrix} M_x \\ M_y \\ 0 \end{Bmatrix} = \begin{Bmatrix} -M_y\phi \\ M_x\phi \\ -M_x u' - M_y v' \end{Bmatrix} \tag{5.4.9}$$

must be subtracted from the appropriate natural boundary conditions.

5.5. Shear Deformable Thin-Walled Beam

If the thin-walled beam is shear deformable, the displacement field along the coordinate (centroidal) axes is

$$\{\mathbf{u}\} = \begin{Bmatrix} u_1 \\ u_2 \\ u_3 \end{Bmatrix} = \begin{bmatrix} 1 & 0 & 0 & 0 & 0 & -\bar{y} & 0 & 0 & 0 & 0 \\ 0 & 1 & 0 & 0 & 0 & \bar{x} & 0 & 0 & 0 & 0 \\ 0 & 0 & 1 & y & -x & 0 & -\Omega & \xi & \eta & \zeta \end{bmatrix} \{\boldsymbol{\alpha}\} = [\mathbf{N}_0]\{\boldsymbol{\alpha}\} \tag{5.5.1}$$

where the generalized displacement

$$\{\boldsymbol{\alpha}\}^{\mathrm{T}} = [u, v, w, \phi_x, \phi_y, \phi_z, \phi_\Omega, \chi_x, \chi_y, \chi_\Omega] \tag{5.5.2}$$

in which ϕ_x, ϕ_y, ϕ_z are the rotational displacements about the coordinate axes (x, y, z), ξ, η, ζ are the generalized warping functions ($\psi_x, \psi_y, \psi_\omega$ of Laudiero et al. [1]), ϕ_Ω, χ_x, χ_y, χ_Ω are the respective warping coordinates, $\bar{y} = y - y_s$ and $\bar{x} = x - x_s$, and (x_s, y_s) are the coordinates of the shear centre with respect to the centroid. The linear axial strain is given by

$$\varepsilon_{zz} = \partial u_3/\partial z = [0, 0, D, yD, -xD, 0, -\Omega D, \xi D, \eta D, \zeta D]\{\boldsymbol{\alpha}\} = [\mathbf{B}, D]\{\boldsymbol{\alpha}\} \tag{5.5.3}$$

and the linear shear strain is given by

$$\gamma_{zs} = [x, -y, \Omega, \xi, \eta, \zeta]_{,s}\{\bar{\boldsymbol{\alpha}}\} = [\bar{\mathbf{B}}_0]\{\bar{\boldsymbol{\alpha}}\} \tag{5.5.4}$$

where $\{\bar{\boldsymbol{\alpha}}\} = [u' - \phi_y, v' - \phi_x, \theta' - \phi_\omega, \chi_x, \chi_y, \chi_\omega]$, a prime denotes partial differentiation and s is the arc-length along the thin-walled section. Note γ_{zs} is used instead of γ_{zx} and γ_{zy} to ensure $\varepsilon_{ss} = \varepsilon_{nn} = \varepsilon_{sn} = \varepsilon_{zn} = 0$ for the linear strains, where n is normal to both s and z. The displacement gradients are as usual when the shear buckling is not considered,

$$\{\boldsymbol{\theta}_z\} = \begin{Bmatrix} \partial u_1/\partial z \\ \partial u_2/\partial z \\ \partial u_3/\partial z \end{Bmatrix} = \begin{bmatrix} D & 0 & 0 & 0 & 0 & -\bar{y}D & 0 & 0 & 0 & 0 \\ 0 & D & 0 & 0 & 0 & \bar{x}D & 0 & 0 & 0 & 0 \\ 0 & 0 & D & yD & -xD & 0 & -\Omega D & \xi D & \eta D & \zeta D \end{bmatrix} \{\boldsymbol{\alpha}\}$$

$$= [\mathbf{G}_1 D]\{\boldsymbol{\alpha}\} = [\mathbf{N}_0]\{\boldsymbol{\alpha}'\} \tag{5.5.5}$$

The variation of the kinetic energy, Eq. (5.3.16) is quite straightforward,

$$\delta T = \int \{\dot{\boldsymbol{\alpha}}\}^{\mathrm{T}}[\mathbf{T}_{00}]\{\delta\dot{\boldsymbol{\alpha}}\}\,\mathrm{d}z$$

where

$$[\mathbf{T}_{00}] = \int [\mathbf{N}_0^{\mathrm{T}} \rho \mathbf{N}_0]\,\mathrm{d}A\,\mathrm{d}z$$

$$= \rho\,\mathrm{diag}[A, A, A, I_{yy}, I_{xx}, I_p, I_{\Omega\Omega}, I_{\xi\xi}, I_{\eta\eta}, I_{\zeta\zeta}]$$

$$+ \rho A[y_s(\mathbf{e}_{16} + \mathbf{e}_{61}) - x_s(\mathbf{e}_{26} + \mathbf{e}_{62})] \tag{5.5.6}$$

in which $\mathbf{e}_{ij}$ is zero everywhere except the ijth element, which is one,

$$I_{jj} = \int j^2\,\mathrm{d}A, \qquad j = y, x, \Omega, \xi, \eta, \zeta$$

and

$$I_p = \int (\bar{x}^2 + \bar{y}^2)\,\mathrm{d}A \tag{5.5.7}$$

The generalized coordinates are normalized so that $I_{ij} = 0$, $i \neq j$. The strain energy is given by

$$U = \tfrac{1}{2}\int E\varepsilon_{zz}^2 + G\gamma_{zs}^2 + \sigma_2^0\left[\left(\frac{\partial u_1}{\partial z}\right)^2 + \left(\frac{\partial u_2}{\partial z}\right)^2 + \left(\frac{\partial u_3}{\partial z}\right)^2\right]\mathrm{d}A\,\mathrm{d}z \tag{5.5.8}$$

where

$$\sigma_z^0 = \frac{P}{A} + \frac{M_y}{I_{yy}}y - \frac{M_y}{I_{xx}}x + \frac{M_x}{I_{\Omega\Omega}}\Omega \tag{5.5.9}$$

The first term of the integral gives

$$\int E\varepsilon_{zz}^2\,\mathrm{d}A = \{\boldsymbol{\alpha}'\}^{\mathrm{T}}[\mathbf{C}_1]\{\boldsymbol{\alpha}'\} \tag{5.5.10}$$

where

$$[C_1] = E\,\mathrm{diag}[0, 0, A, I_{yy}, I_{xx}, 0, I_{\Omega\Omega}, I_{\xi\xi}, I_{\eta\eta}, I_{\zeta\zeta}]$$

The second term gives

$$\int G\varepsilon_{zs}^2\,\mathrm{d}A = GJ(\theta')^2 + \{\bar{\boldsymbol{\alpha}}\}^{\mathrm{T}}[\mathbf{C}_2]\{\bar{\boldsymbol{\alpha}}\} \tag{5.5.11}$$

where $\{\bar{\boldsymbol{\alpha}}\}$ is given by Eq. (5.5.4),

$$J = \int (\bar{x} + \Omega_{,y})^2 + (\bar{y} + \Omega_{,x})^2\,\mathrm{d}A \quad \text{and} \quad [\mathbf{C}_2] = [F_{ij}] \tag{5.5.12}$$

in which

$$F_{ij} = \int \left(\frac{\mathrm{d}i}{\mathrm{d}s}\right)\left(\frac{\mathrm{d}j}{\mathrm{d}s}\right)\mathrm{d}A, \qquad i, j = x, -y, \Omega, \xi, \eta, \zeta$$

Finally, for the last integral of Eq. (5.5.8),

$$\begin{aligned}
&\int \sigma_z^0[(u_1')^2 + (u_2')^2 + (u_3')^2]\,\mathrm{d}A \\
&= \{\boldsymbol{\alpha}')^{\mathrm{T}} \int \left(\frac{P}{A} + \frac{M_x}{I_{yy}}y - \frac{M_y}{I_{xx}}x + \frac{M_\Omega}{I_{\Omega\Omega}}\Omega\right)[\mathbf{N}_0]^{\mathrm{T}}[\mathbf{N}_0]\,\mathrm{d}A\{\boldsymbol{\alpha}'\} \\
&= \{\boldsymbol{\alpha}'\}^{\mathrm{T}}[\mathbf{C}_3]\{\boldsymbol{\alpha}'\}
\end{aligned} \tag{5.5.13}$$

where

$$\begin{aligned}
[\mathbf{C}_3] = {} & \frac{P}{\rho}[\mathbf{T}_{00}] + \frac{M_x}{I_{yy}}\int y[\mathbf{N}_0]^{\mathrm{T}}[\mathbf{N}_0]\,\mathrm{d}A - \frac{M_y}{I_{xx}}\int x[\mathbf{N}_0]^{\mathrm{T}}[\mathbf{N}_0]\,\mathrm{d}A \\
& + \frac{M_\Omega}{I_{\Omega\Omega}}\int \Omega[\mathbf{N}_0]^{\mathrm{T}}[\mathbf{N}_0]\,\mathrm{d}A
\end{aligned} \tag{5.5.14}$$

$$
\text{in which } [N]^{\mathrm{T}}[N] = \begin{bmatrix}
1 & & & & & -\bar{y} & & & & \\
& 1 & & & & \bar{x} & & & & \\
& & 1 & y & -x & & -\Omega & \xi & \eta & \zeta \\
& & & y^2 & -xy & & -y\Omega & y\xi & y\eta & y\zeta \\
& & & & x^2 & & x\Omega & -x\xi & -x\eta & -x\zeta \\
& & & & & \bar{x}^2+\bar{y}^2 & 0 & 0 & 0 & 0 \\
& & & & & & \Omega^2 & -\Omega\xi & -\Omega\eta & -\Omega\zeta \\
& & & & & & & \xi^2 & \xi\eta & \xi\zeta \\
& & \text{sym.} & & & & & & \eta^2 & \eta\zeta \\
& & & & & & & & & \zeta^2
\end{bmatrix}
$$

In general, the integration of Eq. (5.5.14) is difficult because the higher order integration terms like $\int x^3 \, \mathrm{d}A$ may not vanish. However, if certain symmetry of the cross-sectional area exists, many integration terms will become zero.

With the given area property matrices $[\mathbf{T}_{ij}]$ and $[\mathbf{C}_i]$, the governing equations and the natural boundary conditions can be derived in a straightforward manner.

5.6. Analytical Dynamic Stiffness

It is well known that the equilibrium configuration of a structural member is governed by a partial differential equation, which may be reduced to a system of ordinary differential equations depending on one spatial parameter alone, by means of the Kantorovich method [2], due to a certain regularity of the member. It is not so well known that the eigenproblem of the resulting boundary value problem is often defective [3], having repeated eigenvalues and repeated eigenvectors, due to the same regularity of the member. For conservative systems, the eigenproblem associated with the time variable is always non-defective. The natural modes are always distinct, even for multiple natural frequencies [4]. However, the eigenproblem associated with the spatial variable is often defective, even for conservative systems. For example, the eigenproblem associated with a Euler beam is fourfold degenerated, having multiple zero eigenvalues of order four. The exact solutions are just a polynomial up to the cubic power. The eigenproblem associated with a uniform helix [5] is of order 12, having a four-fold zero eigenvalue and two fourfold conjugate complex eigenvalues. The exact solutions having 12 free parameters to be determined by the boundary conditions form one of the cases to be discussed. These exact solutions can be taken as shape functions in a finite element formulation. The resulting finite element is free from all difficulties associated with the assumed shape function approach, e.g. rigid body modes, constant strains, spurious zero-energy modes, the necessity of reduced integration, and slow convergence. If the eigensolutions are found to be in terms of elementary functions, e.g. products of polynomial, exponential and trigonometrical functions, the complete process can easily be automated.

5.6.1. Self-Adjoint Governing Equation

Consider a system of uniform beams or one-dimensional structures with arbitrary cross-section subjected to external loads including static or dynamic excitation. The governing equation can eventually be written in the spatial domain in the general form

$$\mathscr{L}\{\mathbf{u}(x)\} = ([\mathbf{A}_0] + [\mathbf{A}_1]D^{(1)} + \ldots + [\mathbf{A}_n]D^{(n)})\{\mathbf{u}(x)\} = \{\mathbf{f}(x)\} \tag{5.6.1}$$

with boundary conditions

$$\left.\begin{aligned} D^{(i)}\{\mathbf{u}(x)\}|_{x=-l/2} &= D^{(i)}\{\mathbf{u}(-l/2)\} \\ D^{(i)}\{\mathbf{u}(x)\}|_{x=l/2} &= D^{(i)}\{\mathbf{u}(l/2)\} \end{aligned}\right\} \quad i = 0, 1, \ldots, n/2 - 1 \tag{5.6.2}$$

Here $D^{(i)}(\cdot)$ denotes derivatives with respect to the position variable x; l is the length of the element; $\{\mathbf{f}(x)\}$ and $\{\mathbf{u}(x)\}$ are the excitation and response vectors respectively; $[\mathbf{A}_0]$, $[\mathbf{A}_1]$, ..., $[\mathbf{A}_n]$ are real square matrices of order m. The Eq. (5.6.1) is self-adjoint with the condition that $[\mathbf{A}_i]$ is symmetrical or skew-symmetrical when i is even or odd, respectively. The highest differential order n is assumed to be even.

Instead of specifying the elements of the $[\mathbf{A}_i]$, the matrix order m and the differential order n, the self-adjoint Eq. (5.6.1) will be considered here in general form so as to cover various practical structures with all kinds of deformation such as Timoshenko beam, spinning frame [5], open thin-walled elastic beam [7] and helix [8–11]. The homogeneous solution can be obtained by letting $\{\mathbf{u}(x)\} = e^{\lambda x}\{\mathbf{p}\}$ in Eq. (5.6.1) giving

$$[\mathbf{F}(\lambda)]\{\mathbf{p}\} = ([\mathbf{A}_0] + \lambda[\mathbf{A}_1] + \ldots + \lambda^n[\mathbf{A}_n])\{\mathbf{p}\} = \{\mathbf{0}\} \tag{5.6.3}$$

which constitutes an eigenproblem for the nm non-trivial solutions of eigenvectors $\{\mathbf{p}_j\}$ and the corresponding nm eigenvalues $\lambda_j, j = 1, 2, \ldots, nm$.

The conventional method to solve the matrix polynomial eigenproblem (5.6.3) is the companion matrix method [12] which gives an equivalent $nm \times nm$ standard eigenproblem. The computational effort is considerably increased due to the fact that the order of the companion matrix is $n \times m$ and the eigensolutions must be treated as complex, even though the matrix is real. An alternative method is presented here to expand the determinant algebraically resulting in a scalar polynomial equation for the eigenvalues. After the eigenvalues are obtained, the corresponding eigenvectors can be directly evaluated from Eq. (5.6.3).

5.6.2. Solution of the Matrix Polynomial Eigenproblem

For a non-trivial solution of Eq. (5.6.3).

$$\det([\mathbf{A}_0] + \lambda[\mathbf{A}_1] + \ldots + \lambda^n[\mathbf{A}_n]) = 0 \tag{5.6.4}$$

a scalar algebraic equation of degree nm results for the determination of the nm roots λ_j, $j = 1, 2, \ldots, nm$. If, by means of Gauss elimination, the matrix $[\mathbf{F}(\lambda)]$ were transformed to an upper triangular form $[\mathbf{U}(\lambda)]$, then, the algebraic equation would be given by

$$\prod_{i=1}^{n} u_{ii}(\lambda) = 0$$

where $u_{ii}(\lambda)$ is the ith element on the main diagonal of $[\mathbf{U}(\lambda)]$. However, the elements of $[\mathbf{U}(\lambda)]$ would no longer be polynomials in λ and would be very difficult to implement. Alternatively, after eliminating $f_{i1}(\lambda)$, $i = 2, 3, \ldots, m$, to zero, instead of the usual form

$$\left[0, f_{i2} - \frac{f_{i1}f_{12}}{f_{11}}, f_{i3} - \frac{f_{i1}f_{13}}{f_{11}}, \ldots, f_{im} - \frac{f_{i1}f_{1m}}{f_{11}}\right] \tag{5.6.5}$$

of the ith row, equivalently we can write

$$[0, f_{i2}f_{11} - f_{i1}f_{12}, f_{i3}f_{11} - f_{i1}f_{13}, \ldots, f_{im}f_{11} - f_{i1}f_{1m}]/f_{11} \tag{5.6.6}$$

which can easily be obtained explicitly by polynomial multiplication. The corresponding matrix $[\mathbf{F}_1(\lambda)]$ can be written as follows:

$$[\mathbf{F}_1(\lambda)] = \begin{bmatrix} f_{11} & f_{12} & f_{13} & \cdots & f_{1m} \\ 0 & f_{22}^{(1)} & f_{23}^{(1)} & \cdots & f_{2m}^{(1)} \\ 0 & f_{32}^{(1)} & f_{33}^{(1)} & \cdots & f_{3m}^{(1)} \\ \cdot & \cdot & \cdot & \cdots & \cdot \\ 0 & f_{m2}^{(1)} & f_{m3}^{(1)} & \cdots & f_{mm}^{(1)} \end{bmatrix} \tag{5.6.7}$$

where

$$f_{ij}^{(1)}(\lambda) = f_{ij}(\lambda)f_{11}(\lambda) - f_{i1}(\lambda)f_{1j}(\lambda), \qquad i, j = 2, 3, \ldots, m \tag{5.6.8}$$

which has a redundant factor $f_{11}(\lambda)$ when compared with form (5.6.6). It is harmless if $m = 2$ when $f_{22}^{(1)}$ is actually the algebraic equation required.

Eliminating the second column to zero by a similar process, for example, the (3, 3) element is $f_{33}^{(1)}f_{22}^{(1)} - f_{32}^{(1)}f_{23}^{(1)}$ which contains the redundant factor $f_{11}^2(\lambda)f_{22}^{(1)}(\lambda)$ since all $f_{ij}^{(1)}$ contain the redundant factor $f_{11}(\lambda)$, and which is exactly divisible by $f_{11}(\lambda)$. Therefore,

$$f_{ij}^{(2)}(\lambda) = (f_{ij}^{(1)}f_{22}^{(1)} - f_{i2}^{(1)}f_{2j}^{(1)})/f_{11}(\lambda), \qquad i, j = 3, 4, \ldots, m \tag{5.6.9}$$

are polynomials in λ. If $m = 3$, $f_{33}^{(2)}$ is the algebraic equation required.

In general, after $m - 1$ steps,

$$[\mathbf{F}_{m-1}(\lambda)] = \begin{bmatrix} f_{11} & f_{12} & f_{13} & \cdots & f_{1m} \\ 0 & f_{22}^{(1)} & f_{23}^{(1)} & \cdots & f_{2m}^{(1)} \\ 0 & 0 & f_{33}^{(2)} & \cdots & f_{3m}^{(2)} \\ \cdot & \cdot & \cdot & \cdots & \cdot \\ 0 & 0 & 0 & \cdots & f_{mm}^{(m-1)} \end{bmatrix} \tag{5.6.10}$$

is an upper triangular matrix and the polynomial elements in Eq. (5.6.10) are given by

$$f_{ij}^{(k)}(\lambda) = (f_{ij}^{(k-1)}f_{kk}^{(k-1)} - f_{ik}^{k-1)}f_{kj}^{(k-1)})/f_{k-1,k-1}^{(k-2)} \tag{5.6.11}$$

with $f_{ij}^{(0)} = f_{ij}$ and $f_{ij}^{(-1)} = 1$, $k = 1, 2, \ldots, m - 1$; $i, j = k + 1, k + 2, \ldots, m$. The last item

$$f_{mm}^{(m-1)}(\lambda) = 0 \tag{5.6.12}$$

is the required algebraic equation for the determination of eigenvalue λ. By substituting λ into Eq. (5.6.3), one obtains the corresponding eigenvector.

5.6.3. Shape Function

Suppose that $\lambda_1, \lambda_2, \ldots, \lambda_{nm}$ are all distinct roots of Eq. (5.6.12), then the homogeneous solutions of Eq. (5.6.1) have the following form:

$$\begin{aligned}\{\mathbf{u}(x)\} &= \sum_{j=1}^{nm} c_j e^{\lambda_j x}\{\mathbf{p}_j\} \\ &= [\mathbf{P}]\,\mathrm{diag}[e^{\lambda_1 x}, e^{\lambda_2 x}, \ldots, e^{\lambda_{nm} x}]\{\mathbf{C}\}\end{aligned} \tag{5.6.13}$$

where the $[\mathbf{P}]$ is an $m \times nm$ matrix composed of $\{\mathbf{p}_j\}$ which is the corresponding eigenvector to λ_j; $\{\mathbf{C}\}$ is a column containing nm constants. We shall discuss the defective case in the next section.

Equation (5.6.13) represents an expression for the displacement amplitude in terms of the nm constants. The nm constants will be determined by substituting Eq. (5.6.13) into the boundary conditions (5.6.2), that is,

$$\{\mathbf{q}\} = [\mathbf{H}]\{\mathbf{C}\} \tag{5.6.14}$$

in which

$$\begin{aligned}\{\mathbf{q}\} = \Bigg[&\left\{\mathbf{u}\left(-\frac{l}{2}\right)\right\}^{\mathrm{T}}, D^{(1)}\left\{\mathbf{u}\left(-\frac{l}{2}\right)\right\}^{\mathrm{T}}, \ldots, D^{((n/2)-1)}\left\{\mathbf{u}\left(-\frac{l}{2}\right)\right\}^{\mathrm{T}}, \\ &\left\{\mathbf{u}\left(\frac{l}{2}\right)\right\}^{\mathrm{T}}, D^{(1)}\left\{\mathbf{u}\left(\frac{l}{2}\right)\right\}^{\mathrm{T}}, \ldots, D^{((n/2)-1)}\left\{\mathbf{u}\left(\frac{l}{2}\right)\right\}^{\mathrm{T}}\Bigg]^{\mathrm{T}}\end{aligned} \tag{5.6.15}$$

$$\begin{aligned}[\mathbf{H}] = \Bigg[&\left[\mathbf{h}_1\left(-\frac{l}{2}\right)\right]^{\mathrm{T}}, \left[\mathbf{h}_2\left(-\frac{l}{2}\right)\right]^{\mathrm{T}}, \ldots, \left[\mathbf{h}_{n/2}\left(-\frac{l}{2}\right)\right]^{\mathrm{T}}, \\ &\left[\mathbf{h}_1\left(\frac{l}{2}\right)\right]^{\mathrm{T}}, \left[\mathbf{h}_2\left(\frac{l}{2}\right)\right]^{\mathrm{T}}, \ldots, \left[\mathbf{h}_{n/2}\left(\frac{l}{2}\right)\right]^{\mathrm{T}}\Bigg]^{\mathrm{T}}\end{aligned} \tag{5.6.16}$$

where

$$[\mathbf{h}_j(x)] = [\mathbf{P}]\,\mathrm{diag}[\lambda_1^{j-1} e^{\lambda_1 x}, \lambda_2^{j-1} e^{\lambda_2 x}, \ldots, \lambda_{nm}^{j-1} e^{\lambda_{nm} x}] \qquad j = 1, 2, \ldots, n/2 \tag{5.6.17}$$

It is evident that the constant $\{\mathbf{C}\}$ can be evaluated in terms of the general modal displacement $\{\mathbf{q}\}$ by inverting Eq. (5.6.14)

$$\{\mathbf{C}\} = [\mathbf{H}]^{-1}\{\mathbf{q}\} \tag{5.6.18}$$

Hence, substituting this into Eq. (5.6.13) leads to

$$\{\mathbf{u}(x)\} = [\mathbf{P}]\,\mathrm{diag}[e^{\lambda_1 x}, e^{\lambda_2 x}, \ldots, e^{\lambda_{nm} x}][\mathbf{H}]^{-1}\{\mathbf{q}\} \tag{5.6.19}$$

The matrix product in front of $\{\mathbf{q}\}$ in Eq. (5.6.19) represents the shape function, by definition, because it expresses the distributed displacements in terms of the nodal displacements. That is,

$$[\mathbf{N}(x)] = [\mathbf{P}]\,\mathrm{diag}[e^{\lambda_1 x}, e^{\lambda_2 x}, \ldots, e^{\lambda_{nm} x}][\mathbf{H}]^{-1} \tag{5.6.20}$$

Note that the shape function matrix is parametrically dependent because all matrices in the right of Eq. (5.6.20) are functions of various parameters including frequencies for dynamics and compressive loads for stability.

5.6.4. Defective Shape Function

Since the eigenvectors corresponding to distinct eigenvalues are linearly independent, the nm vectors $e^{\lambda_j x}\{\mathbf{p}_j\}$, $j = 1, 2, \ldots, nm$, are consequently independent and the expansion relation in the form of (5.6.13) holds when the roots of the characteristic equation (5.6.12) are all distinct. However, this relation and the shape function (5.6.20) might not be always valid when there are repeated roots. The problem of finding eigenvectors associated with a multiple eigenvalue is much more complicated and the general classification [13–14] of the eigensolutions of the eigenproblem (5.6.3) is given in Fig. 5.6.1, where the algebraic multiplicity of an eigensolution is the multiplicity of the eigenvalue and the geometric multiplicity is the number of linearly independent eigenvectors associated with the multiple eigenvalue. If the eigenvalue is distinct, the eigensolution is simple. If the eigenvalue is multiple and the algebraic and geometric multiplicity of the eigensolution are equal, it is semi-simple. If the algebraic and geometric multiplicity are not equal for an eigensolution, it is non-semi-simple or defective. Both simple and semi-simple eigensolutions are non-defective and the non-defective shape functions associated with the semi-simple eigensolutions can also be constructed by formulation (5.6.20) without any additional difficulty.

For the general cases including the defective eigensolutions, there always exists a Jordan pair $(\mathbf{P}, \mathbf{J})$ [12] such that

$$[\mathbf{A}_0][\mathbf{P}] + [\mathbf{A}_1][\mathbf{P}][\mathbf{J}] + \ldots + [\mathbf{A}_n][\mathbf{P}][\mathbf{J}] = [\mathbf{0}] \tag{5.6.21}$$

where the Jordan form

$$[\mathbf{J}] = \text{diag}[\mathbf{J}_i\text{: } i = 1, 2, \ldots, K] \tag{5.6.22}$$

and the ith Jordan block of size n_i is given by

$$[\mathbf{J}_i] = \lambda_i[\mathbf{I}] + [\mathbf{N}_i] \tag{5.6.23}$$

in which λ_i is the ith eigenvalue, and $[\mathbf{N}_i]$ is a nilpotent matrix of order n_i:

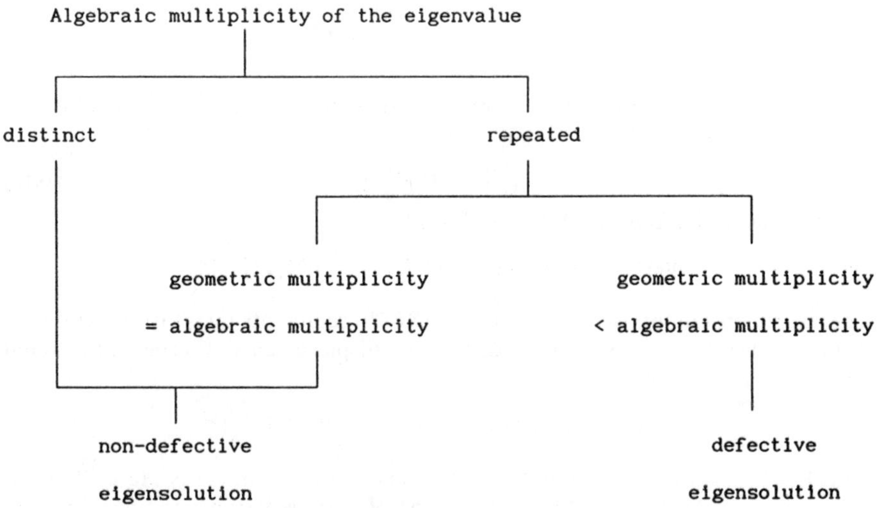

Fig. 5.6.1. Classification of eigensolutions

$$[\mathbf{N}_i] = \begin{bmatrix} 0 & 1 & & & \\ & 0 & 1 & \mathbf{0} & \\ & & 0 & \cdot & \\ & \mathbf{0} & & \cdot & 1 \\ & & & & 0 \end{bmatrix} \tag{5.6.24}$$

in which $\sum_{i=1}^{K} n_i = nm$, the order of the eigenproblem.

When $n_i = 1$, then λ_i is simple and when $n_i = n_j = 1$ and $\lambda_i = \lambda_j$ for $i \neq j$, then λ_i is semi-simple. If $n_i \neq 1$ then λ_i is non-semi-simple or defective. Suppose λ_s is a defective eigenvalue with the Jordan block size $n_i = n_s = k$ ($k > 1$). The columns of $[\mathbf{P}]$ corresponding to the eigenvalue λ_s are made of the Jordan chains $\{\mathbf{p}_1^s\}, \{\mathbf{p}_2^s\}, \ldots, \{\mathbf{p}_k^s\}$ which are obtained using the following k equations

$$\left.\begin{aligned} [\mathbf{F}(\lambda_s)]\{\mathbf{p}_1^s\} &= \{\mathbf{0}\} \\ [\mathbf{F}(\lambda_s)]\{\mathbf{p}_2^s\} + \frac{1}{1!}[\mathbf{F}^{(1)}(\lambda_s)]\{\mathbf{p}_1^s\} &= \{\mathbf{0}\} \\ &\vdots \\ [\mathbf{F}(\lambda_s)]\{\mathbf{p}_k^s\} + \frac{1}{1!}[\mathbf{F}^{(1)}(\lambda_s)]\{\mathbf{p}_{k-1}^s\} + \ldots + \frac{1}{(k-1)!}[\mathbf{F}^{(k-1)}(\lambda_s)]\{\mathbf{p}_1^s\} &= \{\mathbf{0}\} \end{aligned}\right\} \tag{5.6.25}$$

and then the k functions

$$\left.\begin{aligned} \{\mathbf{u}_1^s(x)\} &= \{\mathbf{p}_1^s\}e^{\lambda_s x} \\ \{\mathbf{u}_2^s(x)\} &= (x\{\mathbf{p}_1^s\} + \{\mathbf{p}_2^s\})e^{\lambda_s x} \\ &\vdots \\ \{\mathbf{u}_k^s(x)\} &= \left(\sum_{j=0}^{k-1} \frac{x^j}{j!}\{\mathbf{p}_{k-j}^s\}\right)e^{\lambda_s x} \end{aligned}\right\} \tag{5.6.26}$$

are proved [12] to be linearly independent solutions of Eq. (5.6.1).

Similarly, the corresponding independent solutions for other eigenvalues λ_i can also be obtained by equations (5.6.26) in conjunction with formula (5.6.25) with different block sizes n_i. Then the homogeneous solutions of Eq. (5.6.1) for the general case are of the form, instead of the form (5.6.13), as follows:

$$\begin{aligned} \{\mathbf{u}(x)\} &= \sum_{i=1}^{K} \sum_{j=1}^{n_i} c_{ij}\{\mathbf{u}_j^i(x)\} \\ &= [\mathbf{P}][\mathbf{G}(x)]\{\mathbf{C}\} \end{aligned} \tag{5.6.27}$$

where

$$[\mathbf{P}] = \mathrm{col}[\mathbf{p}_{ij}\colon i = 1, \ldots, K, j = 1, \ldots, n_i] \tag{5.6.28}$$

$$[\mathbf{G}(x)] = \mathrm{diag}[\mathbf{G}_i(x)\colon i = 1, \ldots, K] \tag{5.6.29}$$

in which the $[\mathbf{G}_i(x)]$ is a square matrix of size of n_i and its (s, t)-element is given by

$$g_{st}^i(x) = \begin{cases} 0 & s > t \\ \dfrac{1}{(t-s)!}x^{t-s}e^{\lambda_i x} & s \leq t \end{cases} \tag{5.6.30}$$

By the same process presented in the previous section, the corresponding defective shape function, which treats the non-defective shape function (5.6.20) as a particular case, is of the form

$$[\mathbf{N}(x)] = [\mathbf{P}][\mathbf{G}(x)][\mathbf{H}]^{-1} \tag{5.6.31}$$

with the submatrices of $[\mathbf{H}]$

$$[\mathbf{h}_j(x)] = [\mathbf{P}]\,\mathrm{diag}[\mathbf{D}_i^j(x)\colon i = 1,\ldots,K], \qquad j = 1, 2, \ldots, n/2 \tag{5.6.32}$$

and the (s, t)-element of $[\mathbf{D}_i^j(x)]$

$$d_{st}^{ij}(x) = \begin{cases} 0 & s > t \\ \displaystyle\sum_{k=0}^{j-1} \binom{k}{j-1} \frac{1}{(t-s-k)!} \lambda_i^{j-k-1} x^{t-s-k} e^{\lambda_i x} & s \le t \end{cases} \tag{5.6.33}$$

5.6.5. Stiffness Matrix

For the matrix differential Eq. (5.6.1), the inner product is defined as

$$\langle \mathscr{L}\{\mathbf{u}(x)\}, \{\mathbf{w}(x)\}\rangle = \int_{-l/2}^{l/2} \{\mathbf{w}(x)\}^{\mathrm{T}} \mathscr{L}\{\mathbf{u}(x)\}\,\mathrm{d}x \tag{5.6.34}$$

which can be integrated by parts until all derivatives in $\{\mathbf{u}(x)\}$ have been eliminated. This leads to the transposed form of the inner product and produces a series of boundary items. In general

$$\int_{-l/2}^{l/2} \{\mathbf{w}\}^{\mathrm{T}} \mathscr{L}\{\mathbf{u}\}\,\mathrm{d}x = \int_{-l/2}^{l/2} \{\mathbf{u}\}^{\mathrm{T}} \mathscr{L}\{\mathbf{w}\}\,\mathrm{d}x + [\mathscr{C}^*\{\mathbf{w}\}\mathscr{Y}\{\mathbf{u}\} - \mathscr{Y}^*\{\mathbf{w}\}\mathscr{C}\{\mathbf{u}\}]\big|_{x=-l/2}^{x=l/2} \tag{5.6.35}$$

where $\mathscr{C}(\cdot)$ and $\mathscr{Y}(\cdot)$ are differential operators results from integration by parts; $\mathscr{L}^*(\cdot)$, $\mathscr{C}^*(\cdot)$ and $\mathscr{Y}^*(\cdot)$ are adjoint operators of $\mathscr{L}(\cdot)$, $\mathscr{C}(\cdot)$ and $\mathscr{Y}(\cdot)$, respectively.

Substituting Eq. (5.6.1) into Eq. (5.3.34) gives

$$\begin{aligned}
&\int_{-l/2}^{l/2} \{\mathbf{w}\}^{\mathrm{T}} \mathscr{L}\{\mathbf{u}\}\,\mathrm{d}x \\
&= \int_{-l/2}^{l/2} \{\mathbf{u}\}^{\mathrm{T}} \mathscr{L}\{\mathbf{w}\}\,\mathrm{d}x + (\{\mathbf{u}\}^{\mathrm{T}}(\tfrac{1}{2}[\mathbf{A}_1] + [\mathbf{A}_2]D \\
&\quad + [\mathbf{A}_3]D^{(2)} + \ldots + [\mathbf{A}_n]D^{(n-1)})\{\mathbf{w}\} \\
&\quad - D\{\mathbf{u}\}^{\mathrm{T}}(\tfrac{1}{2}[\mathbf{A}_3]D + [\mathbf{A}_4]D^{(2)} + [\mathbf{A}_5]D^{(3)} + \ldots + [\mathbf{A}_n]D^{(n-2)})\{\mathbf{w}\} + \ldots \\
&\quad + D^{((n/2)-1)}\{\mathbf{u}\}^{\mathrm{T}}(\tfrac{1}{2}[\mathbf{A}_{n-1}]D^{((n/2)-1)} + [\mathbf{A}_n]D^{(n/2)})\{\mathbf{w}\})_{x=l/2}^{x=-l/2} \\
&\quad - (\{\mathbf{w}\}^{\mathrm{T}}(\tfrac{1}{2}[\mathbf{A}_1] + [\mathbf{A}_2]D + [\mathbf{A}_3]D^{(2)} + \ldots + [\mathbf{A}_n]D^{(n-1)})\{\mathbf{u}\} \\
&\quad - D\{\mathbf{w}\}^{\mathrm{T}}(\tfrac{1}{2}[\mathbf{A}_3]D + [\mathbf{A}_4]D^{(2)} + [\mathbf{A}_5]D^{(3)} + \ldots + [\mathbf{A}_n]D^{(n-2)})\{\mathbf{u}\} + \ldots \\
&\quad + D^{((n/2)-1)}\{\mathbf{w}\}^{\mathrm{T}}(\tfrac{1}{2}[\mathbf{A}_{n-1}]D^{((n/2)-1)} + [\mathbf{A}_n]D^{(n/2)})\{\mathbf{u}\})_{x=l/2}^{x=-l/2}
\end{aligned} \tag{5.6.36}$$

Note that all operations upon $\{\mathbf{u}\}$ are the same as those upon $\{\mathbf{w}\}$; that is, $\mathscr{L}^*(\cdot) = \mathscr{L}(\cdot)$, $\mathscr{C}^*(\cdot) = \mathscr{C}(\cdot)$ and $\mathscr{Y}^*(\cdot) = \mathscr{Y}(\cdot)$ because the differential equation considered here is self-adjoint.

After removing the last item on the right of Eq. (5.6.36) to the left and treating the $\{\mathbf{u}\}$ and $\{\mathbf{w}\}$ as factual state and virtual state, respectively, Eq. (5.6.36) is just Betti's reciprocal theorem, which states that the work done by the forces of factual state acting through the displacements of virtual state is equal to the work done by the forces of virtual state acting through the displacements of factual state. If $\{\mathbf{w}\}$ is taken to satisfy the homogeneous form of Eq. (5.6.1) and $\{\mathbf{w}(-l/2)\}, D^{(1)}\{\mathbf{w}(-l/2)\}, \ldots, D^{((n/2)-1)}\{\mathbf{w}(-l/2)\}, \{\mathbf{w}(l/2)\}, D^{(1)}\{\mathbf{w}(l/2)\}, \ldots, D^{((n/2)-1)}\{\mathbf{w}(l/2)\}$ are treated as generalized virtual displacements at the boundaries, the corresponding items which multiply them are generalized forces at the boundaries due to the natural boundary conditions; that is,

$$\left\{\mathbf{s}_j\left(-\frac{l}{2}\right)\right\} = \sum_{k=2j+1}^{n} (-1)^j \alpha_{kj} [\mathbf{A}_k] D^{(k-j-1)} \left\{\mathbf{u}\left(-\frac{l}{2}\right)\right\} \tag{5.6.37}$$

and

$$\left\{\mathbf{s}_j\left(\frac{l}{2}\right)\right\} = \sum_{k=2j+1}^{n} (-1)^{j+1} \alpha_{kj} [\mathbf{A}_k] D^{(k-j-1)} \left\{\mathbf{u}\left(\frac{l}{2}\right)\right\} \qquad j = 0, 1, 2, \ldots, (n/2) - 1 \tag{5.6.38}$$

where

$$\alpha_{kj} = \begin{cases} 1/2 & \text{if } k = 2j + 1 \\ 1 & \text{if } k \neq 2j + 1 \end{cases} \tag{5.6.39}$$

Rewriting Eq. (5.6.27) in terms of the analytical shape function (5.6.31) gives

$$\{\mathbf{u}(x)\} = [\mathbf{N}(x)]\{\mathbf{q}\} \tag{5.6.40}$$

In particular, letting $x = -l/2$ and $x = l/2$ in this equation and then substituting this into Eqs (5.6.37) and (5.6.38), we have

$$\{\mathbf{S}\} = [\mathbf{K}]\{\mathbf{q}\} \tag{5.6.41}$$

in which

$$\begin{aligned}\{\mathbf{S}\} = \Bigg[& \left\{s_0\left(-\frac{l}{2}\right)\right\}^{\mathrm{T}}, \left\{s_1\left(-\frac{l}{2}\right)\right\}^{\mathrm{T}}, \ldots, \left\{s_{(n/2)-1}\left(-\frac{l}{2}\right)\right\}^{\mathrm{T}}, -\left\{s_0\left(-\frac{l}{2}\right)\right\}^{\mathrm{T}}, \\ & -\left\{s_1\left(\frac{l}{2}\right)\right\}^{\mathrm{T}}, \ldots, -\left\{s_{(n/2)-1}\left(\frac{l}{2}\right)\right\}^{\mathrm{T}}\Bigg]^{\mathrm{T}}\end{aligned} \tag{5.6.42}$$

$$\begin{aligned}[\mathbf{K}] = \Bigg[& \left\{\mathbf{k}_0\left(-\frac{l}{2}\right)\right\}^{\mathrm{T}}, \left\{\mathbf{k}_1\left(-\frac{l}{2}\right)\right\}^{\mathrm{T}}, \ldots, \left\{\mathbf{k}_{(n/2)-1}\left(-\frac{l}{2}\right)\right\}^{\mathrm{T}}, \left\{\mathbf{k}_0\left(-\frac{l}{2}\right)\right\}^{\mathrm{T}}, \\ & \left\{\mathbf{k}_1\left(\frac{l}{2}\right)\right\}^{\mathrm{T}}, \ldots, \left\{\mathbf{k}_{(n/2)-1}\left(\frac{l}{2}\right)\right\}^{\mathrm{T}}\Bigg]^{\mathrm{T}}\end{aligned} \tag{5.6.43}$$

where

$$\begin{aligned}[\mathbf{k}_j(x)] &= \sum_{k=2j+1}^{n} (-1)^j \alpha_{kj} [\mathbf{A}_k][\mathbf{N}(x)]^{(k-j-1)} \\ &= \sum_{k=2j+1}^{n} (-1)^j \alpha_{kj} [\mathbf{A}_k][\mathbf{h}_{k-j}(x)][\mathbf{H}]^{-1} \qquad j = 0, 1, 2, \ldots, n/2 - 1\end{aligned} \tag{5.6.44}$$

Thus, the $nm \times nm$ matrix $[\mathbf{K}]$ is the required stiffness matrix which represents the generalized boundary forces resulting from the application of unit generalized boundary displacements. Once the initial matrices $[\mathbf{A}_0]$, $[\mathbf{A}_1]$, ..., $[\mathbf{A}_n]$ are available, then the stiffness matrix can be formed explicitly by the following steps:

1. Evaluate the eigenvalues λ_j and the corresponding eigenvectors $\{\phi_j\}$, $j = 1, 2, \ldots, nm$ from the eigenproblem (5.6.3).
2. Calculate the matrices $[\mathbf{h}_j(l/2)]$ and $[\mathbf{h}_j(-l/2)]$, $j = 1, 2, \ldots, n/2$ by using (5.6.32) together with (5.6.33) and then inverting the matrix $[\mathbf{H}]$.
3. Forming the stiffness matrix $[\mathbf{K}]$ with the aid of (5.6.43) and (5.6.44).

The procedure can be easily implemented with the aid of a microcomputer and is valid for any structure whose governing equations are expressible in the form of Eq. (5.6.1).

Example 5.6.1

A uniform beam with the open thin-walled semicircular cross-section shown in Fig. 5.6.2 is considered. The beam is clamped at $x = -l/2$ and free at $x = l/2$. When rotary and warping inertia terms are included, the free vibration is governed by the following equations [15]:

$$
\begin{aligned}
EI_\eta w^{iv} - \rho I_\eta \ddot{w}'' + P(w'' + y_G\phi'') + m(\ddot{w} + y_G\ddot{\phi}) &= 0 \\
EI_\xi v^{iv} - \rho I_\xi \ddot{v}'' + P(v'' - z_G\phi'') + m(\ddot{v} - z_G\ddot{\phi}) &= 0 \qquad (5.6.45) \\
EI_w\phi^{iv} - \rho I_w\ddot{\phi}'' - GI_t\phi'' + P(r_0^2\phi'' + y_G w'' - z_G v'') + j_0\ddot{\phi} + m(y_G\ddot{w} - z_G\ddot{v}) &= 0
\end{aligned}
$$

Primes and dots indicate differentiations with respect to position x and time t, respectively; P is a central static axial load applied to the beam ends. The cross-sectional data are taken as

$$a = 24.5\,\text{mm} \quad t = 4.0\,\text{mm} \quad l = 820\,\text{mm} \quad A = 308 \times 10^{-5}\,\text{m}^2$$

$$y_G = 15.5\,\text{mm} \quad z_G = 0.0\,\text{mm} \quad m = 0.835\,\text{kg}\,\text{m}^{-1} \quad j_0 = 501 \times 10^{-6}\,\text{kg}\,\text{m}$$

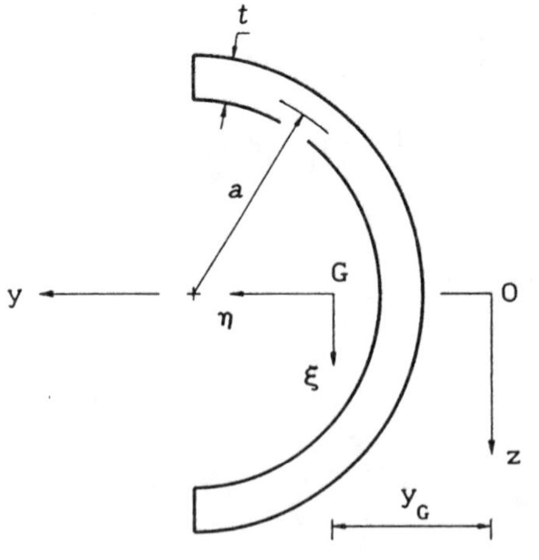

Fig. 5.6.2. The beam cross-section

$$I_\eta = 92.6 \times 10^{-9}\,\mathrm{m}^4 \qquad I_\xi = 17.7 \times 10^{-9}\,\mathrm{m}^4$$
$$I_\mathrm{w} = 1.52 \times 10^{-12}\,\mathrm{m}^6 \qquad I_\mathrm{t} = 1.64 \times 10^{-9}\,\mathrm{m}^4$$
$$E = 68.9\,\mathrm{GPa} \qquad G = 26.5\,\mathrm{GPa}$$

When harmonic conditions are assumed, the matrices of Eq. (5.6.1) are in the forms of

$$[\mathbf{A}_0] = \begin{bmatrix} -m\omega^2 & 0 & -m\omega^2 y_\mathrm{G} \\ & -m\omega^2 & m\omega^2 z_\mathrm{G} \\ \text{sym.} & & -j_0\omega^2 \end{bmatrix}$$

$$[\mathbf{A}_2] = \begin{bmatrix} \rho I_\eta \omega^2 + P & 0 & P y_\mathrm{G} \\ & \rho I_\xi \omega^2 + P & -P z_\mathrm{G} \\ \text{sym.} & & \rho I_\mathrm{w} \omega^2 - G I_\mathrm{t} + p r_0^2 \end{bmatrix} \qquad (5.6.46)$$

$$[\mathbf{A}_4] = \begin{bmatrix} E I_\eta & 0 & 0 \\ & E I_\xi & 0 \\ \text{sym.} & & E I_\mathrm{w} \end{bmatrix}$$

and $[\mathbf{A}_1]$, $[\mathbf{A}_3]$ are zero matrices. The notation used in Eqs (5.6.45) and (5.6.46) is as follows:

A	Cross-sectional area
E	Elastic modulus
G	Shear modulus
I_ξ, I_η	Principal moments of inertia about ξ- and η-axes
I_w	Principal sectorial moment of inertia
I_t	Cross-sectional factor in torsion
j_0	Polar moment of mass inertia per unit beam length with respect to shear centre O
ρ	Mass density
m	Mass per unit beam length
r_0	Polar radius of inertia with respect to shear centre O
v, w	Translations of cross-section at shear centre axis in y- and z-directions
ϕ	Rotation of cross-section about x-axis
y_G, z_G	Coordinates of geometric centre G

The natural frequencies are determined by equating the determinant of the dynamic stiffness of the structure to zero,

$$\det[\mathbf{D}(\omega)] = 0 \qquad (5.6.47)$$

Here $[\mathbf{D}(\omega)]$ is obtained from the dynamic element stiffness matrix $[\mathbf{K}(\omega)]$, which can be explicitly formed by (5.6.43) and (5.6.44) by deleting the rows and columns which correspond to the locked d.o.f. of the clamped–free beam studied. $[\mathbf{K}(\omega)]$ and $[\mathbf{D}(\omega)]$ are real square matrices of order 12 and 6, respectively.

All natural frequencies can be isolated and evaluated by solving Eq. (5.6.47) with the aid of the Wittrick–Williams algorithm [16]. The lowest four frequencies are shown in Table 5.6.1 and compared with the results given by Friberg [17] under the following conditions:

Table 5.6.1. Natural frequencies of a clamped–free beam. Comparison of present results with those of Friberg ([17], given in parentheses)

Mode n	Natural frequency f_n (Hz)		
	Condition 1	Condition 2	Condition 3
1	31.8097(31.81)	31.7987(31.80)	25.0087(25.01)
2	63.7903(63.79)	63.7562(63.76)	61.2839(61.28)
3	137.7362(137.7)	137.5278(137.5)	136.0014(136.0)
4	199.2671(199.3)	198.8706(199.0)	192.3590(192.4)

Condition 1: excluding inertia effects.
Condition 2: including inertia effects.
Condition 3: including inertia effects and the effect of a central compressive load.

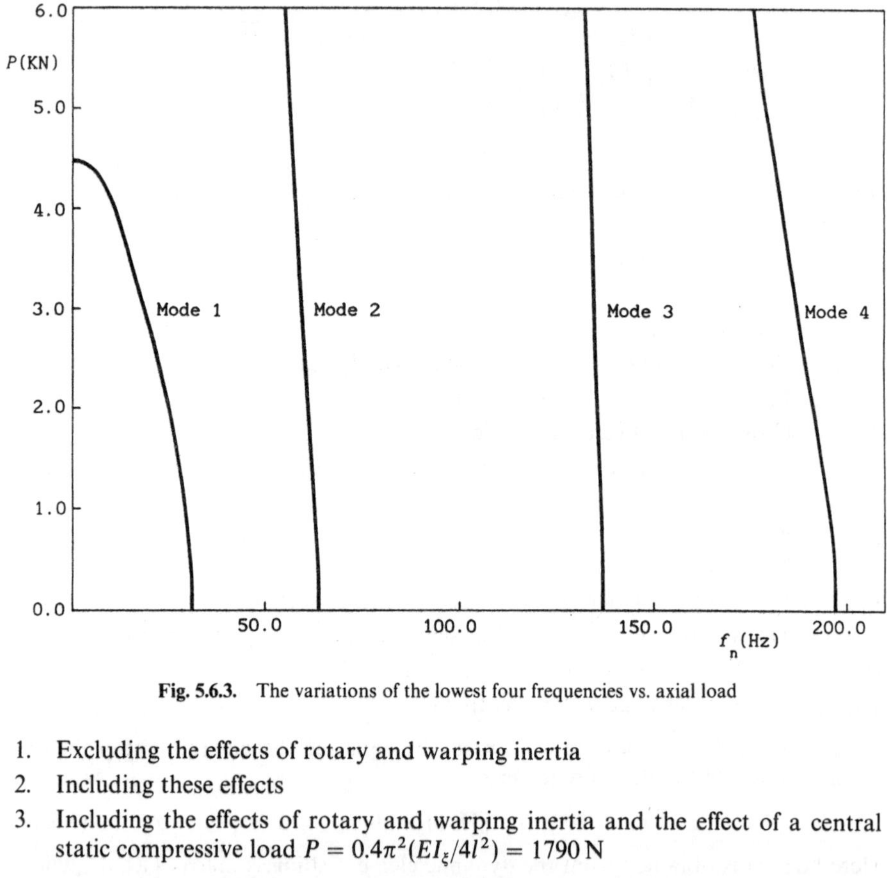

Fig. 5.6.3. The variations of the lowest four frequencies vs. axial load

1. Excluding the effects of rotary and warping inertia
2. Including these effects
3. Including the effects of rotary and warping inertia and the effect of a central static compressive load $P = 0.4\pi^2(EI_\xi/4l^2) = 1790\,\mathrm{N}$

Since the dynamic stiffness matrix formed by the present method is general, the effects of rotary inertia, warping inertia and a central compressive load are considered easily in the matrices $[\mathbf{A}_0]$, $[\mathbf{A}_2]$ and $[\mathbf{A}_4]$ of (5.6.44). In order to investigate the influence of the central compressive load upon the natural frequency, the variation of frequency is plotted against the compressive load P in Fig. 5.6.3.

It is shown in Table 5.6.1 that the agreement between the results calculated by the present method and those obtained by Friberg [17] is considerably close, which proves the validity of the present method. When the thin-walled beam is subjected to a central compressive load $P = 1790\,\mathrm{N}$, the lowest four natural frequencies decrease. More details about this phenomenon can be observed in Fig. 5.6.2. The existence of the compressive load has the strongest effects on the first mode and then the fourth mode, which correspond to bending vibrations in the xy-plane. When the central compressive load P exceeds 4.47 kN, the first mode will vanish.

Example 5.6.2

A cylindrical helical rod with radius R and angle α shown in Fig. 5.6.4 is considered. Let the centre line of the helix be measured by its arc length and the unit tangent,

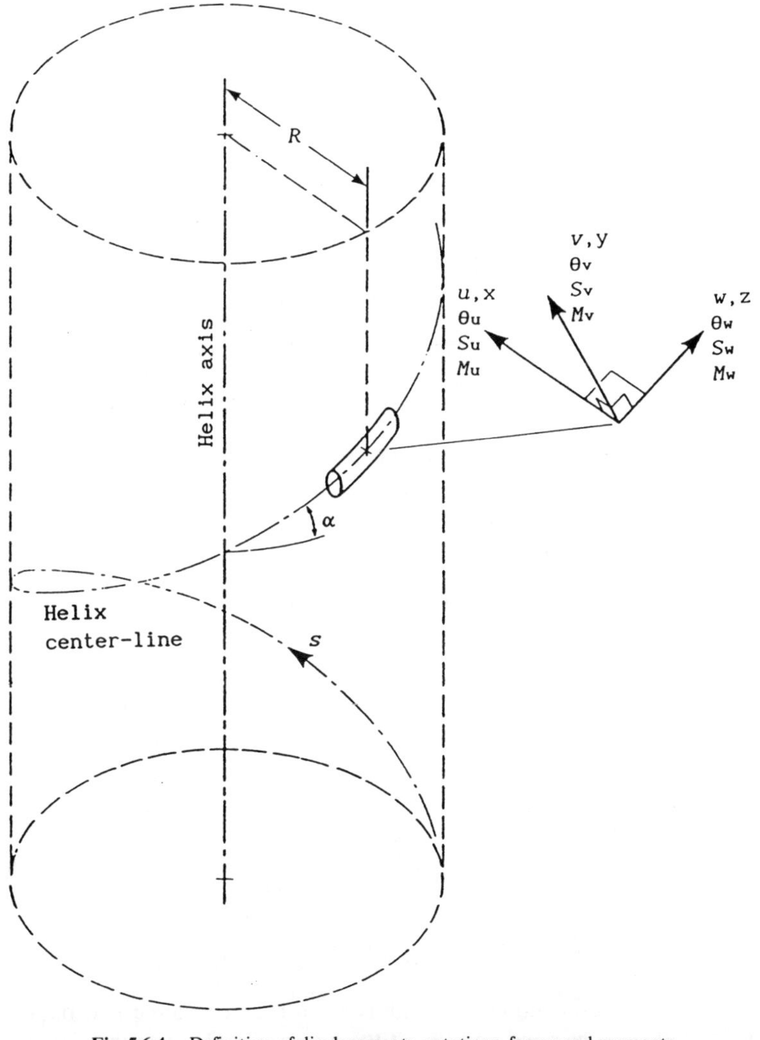

Fig. 5.6.4. Definition of displacements, rotations, forces and moments

normal and binormal vectors along s be $\mathbf{t}$, $\mathbf{n}$ and $\mathbf{b}$, respectively. Then the Frenet–Serret formulae [18] read

$$\frac{\mathrm{d}\mathbf{t}}{\mathrm{d}s} = k\mathbf{n}, \qquad \frac{\mathrm{d}\mathbf{n}}{\mathrm{d}s} = -k\mathbf{t} + \tau\mathbf{b}, \qquad \frac{\mathrm{d}\mathbf{b}}{\mathrm{d}s} = -\tau\mathbf{n} \tag{5.6.48}$$

which define the curvature k and the torsion τ by

$$k = \frac{1}{R}\cos^2\alpha \qquad \text{and} \qquad \tau = \frac{1}{R}\sin\alpha\cos\alpha \tag{5.6.49}$$

Let the displacement vector $\{\mathbf{u}(s)\} = [u, v, w]^{\mathrm{T}}$ and the angular displacement vector $\{\boldsymbol{\theta}(x)\} = [\theta_u, \theta_v, \theta_w]^{\mathrm{T}}$ be defined at s along the local $\mathbf{n}$, $\mathbf{b}$ and $\mathbf{t}$ axes. Then it can be proved [19] that the governing equilibrium equations, in the absence of internal forces, are given by

$$\begin{bmatrix} \mathbf{I}\frac{\mathrm{d}}{\mathrm{d}s} - \mathbf{G} & \mathbf{0} \\ -\mathbf{J} & \mathbf{I}\frac{\mathrm{d}}{\mathrm{d}s} - \mathbf{G} \end{bmatrix} \begin{bmatrix} \mathbf{E}_u & \\ & \mathbf{E}_\theta \end{bmatrix} \begin{bmatrix} \mathbf{I}\frac{\mathrm{d}}{\mathrm{d}s} - \mathbf{G} & -\mathbf{J} \\ \mathbf{0} & \mathbf{I}\frac{\mathrm{d}}{\mathrm{d}s} - \mathbf{G} \end{bmatrix} \begin{Bmatrix} \mathbf{u} \\ \boldsymbol{\theta} \end{Bmatrix} = \{\mathbf{0}\} \tag{5.6.50}$$

where

$$[\mathbf{G}] = \begin{bmatrix} 0 & \tau & -k \\ -\tau & 0 & 0 \\ k & 0 & 0 \end{bmatrix}, \qquad [\mathbf{J}] = \begin{bmatrix} 0 & 1 & 0 \\ -1 & 0 & 0 \\ 0 & 0 & 0 \end{bmatrix}$$

$$[\mathbf{E}_u] = \mathrm{diag}[GA_x, GA_y, EA_z] \qquad \text{and} \qquad [\mathbf{E}_\theta] = \mathrm{diag}[EI_x, EI_y, GI_z]$$

in which $[\mathbf{I}]$ is the identity matrix, A_z is the cross-sectional area of the beam, A_x and A_y the effective shear areas: $A_x = A_z/\gamma_x$ and $A_y = A_z/\gamma_y$, I_z the torsional constant, I_x and I_y the second moments of area about the x and y-axes respectively, and E and G denote Young's modulus and the shear modulus respectively.

For this particular structure, the matrices of the general equation (5.6.1) emerging from Eq. (5.6.47) are specified as

$$[\mathbf{A}_0] = \begin{bmatrix} -GA_y\tau^2 - EA_zk^2 & 0 & 0 & -GA_y\tau & 0 & 0 \\ 0 & -GA_x\tau^2 & GA_xk\tau & 0 & -GA_x\tau & 0 \\ 0 & GA_xk\tau & -GA_xk\tau & 0 & GA_xk & 0 \\ -GA_y\tau & 0 & 0 & -GA_y - EI_y\tau^2 - GI_zk^2 & 0 & 0 \\ 0 & -GA_x\tau & GA_xk & 0 & -GA_x - EI_x\tau^2 & EI_xk\tau \\ 0 & 0 & 0 & 0 & EI_xk\tau & -EI_xk^2 \end{bmatrix}$$

$$[\mathbf{A}_1] = \begin{bmatrix} 0 & -(GA_x + GA_y)\tau & (GA_y + EA_z)k & 0 & -GA_x & 0 \\ (GA_x + GA_y)\tau & 0 & 0 & GA_y & 0 & 0 \\ -(GA_x + EA_z)k & 0 & 0 & 0 & 0 & 0 \\ 0 & -GA_y & 0 & 0 & -(EI_x + EI_y)\tau & (EI_x + GI_z)k \\ GA_x & 0 & 0 & (EI_x + EI_y)\tau & 0 & 0 \\ 0 & 0 & 0 & -(EI_x + GI_z)k & 0 & 0 \end{bmatrix}$$

$$[\mathbf{A}_2] = \mathrm{diag}[GA_x, GA_y, EA_z, EI_x, EI_y, GI_z]$$

Here the circular cross-section with radius $r = 0.1$ is considered and the one-turn helix is fixed at the lower end and free at the upper end. Other properties are

Table 5.6.2. Defective eigensolutions for a helical spring

Mode	Eigenvalues	Generalized eigenvectors (Jordan chains)					
1	0.0000	0.00000 + 0.00000i	0.56713 + 0.00000i	0.10000 + 0.00000i	0.00000 + 0.00000i	0.00000 + 0.00000i	0.00000 + 0.00000i
2		0.10311 + 0.00000i	0.05671 + 0.00000i	0.01000 + 0.00000i	−0.05847 + 0.00000i	0.00000 + 0.00000i	0.00000 + 0.00000i
3		0.00000 + 0.00000i	−3.31634 + 0.00000i	0.00000 + 0.00000i	0.00000 + 0.00000i	0.05671 + 0.00000i	0.01000 + 0.00000i
4		0.00011 + 0.00000i	−3.31634 + 0.00000i	0.00000 + 0.00000i	3.31615 + 0.00000i	0.05671 + 0.00000i	0.01000 + 0.00000i
5	0.9848i	0.00000 + 0.10154i	−0.01763 + 0.00000i	0.10000 + 0.00000i	0.00000 + 0.00000i	0.00000 + 0.00000i	0.00000 + 0.00000i
6		0.10311 + 0.01015i	−0.00176 + 0.59378i	0.01000 + 0.00000i	0.05847 + 0.00000i	0.00000 + 0.01015i	0.00000 − 0.05759i
7		0.10311 + 0.01015i	0.41077 + 0.59378i	0.01000 + 0.00000i	0.05847 − 0.63441i	0.09605 + 0.01015i	−0.70976 − 0.05759i
8		0.00110 + 0.00010i	0.00411 + 8.57774i	0.00010 + 0.00000i	0.84062 − 0.00634i	0.00096 − 0.14646i	−0.00710 − 0.88131i
9	−0.09848i	0.00000 − 0.10154i	−0.01763 + 0.00000i	0.10000 + 0.00000i	0.00000 + 0.00000i	0.00000 + 0.00000i	0.00000 + 0.00000i
10		0.10311 − 0.01015i	−0.00176 − 0.59378i	0.01000 + 0.00000i	0.05847 + 0.00000i	0.00000 − 0.01015i	0.00000 + 0.05759i
11		0.10311 − 0.01015i	0.41077 − 0.59378i	0.01000 + 0.00000i	0.05847 + 0.63441i	0.09605 − 0.01015i	−0.70976 + 0.05759i
12		0.00110 − 0.00010i	0.00411 − 8.57774i	0.00010 + 0.00000i	0.84062 + 0.00634i	0.00096 + 0.14646i	−0.00710 + 0.88131i

Table 5.6.3. Stiffness matrix ($\times 10^3$)

Eight finite elements					
1.502427					
0.000000	1.658459				
0.000000	−0.311238	0.802962			
0.544023	9.187147	−1.724124	126.287064		
−8.322798	−2.955522	8.000187	−19.385968	201.024294	
−0.000001	−16.591698	2.723940	−91.910846	25.670158	252.392436
Sixteen finite elements					
1.386589					
0.000000	1.520920				
0.000000	−0.272404	0.718410			
0.501445	8.423818	−1.508753	120.622759		
−7.679802	−2.550245	7.152973	−16.902250	187.558867	
0.000014	−15.210700	2.358800	−84.246488	21.849794	236.949272
Thirty-two finite elements					
1.358871					
0.000000	1.488134				
0.000000	−0.263387	0.698799			
0.491483	8.242230	−1.458818	119.265964		
−7.526281	−2.456773	6.956646	−16.329430	184.396843	
0.000027	−14.881703	2.274322	−82.424285	20.972024	233.259288
Sixty-four finite elements					
1.352020					
0.000000	1.480042				
−0.000001	−0.261176	0.693988			
0.489035	8.197386	−1.446527	118.930129		
−7.488340	−2.433888	6.908503	−16.188695	183.618758	
0.000027	−14.800525	2.253623	−81.974493	20.757826	232.348262
One hundred and twenty-eight finite elements					
1.350311					
0.000000	1.478020				
0.000000	−0.260624	0.692790			
0.488422	8.186214	−1.443522	118.846493		
−7.478870	−2.428178	6.896496	−16.154176	183.424445	
0.000041	−14.780226	2.248459	−81.862251	20.703724	232.120162
Two hundred and fifty-six finite elements					
1.349793					
−0.000001	1.477447				
0.000001	−0.260514	0.692501			
0.488303	8.184322	−1.443183	118.837393		
−7.477145	−2.427004	6.893581	−16.150010	183.386221	
0.000141	−14.774456	2.247393	−81.843080	20.691822	232.055408
Five hundred and twelve finite elements					
1.349688					
0.000000	1.473326				
0.000000	−0.260581	0.692428			
0.488270	8.183612	−1.442928	118.831964		
−7.476562	−2.426671	6.892866	−16.147244	183.374685	
0.000000	−14.773271	2.247089	−81.836176	20.689496	232.042382
One helix element					
1.349621					
0.000000	1.477276				
0.000000	−0.260468	0.692322			
0.488247	8.183372	−1.442817	118.824771		
−7.476351	−2.426553	6.892620	−16.145032	183.355992	
0.000000	−14.772869	2.246979	−81.831974	20.686070	232.036403

$R = 10.0$
$\alpha = 10^\circ$
$\gamma_x = \gamma_y = 1.2$
$E = 89.0 \times 10^9$
$G = 26.0 \times 10^9$

The eigenvalues for the helix are $\lambda_1 = 0$, $\lambda_2 = ai$, and $\lambda_3 = -ai$, where $a = (k^2 + \tau^2)^{1/2}$ and $i = \sqrt{-1}$. The algebraic multiplicities are found to be four for all three eigenvalues, but the geometric multiplicities are two for λ_1 and one for λ_2 and λ_3, which indicates that the eigensolutions associated with the helical spring considered here are all defective. The corresponding generalized eigenvectors or Jordan chains are calculated by means of formula (5.6.25) and the results are listed in Table 5.6.2. The stiffness matrix using the defective shape function is checked by straight finite element segments using 8, 16, 32, 64, 128, 256 and 512 elements, and the results are presented in Table 5.6.3. It is found that one helix element is as good as 512 conventional finite elements.

5.7. Curved Thin-Walled Beam

Consider a horizontally curved thin-walled beam. Some assumptions which must be made here are:

1. The material is elastic and homogeneous.
2. The length of the beam is very large compared with the cross-sectional dimensions.
3. Every cross-section is rigid in its own plane.
4. Shearing deformation of the middle surface of the member is negligible.
5. Transverse displacements are much larger than the longitudinal displacement.

For a circularly curved member with I-section as shown in Fig. 5.7.1, when the effect of curvature is considered, the cross-sectional displacements at an arbitrary point p (Fig. 5.7.1) are derived according to Vlasov's thin-walled beam theory [20],

$$u_p = u - yv' - z\left(w' - \frac{u}{R}\right) - \Omega\left(\theta' + \frac{v'}{R}\right) \tag{5.7.1}$$

$$v_p = v - z\theta \tag{5.7.2}$$

$$w_p = w + y\theta \tag{5.7.3}$$

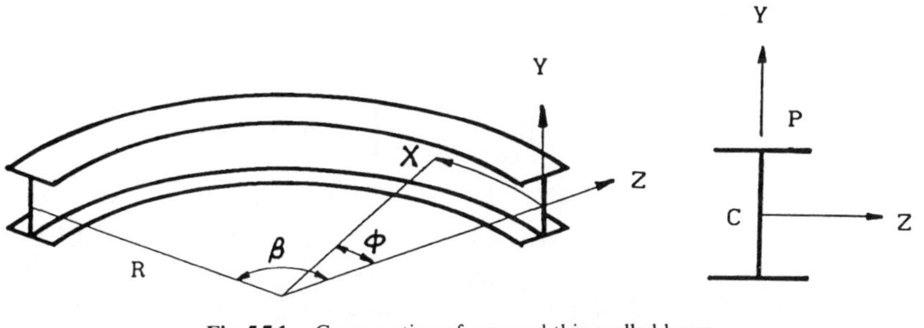

Fig. 5.7.1. Cross-section of a curved thin-walled beam

where R is the curvature radius, θ is the twist angle, v and w denote the transverse displacements of the centroid C from the original position, and u is the longitudinal displacement of C. The displacements u, v, w and θ are functions of coordinate x which is tangent to the curved axis of the member. In Eq. (5.7.1), a prime denotes differentiation with respect to coordinate x, and Ω the normalized sectorial area. Coordinate axes x, y, z form a right-hand frame. The effect of curvature is considered through the expressions of various quantities, such as strains and volumes in terms of curvature radius R.

From the finite displacement theory, the first order linear and the second order non-linear components of the strains can be expressed in terms of the displacements $[u_p, v_p, w_p, \theta]^{\mathrm{T}}$ and hence in terms of the displacements of the centroid C, $[u, v, w, \theta]^{\mathrm{T}}$, from Eqs (5.7.1)–(5.7.3). Therefore we can get a set of strain–displacement relations and hence stress–displacement relations from Hooke's law. According to the principle of virtual displacements, the dynamic stability of a deformed body can be described in a Lagrangian form

$$\int_v s_{ij}\delta\varepsilon_{ij}\,\mathrm{d}\,vol + \delta T = \text{E.V.W.} \tag{5.7.4}$$

where $\delta T = \int_{vol} \rho\{\ddot{u}\}\delta\{u\}\,\mathrm{d}\,vol$ is the volume integral of the virtual work done by inertia forces; s_{ij} is the second Piola–Kirchhoff stress tensor; $\delta\varepsilon_{ij}$ is the variation of the Green–Lagrange strain tensor; E.V.W. is the external virtual work; vol denotes the initial volume of the body; and the differential $\mathrm{d}\,vol$ equals $(R + z)/R\,\mathrm{d}y\,\mathrm{d}z\,\mathrm{d}x$ in Cartesian coordinates.

A set of governing differential equations of motion for the curved thin-walled beam can be obtained by performing the following steps in Eqs (5.7.5)–(5.7.8):

1. Substitute the expression for ε_{ij} and s_{ij} in terms of $[u, v, w, \theta]$.
2. Integrate each term by parts to obtain the virtual displacements δu, δv, δw and $\delta\theta$.
3. Admit the arbitrary nature of virtual displacements.
4. Neglect the diminishing terms beyond second order.

Finally, we get a set of equations of motion as follows:

$$\begin{aligned} &EA\left(u'' + \frac{w'}{R}\right) + \frac{1}{R}F_x\left(w' - \frac{u}{R}\right) - \frac{1}{R}\left(M_z + \frac{B}{R}\right)\left(\theta' + \frac{v'}{R}\right) + \frac{1}{2R}T'_{sv}v' \\ &\quad - \left(F_y - \frac{M_x}{R}\right)\left(v'' - \frac{\theta}{R}\right) - F'_z\left(w' - \frac{u}{R}\right) - F_z\left(w'' + \frac{w}{R^2}\right) - m\left(A + \frac{3I_y}{R^2}\right)\ddot{u} \\ &\quad + \frac{2}{R}mI_y\ddot{w}' = 0 \end{aligned} \tag{5.7.5}$$

$$\begin{aligned} &EI_y\left(w^{\mathrm{iv}} + 2\frac{w''}{R^2} + \frac{w}{R^4}\right) + \frac{EA}{R}\left(u' + \frac{w}{R}\right) - \left[F_x\left(w' - \frac{u}{R}\right)\right]' \\ &\quad + \left[\left(M_z + \frac{B}{R}\right)\left(\theta' + \frac{v'}{R}\right)\right]' - \frac{1}{2}(T'_{sv}v')' - F_y\left(\theta' + \frac{v'}{R}\right) + F_z\left(u'' + \frac{u}{R^2}\right) \\ &\quad + F'_z\left(u' + \frac{w}{R}\right) - \left[M_x\left(v'' - \frac{\theta}{R}\right)\right]' + mA\ddot{w} - mI_y\ddot{w}'' + \frac{2}{R}mI_y\ddot{u}' = 0 \end{aligned} \tag{5.7.6}$$

$$EI_z\left(v^{\text{iv}} - \frac{\theta''}{R}\right) - \frac{GJ}{R}\left(\theta'' + \frac{v''}{R}\right) - (F_x v')' + \left[M_y\left(\theta' + \frac{v'}{R}\right)\right]' + (F_z\theta)'$$
$$+ F_y\left(u'' + \frac{w'}{R}\right) + \frac{1}{R}\left[\left(M_z + \frac{B}{R}\right)\left(w' - \frac{u}{R}\right)\right]' - \frac{r^2}{R}\left[\left(F_x + \frac{M_y}{R}\right)\left(\theta' + \frac{v'}{R}\right)\right]'$$
$$+ \left[(M_x - \tfrac{1}{2}T_{\text{sv}})'\left(w' - \frac{u}{R}\right)\right]' + \left[M_x\left(w'' - \frac{u'}{R}\right)\right]' + mA\ddot{v}$$
$$- m\left(I_z + \frac{3I_\Omega}{R^2}\right)\ddot{v}'' - m\frac{I_y}{R}\ddot{\theta} - m\frac{2I_\Omega}{R}\ddot{\theta}'' = 0 \qquad (5.7.7)$$

$$EI_\Omega\left(\theta^{\text{iv}} + \frac{2\theta''}{R^2} + \frac{\theta}{R^4}\right) - \frac{EI_z}{R}\left(v'' - \frac{\theta}{R}\right) - GJ\left(\theta'' + \frac{v''}{R}\right) + M_y\left(v'' - \frac{\theta}{R}\right)$$
$$+ \left[\left(M_z + \frac{B}{R}\right)\left(w' - \frac{u}{R}\right)\right]' - r^2\left[\left(F_x + \frac{M_y}{R}\right)\left(\theta' + \frac{v'}{R}\right)\right]' + F_y\left(w' - \frac{u}{R}\right)$$
$$- \frac{M_x}{R}\left(w' - \frac{u}{R}\right) + m(I_y + I_z + r^2A)\ddot{\theta} - mI_\Omega\ddot{\theta}'' - \frac{1}{R}mI_y\ddot{v} - \frac{2}{R}mI_\Omega\ddot{v}'' = 0 \quad (5.7.8)$$

where F_x, F_y, F_z, M_x, M_y, M_z are the stress resultants on a cross-section of the member; T_{sv} is the St. Venant torque; B is the bimoment; I_y and I_z are the moments of inertia about the y- and z-axes, respectively; I_Ω is the warping constant; J is the torsional constant; m is the mass per unit volume; r is the polar radius of gyration; and A is the area of the cross-section.

Equations (5.7.5)–(5.7.8) represent the dynamic stability of a curved thin-walled beam. The different deformation modes, such as axial, flexural and torsional modes (including possible warping), are coupled as the result of including the effect of curvature.

Example 5.7.1

For the present purpose, a horizontally curved member of an I-section as shown in Fig. 5.7.1 will be considered. The circular beam is subjected to a constant bending moment M_y with different boundary conditions in the present studies. Figure 5.7.2 is an in-plane diagram of the beam. The section properties adopted from reference

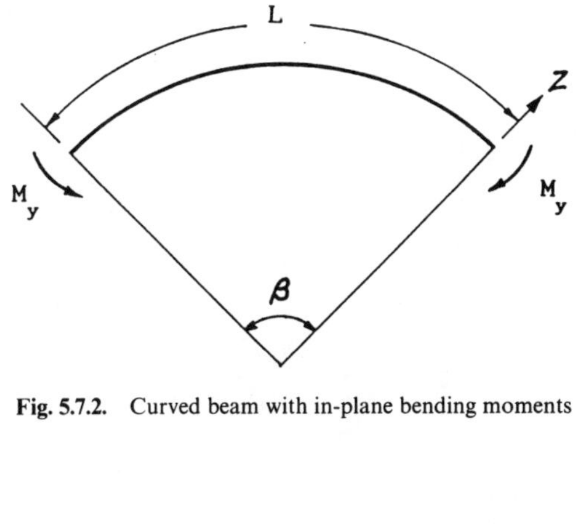

Fig. 5.7.2. Curved beam with in-plane bending moments

[20] are: $A = 92.9\,\text{cm}^2$; $I_z = 11\,360\,\text{cm}^4$; $I_y = 3870\,\text{cm}^4$; $I_\omega = 55\,590\,\text{cm}^6$, $J = 58.9\,\text{cm}^6$, $r = 12.81$ cm; and $L = 1024$ cm. The moduli of elasticity are $E = 200$ GPa, $G = 77.2$ GPa.

If the condition of inextensibility, $u' + w/R = 0$, and the harmonic conditions are assumed, u is eliminated from Eq. (5.7.5) and Eqs (5.7.6)–(5.7.8) become

$$EI_y\left(w^{\text{iv}} + \frac{2w''}{R^2} + \frac{w}{R^4}\right) - \frac{M_y}{R}\left(w'' + \frac{w}{R^2}\right) - Am\omega^2 w + I_y m\omega^2 w'' + \frac{2I_y}{R^2} m\omega^2 = 0 \tag{5.7.9}$$

$$EI_z\left(v^{\text{iv}} - \frac{\theta''}{R}\right) - \frac{GJ}{R}\left(\theta'' + \frac{v''}{R}\right) + M_y\left(1 - \frac{r^2}{R^2}\right)\left(\theta'' + \frac{v''}{R}\right) - Am\omega^2 v$$
$$+ \left(I_z + \frac{3I_\Omega}{R^2}\right) m\omega^2 v'' + \frac{I_y}{R} m\omega^2\theta + \frac{2I_\Omega}{R} m\omega^2\theta'' = 0 \tag{5.7.10}$$

$$EI_\Omega\left(\theta^{\text{iv}} + \frac{2\theta''}{R^2} + \frac{\theta}{R^4}\right) - \frac{EI_z}{R}\left(v'' - \frac{\theta}{R}\right) - GJ\left(\theta'' + \frac{v''}{R}\right) + M_y v'' - \frac{r^2}{R} M_y\left(\theta'' + \frac{v''}{R}\right)$$
$$-(I_y + I_z + r^2 A)m\omega^2\theta + I_\Omega m\omega^2\theta'' + \frac{I_y}{R} m\omega^2 v + \frac{2I_\Omega}{R} m\omega^2 v'' = 0 \tag{5.7.11}$$

The corresponding natural boundary conditions are

$$[Q_1\delta w + Q_2\delta v + Q_3\delta\theta + Q_4\delta w' + Q_5\delta v' + Q_6\delta\theta']_0^L = 0 \tag{5.7.12}$$

where the generalized forces are

$$Q_1(x) = -EI_y\left(w''' + \frac{w'}{R^2}\right) + \rho w^2 I_y w'$$

$$Q_2(x) = -EI_z\left(v''' - \frac{\theta'}{R}\right) + \frac{GJ}{R}\left(\theta' + \frac{v'}{R}\right) - M_y\left(1 - \frac{r^2}{R^2}\right)\left(\theta' + \frac{v'}{R}\right)$$
$$- m\omega^2\left(I_z + \frac{3I_\Omega}{R^2}\right)v' - 2m\omega^2\frac{I_\Omega}{R}\theta$$

$$Q_3(x) = -EI_\Omega\left(\theta''' + \frac{\theta'}{R^2}\right) + GJ\left(\theta' + \frac{v'}{R}\right) + M_y\frac{r^2}{R}\theta' - M_y\left(1 - \frac{r^2}{R^2}\right)v'$$
$$- m\omega^2 I_\Omega\left(\theta' + \frac{2v'}{R}\right)$$

$$Q_4(x) = EI_y\left(w'' + \frac{w}{R^2}\right) + M_y$$

$$Q_5(x) = EI_z\left(v'' - \frac{\theta}{R}\right)$$

$$Q_6(x) = EI_\Omega\left(\theta'' + \frac{\theta}{R}\right)$$

Equations (5.7.9)–(5.7.11) can be written in the form of Eqs (5.6.1), and the matrices of Eq. (5.6.1) become

$$\mathbf{A}_0 = \begin{bmatrix} \frac{EI_y}{R^4} - \frac{M_y}{R^3} - m\omega^2\left(A - \frac{2I_y}{R^2}\right) & 0 & 0 \\ & -m\omega^2 A & m\omega^2 \frac{I_y}{R} \\ \text{sym.} & & \frac{EI_\Omega}{R^4} + \frac{EI_z}{R^2} - m\omega^2(I_y + I_z + r^2 A) \end{bmatrix}$$

$$\mathbf{A}_2 = \begin{bmatrix} \frac{2EI_y}{R^4} - \frac{M_y}{R} + m\omega^2 I_y & 0 & 0 \\ & -\frac{GJ}{R} + \frac{M_y}{R}\left(1 - \frac{r^2}{R^2}\right) + m\omega^2\left(I_z + \frac{3I_\Omega}{R^2}\right) & -\frac{EI_z}{R} - \frac{GJ}{R} + M_y\left(1 - \frac{r^2}{R^2}\right) + m\omega^2 \frac{2I_\Omega}{R} \\ \text{sym.} & & \frac{2EI_\Omega}{R^2} - GJ - \frac{r^2}{R} M_y + m\omega^2 I_\Omega \end{bmatrix}$$

$$\mathbf{A}_4 = \begin{bmatrix} EI_y & 0 & 0 \\ & EI_z & 0 \\ & & EI_\Omega \end{bmatrix} \tag{5.7.13}$$

and $[\mathbf{A}_1]$, $[\mathbf{A}_3]$ are zero matrices in this case. Thus, the dynamic stiffness matrix can be formed explicitly by the following procedure with the help of a microcomputer:

1. Solve for the eigenvalues λ_j and the corresponding eigenvectors $\{\boldsymbol{\phi}_j\}$, $j = 1, 2, \ldots, nm$ from the eigenproblem (5.7.3).
2. Calculate the matrix $[\mathbf{H}]$ and then invert it to obtain the shape function.
3. Form the dynamic stiffness matrix $[\mathbf{K}]$. The natural frequencies of the system are then determined by equating the determinant of the dynamic stiffness of the structure to zero,

$$\det[\mathbf{D}(\omega)] = 0 \tag{5.7.14}$$

Here $[\mathbf{D}(\omega)]$ is the system dynamic stiffness matrix which is obtained from the element matrices $[\mathbf{K}(\omega)]$. $[\mathbf{K}(\omega)]$ is a real square matrix of order 12. $[\mathbf{D}(\omega)]$ is one having order of 12, 6, 6, 3 when the natural boundary conditions are free–free, clamped–free, pinned–pinned and clamped–pinned, respectively.

The lowest six natural frequencies are plotted in Fig. 5.7.3 for various boundary conditions with different end bending moment M_y against the subtended angles. Figure 5.7.4 represents the relations between frequencies and bending moment M_y for different boundary conditions and subtended angles. Figure 5.7.5, representing the variation of frequencies against subtended angles, is plotted by adopting the boundary conditions as pinned–pinned and using the analytical solution of Eqs (5.7.9)–(5.7.11) as,

$$\left.\begin{aligned} w &= a_1 \sin \lambda x \\ v &= a_2 \sin \lambda x \\ \theta &= a_3 \sin \lambda x \end{aligned}\right\} \tag{5.7.15}$$

where $\lambda = n\pi/L$, $n = 1, 2, 3, \ldots$. Substitution of Eq. (5.7.15) in Eqs (5.7.9)–(5.7.11) generates a characteristic problem

$$([\mathbf{A}] - \omega^2[\mathbf{B}])\{\boldsymbol{\phi}\} = \{\mathbf{0}\} \qquad \text{or} \qquad \det[\mathbf{A} - \omega^2\mathbf{B}] = 0 \tag{5.7.16}$$

where

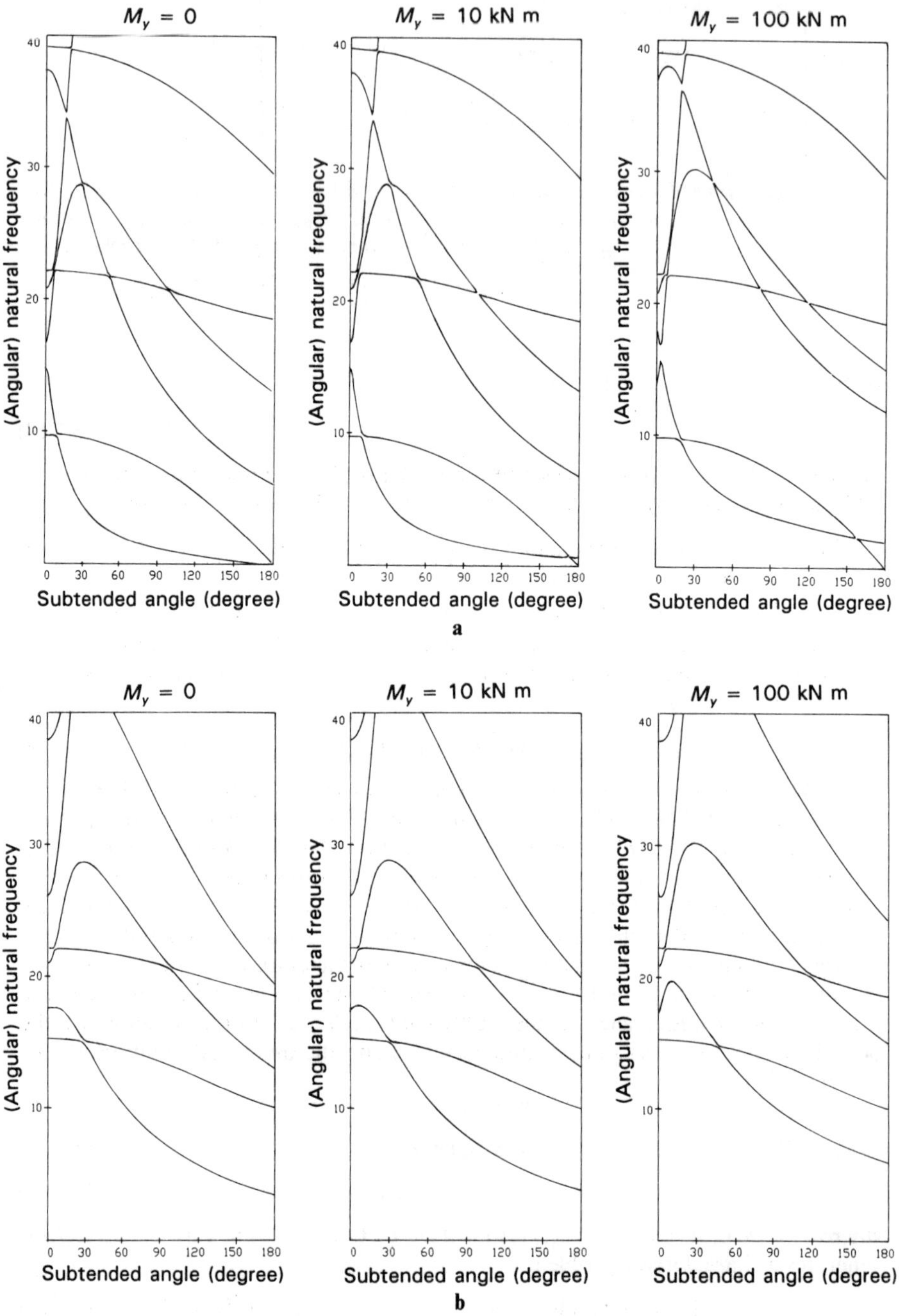

Fig. 5.7.3. Frequency diagrams (dynamic stiffness solutions) of a curved thin-walled beam: **a** pinned–pinned; **b** clamped–pinned; **c** clamped–free; **d** free–free

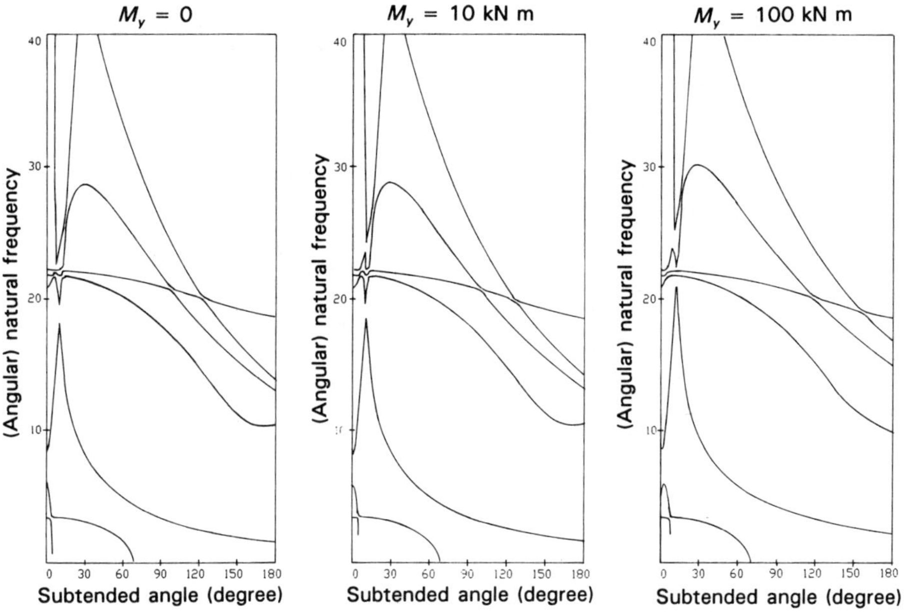

c

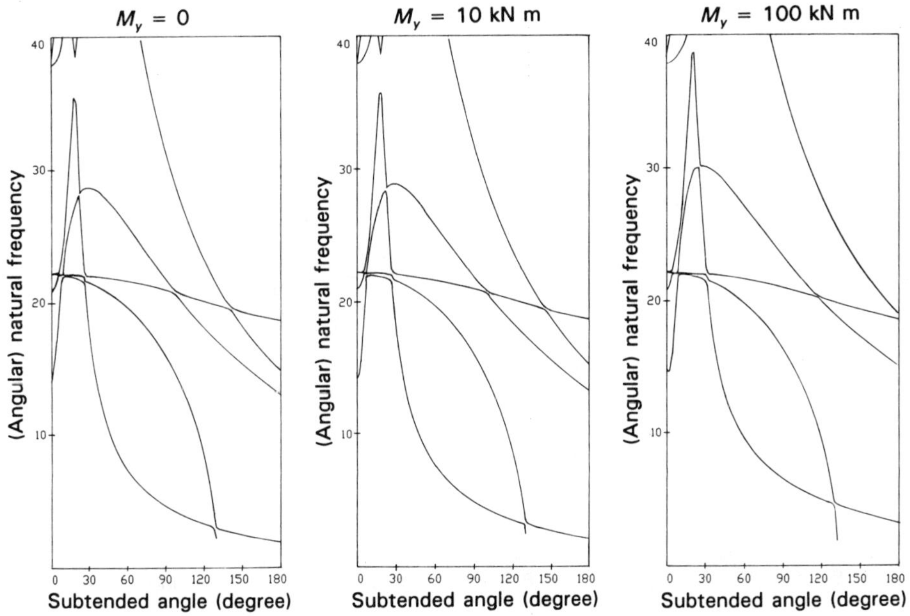

d

Fig. 5.7.3 *(continued)*

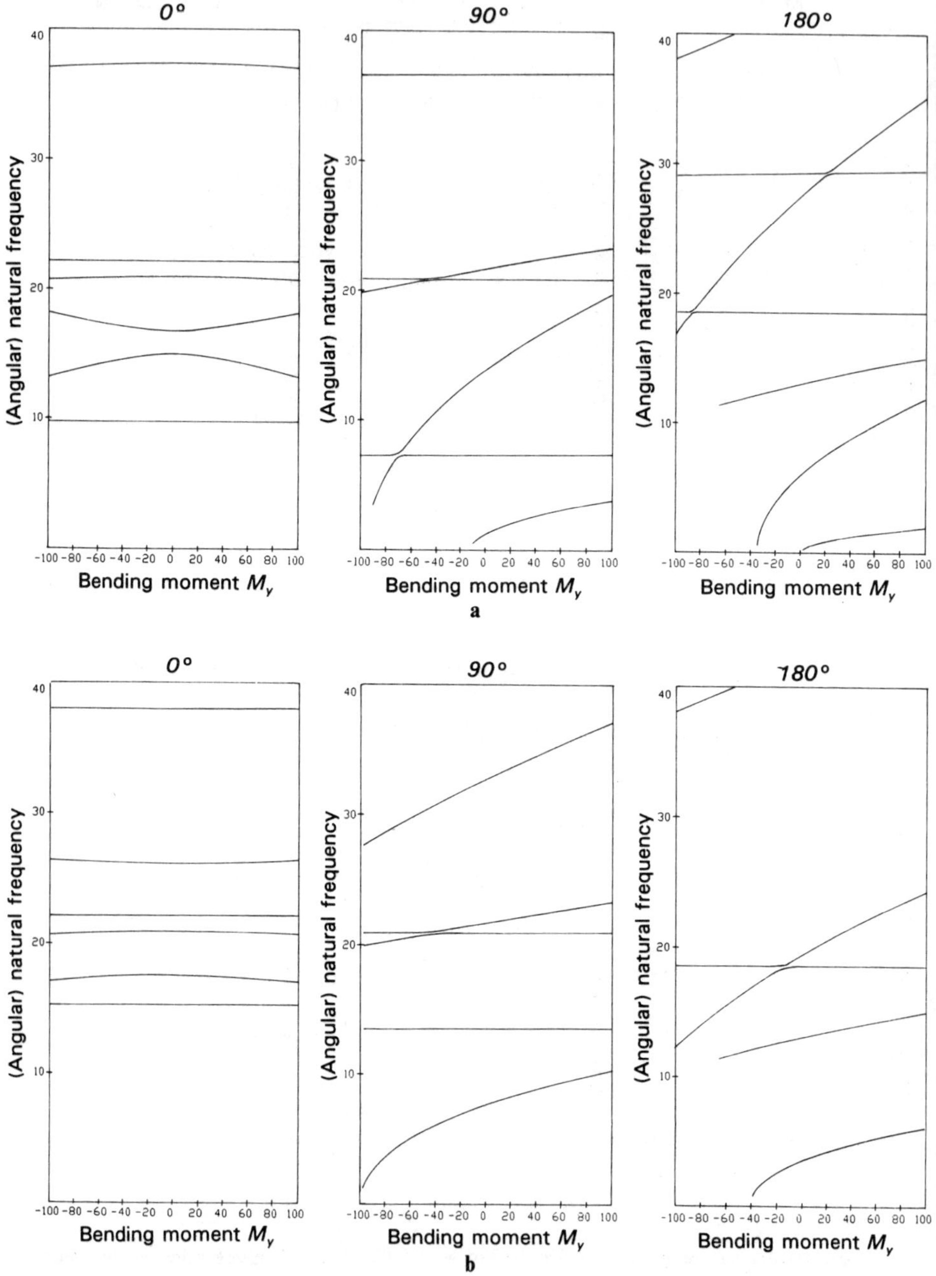

Fig. 5.7.4. Frequency diagrams of a curved thin-walled beam against bending moments M_y: **a** pinned–pinned; **b** clamped–clamped; **c** clamped–free; **d** free–free

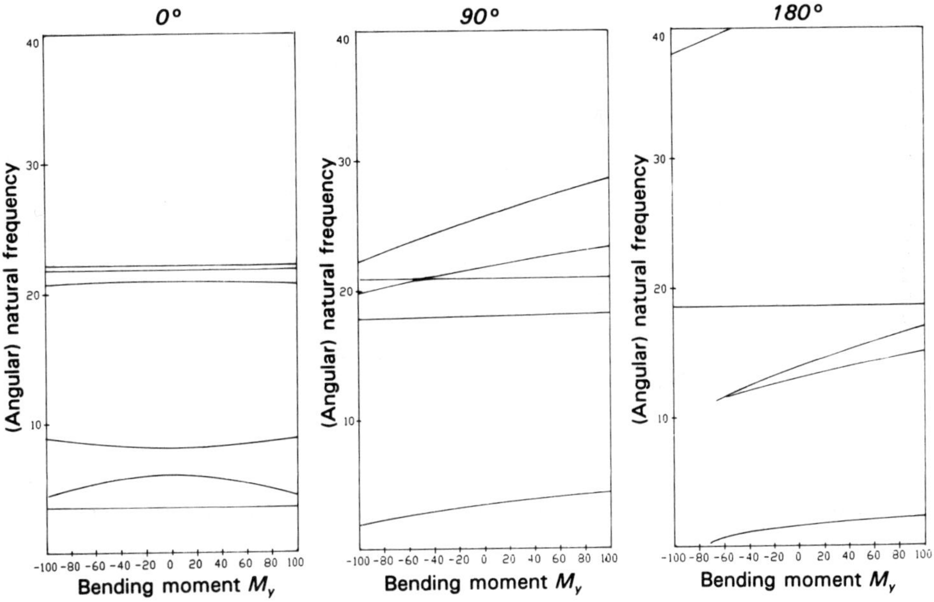

c

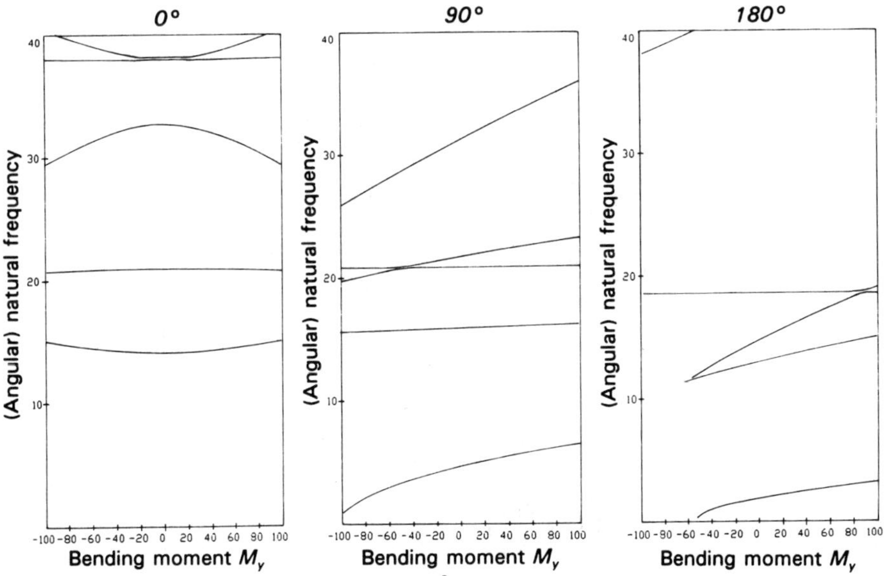

d

Fig. 5.7.4 (*continued*)

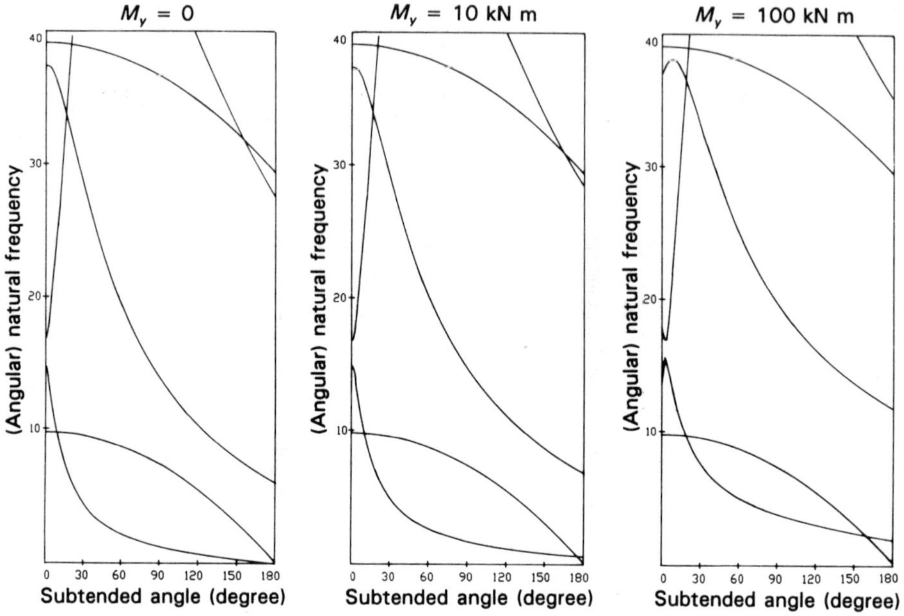

Fig. 5.7.5. Frequency diagram of a pinned curved thin-walled beam (analytical solutions)

$$[\mathbf{A}] = \begin{bmatrix} a & 0 & 0 \\ 0 & b & c \\ 0 & c & d \end{bmatrix}, \qquad [\mathbf{B}] = \begin{bmatrix} e & 0 & 0 \\ 0 & f & g \\ 0 & g & h \end{bmatrix}, \qquad \{\boldsymbol{\phi}\} = \begin{Bmatrix} w \\ v \\ \theta \end{Bmatrix}$$

$$\left.\begin{aligned}
a &= EI_y\left(\lambda^4 - \frac{2}{R^2}\lambda^2 + \frac{1}{R^4}\right) + \frac{M_y}{R}\left(\lambda^2 - \frac{1}{R^2}\right) \\
b &= EI_z\lambda^4 + \frac{GJ}{R^2}\lambda^2 - \frac{M_y}{R}\left(1 - \frac{r^2}{R^2}\right)\lambda^2 \\
c &= \frac{EI_z}{R}\lambda^2 + \frac{GJ}{R}\lambda^2 - M_y\left(1 - \frac{r^2}{R^2}\right)\lambda^2 \\
d &= EI_\Omega\left(\lambda^4 - \frac{2}{R^2}\lambda^2 + \frac{1}{R^4}\right) + \frac{EI_z}{R^2} + GJ\lambda^2 + M_y\frac{r^2}{R}\lambda^2 \\
e &= mA - 2m\frac{I_y}{R^2} + mI_y\lambda^2 \\
f &= mA + m\left(I_z + \frac{3I_\Omega}{R^2}\right)\lambda^2 \\
g &= -m\frac{I_y}{R} + m\frac{2I_\Omega}{R}\lambda^2 \\
h &= m(I_y + I_z + r^2A) + mI_\Omega\lambda^2
\end{aligned}\right\} \qquad (5.7.17)$$

Equation (5.7.16) gives a relation between frequency ω and subtended angle β.

Comparing Fig. 5.7.5 and Fig. 5.7.3a, it is apparent that they fit completely. In Fig. 5.7.3a two extra curves occur which represent the two lowest frequencies with the clamped–clamped boundary condition, where the value of the determinant crosses ∞ and $-\infty$. For the clamped–clamped boundary condition, the order of $[\mathbf{D}(\omega)]$ in Eq. (5.7.12) is zero, and the determinant of $[\mathbf{D}(\omega)]$ tends to be infinite. The same phenomenon is found in other cases with various boundary conditions.

Figure 5.7.3 shows that increasing the subtended angles of the beam softens the flexural modes including in-plane and out-of-plane flexural modes but hardens the torsional mode simultaneously. The natural frequencies do not vary in a monotonic way due to the exchange of modes between flexure and torsion. Avoided crossing, or frequency veering, occurs if two or more frequencies approach each other, but then veer off without becoming equal.

5.8. Helix

Let the centre line of the helix shown in Fig. 5.6.4 be measured by its arc length x, $0 \leq x \leq L$. Let the unit vectors of tangent, binormal and normal along x be represented by $\vec{\mathbf{t}}$, $\vec{\mathbf{b}}$ and $\vec{\mathbf{n}}$ respectively and let $\{\vec{\mathbf{p}}\} = [\vec{\mathbf{t}}, \vec{\mathbf{b}}, \vec{\mathbf{n}}]$, then the Frenet–Serret formula [21] gives the explicit derivatives of the unit vectors

$$\frac{\mathrm{d}}{\mathrm{d}x}\{\vec{\mathbf{p}}\} = [\boldsymbol{\kappa}]\{\vec{\mathbf{p}}\} \tag{5.8.1}$$

where

$$[\boldsymbol{\kappa}] = \begin{bmatrix} 0 & 0 & \kappa \\ 0 & 0 & -\tau \\ -\kappa & \tau & 0 \end{bmatrix}$$

is the matrix of curvature containing the torsion τ and curvature κ. Let the displacement vector of a point P, position vector $\vec{\mathbf{p}}$, of coordinates (y, z) relative to the centroidal axis be $\{\mathbf{u}_p(x, y, z)\} = [u_p, v_p, w_p]^{\mathrm{T}}$, and let the reference displacement vector be $\{\mathbf{u}(x)\} = [u(x), v(x), w(x)]$ with respect to the centroid, and the angular displacement vector about $\vec{\mathbf{t}}$, $\vec{\mathbf{b}}$, $\vec{\mathbf{n}}$ respectively be $\{\boldsymbol{\psi}(x)\} = [\psi_x, \psi_y, \psi_z]^{\mathrm{T}}$. The cross-sectional area is assumed to have double symmetry so that shear centre and mass centre of the cross-section coincide. Assume a plane normal to the centroidal axis before deformation remains plane after deformation. It is admissible to write

$$\{\mathbf{u}_p(x, y, z)\} = \{\mathbf{u}(x)\} + [\mathbf{R}(y, z)]\{\boldsymbol{\psi}(x)\} - \Omega\kappa\{\mathbf{e}_1\} \tag{5.8.2}$$

where

$$[\mathbf{R}(y, z)] = \begin{bmatrix} 0 & z & -y \\ -z & 0 & 0 \\ y & 0 & 0 \end{bmatrix}$$

is the rigid body matrix, Ω is the normalized warping coordinate so that $\int \Omega \, \mathrm{d}A = 0$, $\kappa(x)$ is the local warping displacement and $\{\mathbf{e}_1\} = [1, 0, 0]^{\mathrm{T}}$.

The point P with position vector $\vec{\mathbf{p}}$ before deformation moves to the point Q during deformation with position vector $\vec{\mathbf{q}}$, where

$$\vec{\mathbf{q}} = \vec{\mathbf{p}} + \vec{\mathbf{p}}_{,x}\delta x + \vec{\mathbf{p}}_{,y}\delta y + \vec{\mathbf{p}}_{,z}\delta z \tag{5.8.3}$$

in which a comma denotes partial differentiation. Let P' be a point adjacent to P

before deformation. P' moves to point Q' after deformation. The relative position vectors are given by $\vec{\mathbf{r}} = \overrightarrow{pp'}$ and $\vec{\mathbf{r}}^* = \overrightarrow{QQ'}$. If l, m, n are the direction cosines of the undeformed fibre, then

$$\vec{\mathbf{r}} = [\delta x, \delta y, \delta z]\{\vec{\boldsymbol{\rho}}\} = r[l, m, n]\{\vec{\boldsymbol{\rho}}\} \tag{5.8.4}$$

and

$$\vec{\mathbf{r}}^* = \vec{\mathbf{r}} + \vec{\mathbf{q}} - \vec{\mathbf{p}} = [\delta x \quad \delta y \quad \delta z]\{\vec{\boldsymbol{\rho}}\} + [\delta x \quad \delta y \quad \delta z][\vec{\mathbf{p}}_{,x} \quad \vec{\mathbf{p}}_{,y} \quad \vec{\mathbf{p}}_{,z}]^{\mathrm{T}} \tag{5.8.5}$$

Let the displacement vector at point x, y, z be $\{\mathbf{u}_p(x, y, z)\} = [u_p, b_p, w_p]^{\mathrm{T}}$, then

$$\vec{\mathbf{p}}_{,x} = u_{p,x}\vec{\mathbf{n}} + v_{p,x}\vec{\mathbf{b}} + w_{p,x}\vec{\mathbf{t}} = \{\mathbf{u}_{p,x}\}^{\mathrm{T}}\{\vec{\boldsymbol{\rho}}\} + \{\mathbf{u}_p\}^{\mathrm{T}}\{\vec{\boldsymbol{\rho}}_{,x}\} = [\mathbf{u}_{p,x}^{\mathrm{T}} + \mathbf{u}_p^{\mathrm{T}}\boldsymbol{\kappa}]$$
$$\vec{\mathbf{p}}_{,y} = \{\mathbf{u}_{p,y}\}^{\mathrm{T}}\{\vec{\boldsymbol{\rho}}\}$$
$$\vec{\mathbf{p}}_{,z} = \{\mathbf{u}_{p,z}\}^{\mathrm{T}}\{\vec{\boldsymbol{\rho}}\}$$

where the Frenet–Serret formula (Eq. (5.8.1)) has been used. In collective form,

$$\begin{Bmatrix} \vec{\mathbf{p}}_{,x} \\ \vec{\mathbf{p}}_{,y} \\ \vec{\mathbf{p}}_{,z} \end{Bmatrix} = \begin{bmatrix} \mathbf{u}_{p,x}^{\mathrm{T}} + \mathbf{u}_p^{\mathrm{T}}\boldsymbol{\kappa} \\ \mathbf{u}_{p,y}^{\mathrm{T}} \\ \mathbf{u}_{p,z}^{\mathrm{T}} \end{bmatrix} \{\vec{\boldsymbol{\rho}}\} = [\mathbf{F}]\{\vec{\boldsymbol{\rho}}\} \tag{5.8.6}$$

where $[\mathbf{F}]$ is the gradient matrix

$$[\mathbf{F}] = [\mathbf{u}_{p,x} - \boldsymbol{\kappa}\mathbf{u}_p, \mathbf{u}_{p,y}, \mathbf{u}_{p,z}] = [\boldsymbol{\theta}_x \boldsymbol{\theta}_y \boldsymbol{\theta}_z] \tag{5.8.7}$$

since $[\boldsymbol{\kappa}]^{\mathrm{T}} = -[\boldsymbol{\kappa}]$, in which $\{\boldsymbol{\theta}_x\}$, $\{\boldsymbol{\theta}_y\}$, $\{\boldsymbol{\theta}_z\}$ are the gradient vectors. Substituting Eq. (5.8.5) into (5.8.6), one has the deformed fibre,

$$\vec{\mathbf{r}}^* = [\delta x\, \delta y\, \delta z][\mathbf{I} + \mathbf{F}]\{\vec{\boldsymbol{\rho}}\}$$
$$|\vec{\mathbf{r}}^*|^2/r^2 = [l \quad m \quad n][\mathbf{I} + \mathbf{F}][\mathbf{I} + \mathbf{F}^{\mathrm{T}}][l \quad m \quad n]^{\mathrm{T}}$$

Therefore, the unit elongation, e, of the original fibre $\overrightarrow{pp'}$ is given by

$$e = \tfrac{1}{2}[(|\vec{\mathbf{r}}^*|^2/r^2) - 1] = \tfrac{1}{2}[l \quad m \quad n][\mathbf{F} + \mathbf{F}^{\mathrm{T}} + \mathbf{F}\mathbf{F}^{\mathrm{T}}][l \quad m \quad n]^{\mathrm{T}} \tag{5.8.8}$$

from which, the Green–St. Venant strain tensor, e_{ij}, can be defined,

$$[\mathbf{e}] = \tfrac{1}{2}[\mathbf{F} + \mathbf{F}^{\mathrm{T}} + \mathbf{F}\mathbf{F}^{\mathrm{T}}] = [\mathbf{e}^0] + [\mathbf{e}''] \tag{5.8.9}$$

where

$$[\mathbf{e}^0] = \tfrac{1}{2}[\mathbf{F} + \mathbf{F}^{\mathrm{T}}] \tag{5.8.10}$$

$$[\mathbf{e}''] = \tfrac{1}{2}[\mathbf{F} \quad \mathbf{F}^{\mathrm{T}}] \tag{5.8.11}$$

are the linear strain and the non-linear strain components respectively. The present section is concerned with finding the critical initial stress distribution under the assumption that changes in the geometrical configuration of the body remain negligible until the stability occurs. We shall not distinguish the initial coordinate system from the deformed coordinate system.

Evaluating $[\mathbf{F}]$ from Eqs (5.8.7) and (5.8.2), we have

$$[\mathbf{F}] = [\boldsymbol{\theta}_x \quad \boldsymbol{\theta}_y \quad \boldsymbol{\theta}_z] = [\mathbf{u}' + \mathbf{R}\boldsymbol{\psi}' - \boldsymbol{\kappa}\mathbf{u} - \mathbf{R}\boldsymbol{\kappa}\boldsymbol{\psi}, \mathbf{R}_{,y}\boldsymbol{\psi}, \mathbf{R}_{,z}\boldsymbol{\psi}] \tag{5.8.12}$$

Substituting Eq. (5.8.12) into (5.8.10), we have, for the linear strains,

$$e_{11}^0 = e_{12}^0 = e_{13}^0 = 0 \tag{5.8.13}$$

which satisfy most beam, plate and shell theories for vanishing direct strain across the thickness. The linear eigineering strain vector for the non-vanishing strains are obtained from Eqs (5.8.12) and (5.8.10),

$$\{\boldsymbol{\varepsilon}\} = \begin{Bmatrix} e^0_{11} \\ 2e^0_{12} \\ 2e^0_{13} \end{Bmatrix} = \begin{Bmatrix} \varepsilon_x \\ \varepsilon_{xy} \\ \varepsilon_{xz} \end{Bmatrix} = \begin{bmatrix} 1 & 0 & 0 & 0 & z & -y & -\Omega & 0 \\ 0 & 1 & 0 & -z-\Omega_{,y} & 0 & 0 & 0 & \Omega_{,y} \\ 0 & 0 & 1 & y-\Omega_{,z} & 0 & 0 & 0 & \Omega_{,z} \end{bmatrix} \begin{Bmatrix} \varepsilon_1 \\ \varepsilon_2 \\ \vdots \\ \varepsilon_8 \end{Bmatrix}$$

$$= [\mathbf{B}(y,z,\Omega)]\{\boldsymbol{\varepsilon}_R\} \tag{5.8.14}$$

where $\{\boldsymbol{\varepsilon}_R\}$ is a vector of eight generalized strains, ε_1 is the axial stretch, ε_2 and ε_3 the transverse flexural shear strains, ε_4 the St. Venant torsional shear strain, ε_5 and ε_6 the bending curvatures with respect to the y- and z-axes, ε_7 the warping curvature and ε_8 the warping shear strain matrix $[\mathbf{B}]$ relating the linear strains to the generalized strains. The generalized strains relate to the generalized displacements by

$$\{\boldsymbol{\varepsilon}_R\} = \begin{bmatrix} D & & \kappa & & & & \\ & D & -\tau & & & & \\ \kappa & \tau & D & & & & \\ & & & D & & \kappa & \\ & & & & D & -\tau & \\ & & & -\kappa & \tau & D & \\ & & & & & & D \\ & & & D & & \kappa & -1 \end{bmatrix} \begin{Bmatrix} u \\ v \\ w \\ \psi_x \\ \psi_y \\ \psi_z \\ \kappa \end{Bmatrix} = [\mathbf{C}_0 + \mathbf{C}_1 D]\{\mathbf{u}_R\} \tag{5.8.15}$$

which is the compatibility equation. The engineering stresses are given by

$$\{\boldsymbol{\sigma}\} = \begin{Bmatrix} \sigma_x \\ \sigma_{xy} \\ \sigma_{xz} \end{Bmatrix} = \begin{bmatrix} E & & \\ & G & \\ & & G \end{bmatrix} \{\boldsymbol{\varepsilon}\} = [\mathbf{E}]\{\boldsymbol{\varepsilon}\} \tag{5.8.16}$$

According to Eq. (5.8.14), the generalized stresses are defined by

$$\begin{Bmatrix} S_1 \\ S_2 \\ S_3 \end{Bmatrix} = \int \{\boldsymbol{\sigma}\}\,\mathrm{d}A$$

$$\begin{Bmatrix} S_4 \\ S_5 \\ S_6 \end{Bmatrix} = \int \begin{bmatrix} & z & y \\ -z-\Omega_{,y} & & \\ y-\Omega_{,z} & & \end{bmatrix}^{\mathrm{T}} \{\boldsymbol{\sigma}\}\,\mathrm{d}A \tag{5.8.17}$$

$$\begin{Bmatrix} S_7 \\ S_8 \end{Bmatrix} = \int \begin{bmatrix} -\Omega & \\ & \Omega_{,y} \\ & \Omega_{,z} \end{bmatrix}^{\mathrm{T}} \{\boldsymbol{\sigma}\}\,\mathrm{d}A$$

Therefore, the generalized stresses and strains are related by the generalized Hooke's law,

$$\{\boldsymbol{\sigma}_R\} = [S_1, S_2, \ldots, S_8]^{\mathrm{T}} = [\mathbf{E}_R]\{\boldsymbol{\varepsilon}_R\} \tag{5.8.18}$$

where

$$[\mathbf{E}_R] = \mathrm{diag}[EA, GA_y, GA_z, GJ, EI_y, EI_z, EI_w, G(I_p - J)] \tag{5.8.19}$$

in which, the terms in the diagonal matrix are the respective rigidities for axial, y-shear, z-shear, St. Venant torsion, y-bending, z-bending, warping bending and warping torsion. In particular,

$$J = \int (y - \Omega_{,z})^2 + (z + \Omega_{,y})^2 \, \mathrm{d}y \, \mathrm{d}z$$

$$I_p = I_y + I_z, \qquad I_w = \int \Omega^2 \, \mathrm{d}y \, \mathrm{d}z \tag{5.8.20}$$

If the distributed forces are f_1, f_2, f_3, distributed torque and moments are f_4, f_5, f_6 and distributed bimoment is f_7, then the external work done is given by

$$W = \int \{\mathbf{f}_R\}^{\mathrm{T}} \{\mathbf{u}_R\} \, \mathrm{d}x \tag{5.8.21}$$

where the generalized applied force vector

$$\{\mathbf{f}\}^{\mathrm{T}} = [f_1, f_2, \ldots, f_7]$$

For equilibrium, the first variation of the total potential energy $V = U - W$, vanishes, where the strain energy U is given by

$$U = \tfrac{1}{2} \int \{\boldsymbol{\sigma}\}^{\mathrm{T}} \{\boldsymbol{\varepsilon}\} \, \mathrm{d}\,vol = \tfrac{1}{2} \int \{\boldsymbol{\sigma}_R\}^{\mathrm{T}} \{\boldsymbol{\varepsilon}_R\} \, \mathrm{d}x \tag{5.8.22}$$

From Eqs (5.8.14)–(5.8.18)

$$\delta V = \delta U - \delta W = \int ([\mathbf{C}_0 + \mathbf{C}_1 D]\{\mathbf{u}_R\})^{\mathrm{T}} [\mathbf{E}_R] \delta([\mathbf{C}_0 + \mathbf{C}_1 D]\{\mathbf{u}_R\}) \, \mathrm{d}x$$

$$- \int \{\mathbf{f}_R\}^{\mathrm{T}} \{\delta \mathbf{u}_p\} \, \mathrm{d}x = 0 \tag{5.8.23}$$

Therefore, the governing equation for $\{\mathbf{u}_R\}$ is given by

$$[\mathbf{A}_{00}]\{\mathbf{u}_R\} - [\mathbf{A}_{01}]\{\mathbf{u}_R'\} + \{\mathbf{A}_{10}\mathbf{u}_R\}' - \{\mathbf{A}_{11}\mathbf{u}_R'\}' - \{\mathbf{f}_R\} = \{\mathbf{0}\} \tag{5.8.24}$$

with the natural boundary condition,

$$\{\mathbf{R}\}^{\mathrm{T}} \{\delta \mathbf{u}_R\} = 0 \tag{5.8.25}$$

where

$$\{\mathbf{R}\} = [\mathbf{A}_{11}]\{\mathbf{u}_R'\} + [\mathbf{A}_{10}]\{\mathbf{u}_R\} \tag{5.8.26}$$

in which

$$[\mathbf{A}_{11}] = [\mathbf{C}_1]^{\mathrm{T}}[\mathbf{E}_R][\mathbf{C}_1] = \mathrm{diag}[EA, GA_y, GA_z, GJ, EI_y, EI_z, EI_w]$$

$$[\mathbf{A}_{01}]^{\mathrm{T}} = [\mathbf{A}_{10}] = [\mathbf{C}_1]^{\mathrm{T}}[\mathbf{E}_R][\mathbf{C}_0]$$

$$= \begin{bmatrix} & & EA\kappa & & & \\ & & -GA_y\tau & & & -GA_y \\ -GA_z\kappa & GA_z\tau & & & GA_z & \\ & & & & & GJ\kappa \\ & & & & & -EI_y\tau \\ & & & -EI_z\kappa & EI_z\tau & \\ & & & & & 0 \end{bmatrix}$$

$$[\mathbf{A}_{01}] = [\mathbf{C}_0]^{\mathrm{T}}[\mathbf{E}_R][\mathbf{C}_0]$$

$$= \begin{bmatrix} GA_z\kappa^2 & -GA_z\kappa\tau & & & -GA_z\kappa & & \\ & GA_z\tau^2 & & & GA_z\tau & & \\ & & EA\kappa^2+GA_y\tau^2 & & & GA_y\tau & \\ & & & EI_z\kappa^2 & -EI_z\kappa\tau & & \\ & & & & EI_z\tau^2 & & \\ & & & & & GJ\kappa^2+EI_y\tau^2 & -\kappa G(I_p-J) \\ \text{sym.} & & & & & & G(I_p-J) \end{bmatrix}$$

and the generalized force vector is given by

$$\{\mathbf{R}\} = \begin{bmatrix} EAD & & EA\kappa & & & & \\ & GA_yD & -GA_y\tau & & & -GA_y & \\ -GA_z\kappa & GA_z\tau & GA_zD & & GA_z & & \\ & & & GJD & & GJ\kappa & \\ & & & & EI_yD & -EI_y\tau & \\ & & & -EI_z\kappa & EI_y\tau & EI_zD & \\ & & & & & & EI_wD \end{bmatrix} \{\mathbf{u}_R\} \tag{5.8.27}$$

For a uniform cross-section,

$$[\mathbf{A}_{00}]\{\mathbf{u}_R\} + [\mathbf{A}_{10} - \mathbf{A}_{01}]\{\mathbf{u}'_R\} - [\mathbf{A}_{11}]\{\mathbf{u}''_R\} = \{\mathbf{f}_R\} \tag{5.8.28}$$

The kinetic energy of a helix with mass density ρ is given by

$$T = \tfrac{1}{2}\int \rho\{\dot{\mathbf{u}}_p\}^{\mathrm{T}}\{\dot{\mathbf{u}}_p\}\,\mathrm{d}\,vol \tag{5.8.29}$$

When the displacement field $\{\mathbf{u}_p\}$ in Eq. (5.8.2) is substituted into Eq. (5.2.28), we have

$$T = \tfrac{1}{2}\int \{\dot{\mathbf{u}}_R\}[\mathbf{M}_R]\{\dot{\mathbf{u}}_R\}\,\mathrm{d}x \tag{5.8.30}$$

where the generalized inertia matrix is given by

$$[\mathbf{M}_R] = \rho \begin{bmatrix} A+3I_y\kappa^2 & & & & 3I_y\kappa & & \\ & A & & -I_y\kappa & & & \\ & & A & & & & \\ & & & I_p & & & \\ & & & & I_y & & \\ & & & & & I_z+I_w\kappa^2 & \\ \text{sym.} & & & & & & I_w \end{bmatrix} \tag{5.8.31}$$

From Hamilton's principle, we have

$$[\mathbf{A}_{00}]\{\mathbf{u}_R\} + [\mathbf{A}_{10} - \mathbf{A}_{01}]\{\mathbf{u}'_R\} - [\mathbf{A}_{11}]\{\mathbf{u}''_R\} = \{\mathbf{f}_R\} - [\mathbf{M}_R]\{\ddot{\mathbf{u}}_R\} \tag{5.8.32}$$

if inertial force is included.

To investigate the buckling behaviour, the non-linear strains in Eq. (5.8.9) must be considered. Define $[\boldsymbol{\tau}]$ a 9×9 matrix of initial stresses by

$$[\boldsymbol{\tau}] = \begin{bmatrix} \sigma_x \mathbf{I}_3 & \sigma_{xy} \mathbf{I}_3 & \sigma_{xz} \mathbf{I}_3 \\ & \sigma_y \mathbf{I}_3 & \sigma_{yz} \mathbf{I}_3 \\ \text{sym.} & & \sigma_z \mathbf{I}_3 \end{bmatrix} \tag{5.8.33}$$

then the strain energy due to the initial stresses is given by

$$U_0 = \tfrac{1}{2} \int \{\boldsymbol{\theta}\}^{\mathrm{T}} [\boldsymbol{\tau}] \{\boldsymbol{\theta}\} \, \mathrm{d}\, vol \tag{5.8.34}$$

Here, from Eqs (5.8.14)–(5.8.19)

$$\{\boldsymbol{\sigma}\} = \begin{Bmatrix} \sigma_x \\ \sigma_{xy} \\ \sigma_{xz} \end{Bmatrix} = [\mathbf{E}]\{\boldsymbol{\varepsilon}\} = [\mathbf{E}][\mathbf{B}]\{\boldsymbol{\varepsilon}_R\} = [\mathbf{E}][\mathbf{B}][\mathbf{E}_R]^{-1}\{\boldsymbol{\sigma}_R\} \tag{5.8.35}$$

and after evaluation,

$$\{\boldsymbol{\sigma}\} = \begin{bmatrix} 1/A & & & & z/I_y & -y/I_z & -\Omega/I_w & \\ & 1/A_y & & -(z+\Omega_{,y})/J & & & & \Omega_{,y}/(I_p - J) \\ & & 1/A_z & (y - \Omega_{,z})/J & & & & \Omega_{,z}/(I_p - J) \end{bmatrix} \{\boldsymbol{\sigma}_R\} \tag{5.8.36}$$

From Eq. (5.8.12), we can obtain the gradient vector $\{\boldsymbol{\theta}\}$ in terms of the generalized displacement $\{\mathbf{u}_R(x)\}$,

$$\{\boldsymbol{\theta}(x, y, z)\} = [\mathbf{G}(y, z)]\{\mathbf{u}_R(x)\} \tag{5.8.37}$$

where

$$[\mathbf{G}] = \begin{bmatrix} D & & \kappa & y\kappa & zD - y\tau & -z\tau - yD & -\Omega D \\ & D & -\tau & -zD & & -z\kappa & \\ -\kappa & \tau & D & yD & & y\kappa & \\ & & & & & -1 & -\Omega_{,y} \\ & & & 1 & & & \\ & & & & 1 & & -\Omega_{,z} \\ & & & -1 & & & \end{bmatrix}$$

in which the fifth and the ninth rows are zero rows. Substituting Eqs (5.8.36) and (5.8.37) into (5.8.34), we have the strain energy of initial stresses. To integrate U_σ, assume without loss of generality that σ_x is the only important term and obtain

$$U_\sigma = \tfrac{1}{2} \int \begin{Bmatrix} \mathbf{u}' + \boldsymbol{\kappa}\mathbf{u} \\ \boldsymbol{\psi}' + \boldsymbol{\kappa}\boldsymbol{\psi} \\ \boldsymbol{\kappa}' \end{Bmatrix}^{\mathrm{T}} \begin{bmatrix} S_1 A & & & & S_5 I_y & S_6 I_z & S_7 I_w \\ & S_1 A & & -S_5 I_y & & & \\ & & S_1 A & -S_6 I_y & & & \\ & -S_5 I_y & -S_6 I_z & S_1 I_p & & & \\ S_5 I_y & & & & S_1 I_y & & \\ S_6 I_z & & & & & S_1 I_z & \\ S_7 I_w & & & & & & S_1 I_w \end{bmatrix} \times$$

$$\times \begin{Bmatrix} \mathbf{u}' + \boldsymbol{\kappa}\mathbf{u} \\ \boldsymbol{\psi}' + \boldsymbol{\kappa}\boldsymbol{\psi} \\ \boldsymbol{\kappa}' \end{Bmatrix} \mathrm{d}x$$

$$= \tfrac{1}{2}\int ([D\mathbf{I} + \mathbf{k}]\{\mathbf{u}_R\})^{\mathrm{T}}[\boldsymbol{\tau}_R]([D\mathbf{I} + \mathbf{k}]\{\mathbf{u}_R\})\,\mathrm{d}x \tag{5.8.38}$$

where $[\mathbf{k}] = \mathrm{diag}[\boldsymbol{\kappa}, \boldsymbol{\kappa}, 0]$ and $[\boldsymbol{\tau}_R]$ is the 7×7 generalized initial stress matrix. Now, the total potential energy is given by

$$V = U + U_\sigma - W$$

and, after taking the first variation, the governing Eqs (5.8.24)–(5.8.32) have exactly the same form but for buckling problems the matrices $[\mathbf{A}_{ij}]$, $i, j = 0, 1$ must be replaced by $[\mathbf{A}_{00}] + [\boldsymbol{\tau}_R]$, $[\mathbf{A}_{01}] + [\boldsymbol{\tau}_R][\mathbf{k}]$, $[\mathbf{A}_{10}] + [\mathbf{k}]^{\mathrm{T}}[\boldsymbol{\tau}_R]$, and $[\mathbf{A}_{11}] + [\mathbf{k}]^{\mathrm{T}}[\boldsymbol{\tau}_R][\mathbf{k}]$ respectively. Since S_1, S_5, S_6 and S_7 are the axial force, the bending moments about the y- and z-axes and the warping moment respectively, we can analyse axial and moment buckling with the modified equations.

5.9. Curvature Effect

Similar to shell theories, curvature has an important effect on helices if the thickness is comparable to the radius of curvature $1/\kappa$, i.e. $z\kappa$ is not negligible compared with unity. In this case, the displacement field is related to the generalized displacements by

$$\{\mathbf{u}(x, y, z)\} = \begin{Bmatrix} u_0(1 + z\kappa) \\ v_0 \\ w_0 \end{Bmatrix} + \begin{bmatrix} 0 & z & -(y + \Omega\kappa) \\ -z & 0 & 0 \\ y & 0 & 0 \end{bmatrix} \begin{Bmatrix} \psi_x \\ \psi_y \\ \psi_z \end{Bmatrix} - \begin{Bmatrix} \Omega\kappa \\ 0 \\ 0 \end{Bmatrix} \tag{5.9.1}$$

which approaches Eq. (5.8.2) if $1 \geq z\kappa$ and $y \geq \Omega\kappa$. The engineering strain components ε_x, ε_{xy} and ε_{xz} are given by

$$\begin{Bmatrix} \varepsilon_x \\ \varepsilon_{xy} \\ \varepsilon_{xz} \end{Bmatrix} = \begin{Bmatrix} \varepsilon_0/(1 + z\kappa) \\ \gamma_y \\ \gamma_z \end{Bmatrix} + \frac{1}{1 + z\kappa} \begin{bmatrix} 0 & z & -y \\ -z & 0 & 0 \\ y + \Omega\kappa & 0 & 0 \end{bmatrix} \begin{Bmatrix} \gamma_s \\ k_y \\ k_z \end{Bmatrix} - \begin{Bmatrix} \Omega k_w/(1 + z\kappa) \\ \Omega_{,y}(\gamma_s - \gamma_w) \\ \Omega_{,z}(\gamma_s - r_w) \end{Bmatrix} \tag{5.9.2}$$

where the generalized strains

$$\{\boldsymbol{\varepsilon}_R\} = [\varepsilon_0, \gamma_y, \gamma_z, \gamma_s, k_y, k_z, k_w, \gamma_w]^{\mathrm{T}} \tag{5.9.3}$$

have the same meaning as before. The compatibility equation is given by

$$\{\boldsymbol{\varepsilon}\} = \begin{bmatrix} D & & \kappa & & & & \\ & D & -\tau & & & -1 & \\ -\kappa & \tau & D & & 1 & & \\ & & & D & & & \\ & & & & D & & \\ & & & -\kappa & \tau & D & \\ & & & & & & D \\ & & & D & & \kappa & -1 \end{bmatrix} \begin{Bmatrix} u_0 \\ v_0 \\ w_0 \\ \psi_x \\ \psi_y + \kappa u_0 \\ \psi_z \\ \psi + \psi_z \kappa \end{Bmatrix} \tag{5.9.4}$$

or

$$\{\boldsymbol{\varepsilon}_R\} = [\mathbf{C}_0 + \mathbf{C}_1 D]\{\bar{\mathbf{u}}_R\} \tag{5.9.5}$$

where, due to curvature effect, $\bar{\psi}_y = \psi_y + \kappa u_0$, $\bar{\psi} = \psi + \kappa\psi_z$ and $\{\bar{\mathbf{u}}_R\} = [u_0, v_0, w_0, \psi_x, \bar{\psi}_y, \psi_z, \bar{\psi}]^{\mathrm{T}}$. Before the strain energy of Eq. (5.8.22) is integrated over the cross-sectional area, the forms

$$1/(1 + z\kappa) \doteq 1 - z\kappa \quad \text{and} \quad \mathrm{d}A = (1 + z\kappa)\,\mathrm{d}y\,\mathrm{d}z \tag{5.9.6}$$

are assumed. For a bisymmetrical cross-section,

$$\begin{aligned} [\mathbf{E}_R] = {} & \mathrm{diag}[E(A + I_y\kappa^2), GA_y, GA_z, G(J + I_w\kappa^2), EI_y, EI_z, EI_w, G(I_p - J)] \\ & + GI_y\kappa(\mathbf{e}_{24} + \mathbf{e}_{42}) \end{aligned} \tag{5.9.7}$$

and

$$U = \tfrac{1}{2}\int \{\boldsymbol{\varepsilon}_R\}^{\mathrm{T}}[\mathbf{E}_R]\{\boldsymbol{\varepsilon}_R\}\,\mathrm{d}x$$

where $\mathbf{e}_{ij}$ is a zero matrix with ij element equal to one, and $\int y^m z^n\,\mathrm{d}y\,\mathrm{d}z = 0$ if m, $n > 2$. The equilibrium equation (5.8.28) becomes

$$[\mathbf{A}_{00}]\{\bar{\mathbf{u}}_R\} + [\mathbf{A}_{10} - \mathbf{A}_{01}]\{\bar{\mathbf{u}}'_R\} - [\mathbf{A}_{11}]\{\bar{\mathbf{u}}''_R\} = \{\mathbf{f}_R\} \tag{5.9.8}$$

with the natural boundary conditions

$$\{\mathbf{R}\}^{\mathrm{T}}\{\delta\bar{\mathbf{u}}_R\} = 0 \tag{5.9.9}$$

where $\{\mathbf{R}\} = [\mathbf{A}_{11}]\{\bar{\mathbf{u}}'_R\} + [\mathbf{A}_{10}]\{\bar{\mathbf{u}}_R\}$

$$[\mathbf{A}_{11}] = \mathrm{diag}[E(A + \kappa^2 I_y), GA_y, GA_z, G(J + \kappa^2 I_w), EI_y, EI_z, EI_w]$$

$$[\mathbf{A}_{01}]^{\mathrm{T}} = [\mathbf{A}_{10}]$$

$$= \begin{bmatrix} & & \kappa E(A + \kappa^2 I_y) & & & & \\ & & -\tau GA_y & & & GI_y\kappa^2 - GA_y & \\ -\kappa GA_z & \tau GA_z & & & GA_z & & \\ & & -\tau\kappa GI_y & & & \kappa G(J + \kappa^2 I_w - I_y) & \\ & & & & & -\tau EI_y & \\ & & & -\kappa EI_z & \tau EI_z & & \\ & & & & & & 0 \end{bmatrix}$$

$$[\mathbf{A}_{00}] =$$

$$\begin{bmatrix} GA_z\kappa^2 & -GA_z\kappa\tau & & & -GA_z\kappa & & \\ & GA_z\tau^2 & & & GA_z\tau & & \\ & & E\kappa^2 A + GA_y\tau^2 & & & \tau G(A_y+\kappa^2 I_y) & \\ & & & EI_z\kappa^2 & -EI_z\kappa\tau & & \\ & & & & EI_z\tau^2 & & \\ & & & & & \kappa^2 G(J+\kappa^2 I_w - 2GI_y\kappa) + EI_y\tau^2 & -\kappa G(I_p - J) \\ \text{sym.} & & & & & & G(I_p - J) \end{bmatrix}$$

The kinetic energy of the helix with mass density ρ and with curvature effect included is given by

$$T = \tfrac{1}{2}\int \rho\{\dot{\mathbf{u}}\}^{\mathrm{T}}\{\dot{\mathbf{u}}\}\,\mathrm{d}\,vol = \tfrac{1}{2}\int \{\dot{\mathbf{u}}_R\}^{\mathrm{T}}[\mathbf{M}_R]\{\dot{\mathbf{u}}_R\}\,\mathrm{d}x \tag{5.9.10}$$

where the displacement field $\{\mathbf{u}\}$ is related to the generalized displacement $\{\mathbf{u}_R\}$ by Eq. (5.8.1) and

$$[\mathbf{M}_R] = \rho\begin{bmatrix} A + 3I_y\kappa^2 & & & & 3I_y\kappa & & \\ & A & & -I_y\kappa & & & \\ & & A & & & & \\ & & & I_p & & & \\ & & & & I_y & & \\ & & & & & I_z + I_w\kappa^2 & \\ \text{sym.} & & & & & & I_w \end{bmatrix} \tag{5.9.11}$$

From Hamilton's principle, we have

$$[\mathbf{A}_{00}]\{\bar{\mathbf{u}}_R\} + [\mathbf{A}_{10} - \mathbf{A}_{01}]\{\mathbf{u}'_R\} - [\mathbf{A}_{11}]\{\mathbf{u}''_R\} = \{\mathbf{f}_R\} - \{\mathbf{M}_R]\{\ddot{\mathbf{u}}_R\} \tag{5.9.12}$$

5.10. Extensions

In summary, the local Cartesian displacements $\{\mathbf{u}_p(x, y, z)\}$ are expressed in terms of the generalized displacements $\{\mathbf{u}(x)\}$ and $\{\boldsymbol{\psi}(x)\}$ by means of the rigid body matrix $[\mathbf{R}(y, z)]$ of Eq. (5.8.2). The gradient matrix $[\mathbf{F}]$ is found in Eq. (5.9.7) which enables us to obtain the initial strain $[\mathbf{e}^0]$ and the non-linear strain $[\mathbf{e}'']$ of Eqs. (5.8.10) and (5.8.11). It is assumed further that the initial direct stress is the most significant one in the initial buckling analysis as given in Eq. (5.8.33). The integration of the various energy terms over the cross-sectional area and the variational operation are purely algebraic. Finally, the governing equations and the associated boundary conditions are obtained.

This section presents ways to improve the theory. The first improvement is for an asymmetrical thin-walled cross-sectional area whose shear centre is at coordinates (y_s, z_s) with respect to the centroid, the rigid body matrix $[\mathbf{R}]$ in Eq. (5.8.2) being modified to

$$[\mathbf{R}(y,z)] = \begin{bmatrix} 0 & z & -y \\ -(z-z_s) & 0 & 0 \\ y-y_s & 0 & 0 \end{bmatrix} \tag{5.10.1}$$

For moderately thick and short helices, one may expand Eq. (5.8.1) using a Taylor series with higher order terms,

$$\{\mathbf{u}_p\} = \begin{bmatrix} 1 & 0 & 0 & 0 & z & -y & -\Omega & xy & \dots \\ 0 & 1 & 0 & -z & 0 & 0 & 0 & z^2 & \dots \\ 0 & 0 & 1 & y & 0 & 0 & 0 & y^2 & \dots \end{bmatrix} \begin{Bmatrix} \mathbf{u}(x) \\ \boldsymbol{\psi}(x) \\ \boldsymbol{\kappa}(x) \\ \boldsymbol{\phi}(x) \end{Bmatrix} \tag{5.10.2}$$

to establish refined theories, where $\{\boldsymbol{\phi}(x)\}$ is a vector of the additional generalized displacements to be determined.

Finally, for helicoidal thin-walled shells with circular cross-sections, $\Omega = 0$ and we may assume that

$$\{\mathbf{u}_p\} = \begin{Bmatrix} u + z\psi_y - y\psi_z + \sum_j u_j(x)\cos j\pi\theta \\ v - z\psi_x \quad + \sum_j v_j(x)\cos j\pi\theta \\ w + y\psi_x \quad + \sum_j w_j(x)\sin j\pi\theta \end{Bmatrix} \tag{5.10.3}$$

where u_j, v_j and w_j are the additional generalized displacements to be determined.

Although the integration of the various energy terms over the cross-sectional area is purely algebraic, to find the generalized strain vector giving a diagonal constitutive matrix $[\mathbf{E}_R]$ will not be generally possible.

5.11. Symmetry of the Dynamic Stiffness Matrix

It was claimed in the previous sections that the dynamic stiffness matrices derived are symmetric for conservative systems. It will be proved in this section that they are indeed so. Assume the following displacement field

$$\{\mathbf{u}(x,y,z)\} = [\mathbf{N}_0(x,y) + \mathbf{N}_1(x,y)D]\{\boldsymbol{\alpha}(z)\} \tag{5.11.1}$$

where $D \equiv \partial/\partial z$, $\{\boldsymbol{\alpha}(z)\}$ is the generalized displacement vector, and the time variable is implicit. In Eq. (5.11.1), terms $\mathbf{N}_i D^i$, $i > 2$ have been neglected, without loss of generality, for convenience. The strain and gradient fields are

$$\{\boldsymbol{\varepsilon}(x,y,z)\} = [\mathbf{B}_0(x,y) + \mathbf{B}_1(x,y)D + \mathbf{B}_2(x,y)D^2]\{\alpha\} \tag{5.11.2}$$

$$\{\boldsymbol{\theta}(x,y,z)\} = [\mathbf{G}_0(x,y) + \mathbf{G}_1(x,y)D + \mathbf{G}_2(x,y)D^2]\{\alpha\} \tag{5.11.3}$$

For harmonic oscillation with frequency ω, Hamilton's principle requires (cf. Eqs (5.3.10)–(5.3.22))

$$-\delta H = \int \begin{Bmatrix} \boldsymbol{\alpha} \\ \boldsymbol{\alpha}' \\ \boldsymbol{\alpha}'' \end{Bmatrix}^{\mathrm{T}} \begin{bmatrix} \mathbf{A}_{00} & \mathbf{A}_{01} & \mathbf{A}_{02} \\ \mathbf{A}_{10} & \mathbf{A}_{11} & \mathbf{A}_{12} \\ \mathbf{A}_{20} & \mathbf{A}_{21} & \mathbf{A}_{22} \end{bmatrix} \begin{Bmatrix} \beta \\ \beta' \\ \beta'' \end{Bmatrix} - \omega^2 \begin{Bmatrix} \boldsymbol{\alpha} \\ \boldsymbol{\alpha}' \end{Bmatrix}^{\mathrm{T}} \begin{bmatrix} \mathbf{T}_{00} & \mathbf{T}_{01} \\ \mathbf{T}_{10} & \mathbf{T}_{11} \end{bmatrix} \begin{Bmatrix} \boldsymbol{\beta} \\ \boldsymbol{\beta}' \end{Bmatrix} \mathrm{d}z$$

$$+ [\mathbf{S}_0^{\mathrm{T}}(0) + \mathbf{S}_1^{\mathrm{T}}(0) - \mathbf{S}_0^{\mathrm{T}}(l) - \mathbf{S}_1^{\mathrm{T}}(l)][\boldsymbol{\beta}^{\mathrm{T}}(0)\boldsymbol{\beta}'^{\mathrm{T}}(0)\boldsymbol{\beta}^{\mathrm{T}}(l)\boldsymbol{\beta}'^{\mathrm{T}}(l)]^{\mathrm{T}}$$

$$+ \int \{\boldsymbol{\beta}\}^{\mathrm{T}}[\mathbf{E}(D)]\{\boldsymbol{\alpha}\}\,\mathrm{d}z = 0 \tag{5.11.4}$$

where $\{\boldsymbol{\beta}, \boldsymbol{\beta}', \boldsymbol{\beta}''\} = \{\delta\boldsymbol{\alpha}, \delta\boldsymbol{\alpha}', \delta\boldsymbol{\alpha}''\}$, the matrices $[\mathbf{A}_{ij}]$ and $[\mathbf{T}_{ij}]$ being given by Eqs (5.3.12) and (5.3.17) respectively, $0 \le z \le l$ and $[\mathbf{E}(D)]\{\boldsymbol{\alpha}\}$ denotes the left-hand side of Eq. (5.3.20). Now, since the generalized force is defined by

$$\{\mathbf{Q}\} = \begin{Bmatrix} \mathbf{S}_0(0) \\ \mathbf{S}_1(0) \\ \mathbf{S}_0(l) \\ \mathbf{S}_1(l) \end{Bmatrix} = [\mathbf{D}(\omega)] \begin{Bmatrix} \boldsymbol{\alpha}(0) \\ \boldsymbol{\alpha}'(0) \\ \boldsymbol{\alpha}(l) \\ \boldsymbol{\alpha}'(l) \end{Bmatrix} \tag{5.11.5}$$

where $[\mathbf{D}(\omega)]$ is the dynamic stiffness matrix and $[\mathbf{E}(D)]\{\boldsymbol{\alpha}\} = 0$ for equilibrium, Eq. (5.11.4) becomes

$$\int \begin{Bmatrix} \boldsymbol{\alpha} \\ \boldsymbol{\alpha}' \\ \boldsymbol{\alpha}'' \end{Bmatrix}^{\mathrm{T}} \begin{bmatrix} \mathbf{A}_{00} & \mathbf{A}_{01} & \mathbf{A}_{02} \\ \mathbf{A}_{10} & \mathbf{A}_{11} & \mathbf{A}_{12} \\ \mathbf{A}_{20} & \mathbf{A}_{21} & \mathbf{A}_{22} \end{bmatrix} \begin{Bmatrix} \beta \\ \beta' \\ \beta'' \end{Bmatrix} - \omega^2 \begin{Bmatrix} \boldsymbol{\alpha} \\ \boldsymbol{\alpha}' \end{Bmatrix}^{\mathrm{T}} \begin{bmatrix} \mathbf{T}_{00} & \mathbf{T}_{01} \\ \mathbf{T}_{10} & \mathbf{T}_{11} \end{bmatrix} \begin{Bmatrix} \boldsymbol{\beta} \\ \boldsymbol{\beta}' \end{Bmatrix} \mathrm{d}z$$

$$= [\boldsymbol{\alpha}^{\mathrm{T}}(0), \boldsymbol{\alpha}'^{\mathrm{T}}(0), \boldsymbol{\alpha}^{\mathrm{T}}(l), \boldsymbol{\alpha}'^{\mathrm{T}}(l)][\mathbf{D}(\omega)]^{\mathrm{T}}[\boldsymbol{\beta}^{\mathrm{T}}(0)\boldsymbol{\beta}'^{\mathrm{T}}(0)\boldsymbol{\beta}^{\mathrm{T}}(l)\boldsymbol{\beta}'^{\mathrm{T}}(l)]^{\mathrm{T}} \tag{5.11.6}$$

Equation (5.11.6) is scalar. Taking the transpose of the whole equation and comparing, we show that $[\mathbf{D}(\omega)] = [\mathbf{D}(\omega)]^{\mathrm{T}}$.

References

1. F Laudiero, M. Savoia, D Zaccaria 1991. The influence of shear deformation on the stability of thin-walled beams under non-conservative loading. Int J Solids Struct 27, 1351–1370
2. A Libai, JG Simmonds 1988. The nonlinear theory of elastic shells in one spatial dimension. Academic Press
3. I. Gohberg, P. Lancaster, L. Rodman 1982. Matrix polynomials. Academic Press
4. YK Cheung, AYT Leung 1991. Finite element method in dynamic analysis. Kluwer Academic
5. AYT Leung 1991. Exact stiffness matrix for twisted helix beam. Finite Elements Anal Des 9, 23–32
6. AYT Leung, TC Fung 1988. Spinning finite elements. J Sound Vib 125, 523–537
7. PO Friberg 1985. Beam element matrices derived from Vlasov's theory of open thin-walled elastic beams. Int J Num Meth Engng 21, 1205–1228
8. JE Mottershead 1980. Finite elements for dynamical analysis of helical rods. Int J Mech Sci 22, 267–283
9. D Pearson 1982. The transfer matrix method for the vibration of compressed helical springs. Int J Mech Engng Sci 24, 163–171
10. D Pearson, WH Wittrick 1986. An exact solution for the vibration of helical springs using a Bernoulli–Euler model. Int J Mech Sci 28, 83–96
11. B. Tabarrok, AN Sinclair, M Farshad, H Yi 1988. On the dynamics of spatially curved and twisted rods – a finite element formulation. J Sound Vib 123, 315–326
12. P. Lancaster, M. Tismenetsky 1985. The theory of matrices. Academic Press, London
13. RA Frazer, WJ Duncan, AR Collar 1947. Elementary matrices, Cambridge, New York
14. AYT Leung 1990. Perturbed general eigensolutions. Commun Appl Num Meth 6, 401–409

15. VZ Vlasov 1959. Thin-walled elastic beams. Moscow
16. WH Wittrick, FW Williams 1970. A general algorithm for computing natural frequencies of elastic structures. Q J Mech Appl Math 24, 263–284
17. PO Friberg 1985. Beam element matrices derived from Vlasov's theory of open thin-walled elastic beams. Int J Num Meth Engng 21, 1205–1228
18. KL Wardle 1965. Differential geometry. Routledge and Kegan Paul, London
19. WH Wittrick 1966. On elastic wave propagation in helical springs. Int J Mech Sci 8, 25–47
20. YB Yang, SR Kuo 1986. Static stability of curved thin-walled beams. J Engng Mech 112, 821–841
21. HW Guggenheimer 1977. Differential geometry. Dover, New York, pp 238–239

Subject Index